Fluvial Processes in
Geomorphology

SECOND EDITION

Luna B. Leopold | M. Gordon Wolman | John P. Miller

With a new Foreword by Dr. Ellen Wohl

DOVER PUBLICATIONS
Garden City, New York

Bibliographical Note

This Dover edition, first published in 2020, is an unabridged and unaltered republication of the work first published by W. H. Freeman and Company, San Francisco, in 1964. A new Foreword, written by Dr. Ellen Wohl, has been specially prepared for this volume.

Library of Congress Cataloging-in-Publication Data

Names: Leopold, Luna B. (Luna Bergere), 1915–2006, author. | Wolman, M. Gordon, 1924–2010, author. | Miller, John P. (John Preston), 1923–1961, author. | Wohl, Ellen E., 1962– author of introduction, etc.
Title: Fluvial processes in geomorphology / Luna B. Leopold, M. Gordon Wolman, John P. Miller.
Description: Dover ed. | Garden City, New York : Dover Publications, 2020. | "an unbridged and unaltered republication of the work first published by W. H. Freeman and Company, San Francisco, in 1964. A new Foreword, written by Dr. Ellen Wohl, has been specially prepared for this volume"—CIP data view. | Includes bibliographical references and index. | Summary: "A pioneering study that encompasses both field and laboratory research, this text explores the landscapes of mountains, rivers, and seacoasts. Topics include weathering, climate, and erosion"—Provided by publisher.
Identifiers: LCCN 2019060217 | ISBN 9780486845524 (trade paperback)
Subjects: LCSH: Physical geography. | Geomorphology. | River channels.
Classification: LCC GB55 .L4 2020 | DDC 551.3/5—dc23
LC record available at https://lccn.loc.gov/2019060217

Manufactured in the United States of America
84552402   2022
www.doverpublications.com

To Carolyn, Elaine, and Laura

Foreword

IF YOU search for this title online, you may find the comment "This excellent text is a pioneering work in the study of landform development under processes associated with running water." That single sentence effectively summarizes this classic book, but so much more lies behind, and within, the book.

River Textbooks Prior to This Book

An important aspect of *Fluvial Processes in Geomorphology* is that it was essentially unique when first published in 1964. At present, someone looking for a textbook for an advanced undergraduate- or graduate-level course in fluvial geomorphology has several fine textbooks to choose from. In 1964, however, *Fluvial Processes* was really the only choice. Earlier books had dealt with many of the general themes covered in *Fluvial Processes*, but most of these books provided a descriptive, qualitative treatment of the material. *Fluvial Processes* reflected a relatively recent but ongoing *quantitative* revolution in the study of river process and form.

By 1964, multiple books had established a history of written descriptions of rivers. Among the earliest in North America was LeConte's 1878 textbook *Elements of Geology*. This book, which predictably covers all of geology, includes two pages showing the development of proportional equations for the ratio of sediment transporting power of a current of water to the sixth power of velocity. Geikie's 1901 *Text-Book of Geology* also includes a discussion of river process and form in which velocity is related to grain size that can be transported. Between these 1878 and 1901 texts come two books focused on rivers: William Morris Davis's short monograph *The Rivers and Valleys of Pennsylvania* (1889) and I. C. Russell's *Rivers of North America* (1898). Although nicely illustrated, the Davis monograph includes no equations, and the Russell book includes only simple relationships between velocity and the size of sediment that can be transported. The latter two are essentially the same as those in the LeConte and Geikie textbooks.

Subsequent textbooks include Chamberlin and Salisbury's 1905 *Geology*, the first volume of which includes diagrams of drainage network erosion and resulting drainage patterns, as well as illustrations of processes such as turbulent flow, meander geometry, and lateral channel migration. Davis and Snyder's *Physical Geography* (1898) and Davis's *Geographical Essays* (1909), although extensively discussing river process and form, include only drawings and photographs to illustrate the text. Russell's 1909 *River Development* is essentially the same text as his 1898 book and includes the same brief mention of velocity and sediment transport.

The first significant change in the way river processes are explained came with G. K. Gilbert's 1914 U.S. Geological Survey Professional Paper, *The transportation of debris by running water*. Gilbert, who spent his career with the U.S. Geological Survey, is commonly described as the father of modern geomorphology because of his emphasis on process-oriented, quantitative treatments of landforms and landscapes. Gilbert's long paper employs numerous equations and graphs to explain river process and form and is thus a direct intellectual precursor of the quantitative treatment of rivers that characterizes *Fluvial Processes in Geomorphology*.

During and after Gilbert's lifetime, however, many textbooks that included a section on rivers continued to be dominated by the thinking of W. M. Davis, who emphasized change through time as described in the Cycle of Erosion with its youthful, mature, and old landforms. Thornbury's 1954 *Principles of Geomorphology* provides an example. Although the book discusses fluvial processes and illustrates the text with block diagrams, maps, and photographs, no equations or graphs are used to explain river processes. Landscapes are described in a Davisian framework.

Other than Gilbert's much earlier work, the textbook that most closely parallels the quantitative emphasis of *Fluvial Processes in Geomorphology* is Scheidegger's 1961 *Theoretical Geomorphology*. Scheidegger's text covers all types of geomorphic processes. Chapter 4, on river bed processes, contains detailed equations, charts, and graphs dealing with all aspects of the hydraulics of open channel flow (velocity, turbulence, flow resistance, flow classification, energy loss), sediment transport, and bedforms. The subsequent chapter on drainage basin development discusses topics such as channel formation and meander geometry. These two chapters (among a total of nine chapters) provide a more detailed treatment of river mechanics than *Fluvial Processes in*

Geomorphology, but Scheidegger's book did not have nearly the impact of the 1964 fluvial text. This is reflected in part by citations listed in Google Scholar. The three editions of *Theoretical Geomorphology* have 401 citations as of early 2020, whereas *Fluvial Processes* has 6,815 citations. The broader coverage of geomorphic topics in *Theoretical Geomorphology* may have obscured the presence of the river sections for some readers, or the large number of equations may have put off some readers. Whatever the cause, *Theoretical Geomorphology* preceded the quantitative, mechanistic treatment of rivers that characterizes *Fluvial Processes*, but did not have nearly the impact of the subsequent book.

The Men Behind the Book

Behind *Fluvial Processes in Geomorphology* is the story of three men who were particularly influential in shaping the study of rivers during the latter half of the twentieth century: Luna Leopold, "Reds" Wolman, and John Miller. Leopold (1915–2006) grew up with a tradition of research and challenging the status quo in scientific thinking. His father, Aldo Leopold, fundamentally altered the way scientists think about predator-prey relations and wildlife management and gave environmentalists some of the most eloquent writing about ecosystems in the English language.

Luna Leopold followed a meandering path in coming to fluvial geomorphology. He studied civil engineering and biology at the University of Wisconsin and spent the summers working at a forest and range experiment station and on the Navajo Nation reservation in Arizona. Luna's interest in forest and grazing management had been triggered by his father's interest in channel erosion in the western U.S., and after finishing his degree Luna joined the Soil Conservation Service (SCS) in New Mexico. At the time, Harvard geology professor Kirk Bryan (1888–1950) was particularly critical of erosion control efforts because Bryan's observations in the western U.S. suggested that the current widespread episode of alluvial erosion reflected primarily climatic controls rather than land use. Intrigued by this criticism, Luna resigned from the SCS and enrolled as Bryan's graduate student at Harvard. Taking classes at both Harvard and MIT, Luna learned from some of the most prominent scientists of the time, taking classes in geology, ecology, and meteorology. When his funding ran out, Luna returned to New Mexico and began to work with hydrologist Thomas Maddock Jr.

When World War II began, Luna joined the Army and became a meteorologist stationed at the University of California Los Angeles (UCLA), where he worked with the eminent meteorologist Jacob Bjerknes, who emphasized the importance of reasoning from general principles. Luna earned an MS in Physics-Meteorology at UCLA. After the war Luna became head of the Meteorology Department at the Pineapple Research Institute in Hawaii. While at the institute he worked with another eminent meteorologist, Carl Rossby of the University of Chicago. As Luna wrote toward the end of his life, "I could see that my weakness in mathematics meant that I could never achieve real stature in meteorology. This recognition of weakness is an essential aspect of career development." (Leopold, 2004, p. 5) In 1949 Luna contacted Bryan again and finished his degree at Harvard.

During his time at Harvard, Luna became friends with another of Bryan's graduate students, John P. Miller (1923–1961). Miller had also done field research in New Mexico, and Leopold and Miller collaborated on a study of river terraces in eastern Wyoming. After graduation, Leopold joined the Water Resources Division of the U.S. Geological Survey, at that time dominated by engineers. Leopold later remembered, "The word *geomorphology* was quite unknown in the office. There was no research." (Leopold, 2004, p. 6) Leopold collaborated with Maddock to investigate what influences the width of channels, and the resulting 1953 USGS Professional Paper, *The hydraulic geometry of stream channels and some physiographic implications*, is a classic of fluvial geomorphology. As Leopold noted, this paper, along with Robert Horton's 1945 paper on the erosional development of streams, initiated a trend toward more quantitative analyses of landforms and geomorphic processes.

Miller took a faculty position at Harvard and Leopold became Chief Hydraulic Engineer at the U.S. Geological Survey. They collaborated on several projects and began to discuss writing a book on geomorphology. While doing field research in New Mexico together during August 1961, Miller abruptly became ill and died several days later. Most accounts, including that written by Leopold, describe bubonic plague as the cause of death. (Bubonic plague remains present in the western U.S., transmitted by fleas, although it is easily treated with antibiotics if correctly identified in time.) A newspaper article from August 1961, however, quotes Massachusetts health commissioner Alfred Frechette as stating that Miller had died of a blood infection (hemolytic streptococcus poisoning), rather than plague.

Whatever the cause, John Miller's life was abruptly cut short at the age of thirty-eight. Leopold shifted his collaborative focus to Walter Langbein, an engineer in the USGS Water Division. Together, Leopold and Langbein built the division into a strong research center and wrote a series of foundational papers on the basic physical principles underlying river process and form. Much of the research coming from the division was published in the USGS Professional Papers, which became highly respected and influential publications.

Before Miller's death, Leopold also began to collaborate with M. Gordon Wolman (1924–2010), widely known as "Reds." Their first coauthored papers examined river channel patterns (Leopold and Wolman, 1957) and floodplains (Wolman and Leopold, 1957). After serving in the Navy during World War II, Wolman earned an MS in geography from Johns Hopkins University and a PhD in geology from Harvard in 1953. He then joined the U.S. Geological Survey and worked with Leopold and Miller, before joining the faculty at Johns Hopkins in 1958. Wolman's father, Abel Wolman, was an eminent sanitation engineer who standardized the methods used to chlorinate drinking-water supplies. At Hopkins, Reds Wolman combined the departments of geography and sanitary and water resources to create the department of geography and environmental engineering, which he chaired for twenty years.

Understanding the environments in which Leopold, Wolman, and Miller worked provides vital insight into *Fluvial Processes in Geomorphology*. Leopold and Wolman were both influenced by the places where they lived and worked: in Leopold's case, channel erosion in New Mexico; in Wolman's case, soil erosion and its effects on water supplies at a Connecticut dairy farm where he worked in the summer during his childhood.

Both Leopold and Wolman cared deeply about the environment. Wolman's research on sediment yield and channel change caused by changes in land cover (Wolman, 1967) helped lead to new state regulations, and he headed the Oyster Roundtable, a coalition that designed a plan to reverse degradation in the Chesapeake Bay ecosystem. Leopold wrote forcefully of the importance of environmental protection in books such as *Water in Environmental Planning* (Dunne and Leopold, 1978).

The other fundamental characteristics shared by Leopold, Wolman, and Miller are that they were preeminently field scientists of broad disciplinary background and interests. Each of them based his understanding of, and contributions to, science in field observations. In a retrospective of his career, Leopold wrote, "The most engaging and interesting intellectual work on geomorphic forms such as river channels has not come from computer specialists and theoretical models but from field measurements and observations." (Leopold, 2004, p 10). What has made the field observations of these three scientists so long-lasting is that each man used detailed, site-specific measurements as a springboard to understand the basic underlying forces and processes governing landscapes and drew on a broad base of knowledge and a wide network of collaborators in doing so.

The Content Within the Book

Leopold and Wolman drew heavily on their own research and field observations in structuring and writing *Fluvial Processes in Geomorphology*. They were also writing at a time of a disciplinary shift toward more quantitative approaches to understanding rivers and landscapes. The foundations for this shift had been laid in the late nineteenth century by G. K. Gilbert in his publications on landscape development (1877), lakes (1890), and rivers and sediment transport (1914). During the first half of the twentieth century, experimental and analytical investigations of sediment transport in streams (Rubey, 1938) and eolian environments (Bagnold, 1941), the flow of water across landscapes and the development of river networks (Horton, 1945), the longitudinal profile of streams (Mackin, 1948; Hack, 1957), and the development of hillslopes (Strahler, 1950, 1952) and integration with river channels (Schumm, 1956) gradually shifted the way scientists thought about landscapes and created geomorphology as a distinctive discipline within Earth sciences (Scatena and Varrin, 2010).

The effects of this shift in a conceptual framework are now ingrained in thinking about physical process and form in rivers. Twenty-first century courses in geomorphology start with the framework of an open system with multiple variables that interact in a manner that tends to promote dynamic equilibrium. Geomorphologists emphasize flows of material through landscapes, as well as linkages between these flows and landforms and between individual landforms (e.g., channels and hillslopes or channels and floodplains) within a drainage basin. Leopold, Wolman, and Miller each contributed to this fundamental shift in how we conceptualize Earth's surface. The treatment of rivers in *Fluvial Processes in Geomorphology* exemplifies the new conceptual framework.

From the first sentence of the first chapter, the field emphasis of the book is clear. This first chapter briefly explains the book's emphasis on quantitative geomorphology and processes; the second chapter places this emphasis in a field context by describing three disparate field settings and some of the important questions that a geomorphologist might ask in viewing these places.

Part II of *Fluvial Processes*, "Process and Form," begins with a chapter on climate and erosion. This chapter introduces the first equations and quantitative graphs to explain processes of erosion. The next portion of the chapter discusses, with reservations, the idea of morphogenetic regions as geographic entities in which particular geomorphic processes of weathering predominate. The reservations arise from the recognition that, at the time the book was written, only a few processes had been studied in sufficient detail and over broad areas to allow generalization. The chapter then proceeds to discuss examples of such processes, including climate-related sediment yield, stream flow regime, flood frequency, and sediment transport in rivers.

Chapter 4 returns to the topic of weathering, discussing in detail the processes and products of chemical and physical weathering of bedrock. The chapter also describes soil formation, with an emphasis on Jenny's (1941) work on the primary soil-forming factors.

Chapter 5 starts the coverage of what might today be considered "classic" fluvial geomorphology. This chapter discusses drainage basins, including linear metrics such as stream order, the areal metric of drainage density, and vertical metrics such as the hypsometric curve. The following chapter examines the movement of water and sediment within channels, with an emphasis on quantification of hydraulic forces, energy losses, different forms of sediment transport within channels and measurement of sediment transport, and the grain-size distribution of streambed sediment. In a sense, this is the first chapter of the book that is revolutionary in terms of its emphasis on quantification of physical forces and resultant processes relative to previously published books on rivers. Much of the material in this chapter, including the terminology used to describe processes, is readily familiar to contemporary fluvial geomorphologists.

Chapter 7 builds on the discussion of hydraulic forces and sediment transport in the preceding chapter and delves into the resulting channel forms and processes. This chapter examines cross-sectional geometry, downstream and at-a-station hydraulic geometry, channel changes during floods, longitudinal profiles, the effect of changing local base level, and the application of an equilibrium conceptual framework to interpreting changes in channel geometry and longitudinal profile. The chapter continues with a discussion of process and form in straight, braided, and meandering channels, along with a graph that distinguishes among the three planforms as a function of bankfull discharge and channel gradient. This very long chapter (comprising about twenty percent of the book) concludes with a discussion of floodplain morphology and formation.

Chapter 8 moves outside of the river corridor to discuss hillslopes in relation to the influence of lithology, different forms of downslope movements of sediment and water, and slope morphology as a function of climate and rock type. The chapter concludes with several field examples from different geographic regions.

In their preface to *Fluvial Processes*, the authors apologize for not being able to "make a truly satisfactory translation from the dynamics of process to historical interpretation," but Part III of the book, "The Effects of Time," does take up this topic. The first chapter in this section discusses geochronology, one area of the book in which advances since 1964 are particularly noticeable. The discussion of geochronology includes historical records, dendrochronology, archaeological methods, varves, pollen analysis, and radiocarbon dating. Although Leopold, Wolman, and Miller clearly recognized the potential for inferring detailed chronologies of geomorphic processes using techniques such as radiocarbon, there were at that time few studies that they could cite as examples.

Chapter 10 returns to drainage basins, but with an emphasis on development of river networks through time. This chapter is notable for its discussion of probability and the application of mathematical models such as the random walk to understanding the spatial distribution of river channels. Chapter 11 examines channel changes through time, discussing processes and potential causes of both channel aggradation and degradation. This leads into the topics of river terraces and paleosols. The final chapter of the book discusses the evolution of hillslopes over diverse time scales.

In summary, the salient characteristics of *Fluvial Processes in Geomorphology* include breadth of vision and a mastery of the relevant scientific knowledge of the time. General principles of physical and chemical processes relevant to geomorphology are effectively illustrated with specific field examples. The writing is clear and includes rigorous attention to details. Abundant illustrations in the form of diagrams, graphs, and photographs greatly enhance the clarity and information content of the book. As the book was published in 1964, it naturally does not do justice to the state of relevant knowledge more than fifty years later, but the book remains worth reading because of its elegant treatment of complex topics.

In its emphasis on quantification and integration of process and form, *Fluvial Processes in Geomorphology* is the first modern textbook of fluvial geomorphology, and many of the emphases in this book continue to define the research directions of the discipline in the twenty-first century.

Dr. Ellen Wohl
Department of Geosciences
Colorado State University
Fort Collins, Colorado

REFERENCES

Bagnold, R. A. *The Physics of Blown Sand and Desert Dunes*. Mineola, New York: Dover Publications, Inc., 2005.

Chamberlin, T. C., and R. D. Salisbury. *Geology: Geological Processes and Their Results*. 2nd ed., Vol. I. New York: Henry Holt & Co., 1905.

Davis, W. M. "The Rivers and Valleys of Pennsylvania," reprinted from *National Geographic Magazine*, 1889. 71 pp.

Davis, W. M. *Geographical Essays*. Boston: Ginn & Co., 1909.

Davis, W. M., and W. H. Snyder. *Physical Geography*. Boston: Ginn & Co., 1898.

Dunne, T., and L. B. Leopold. *Water in Environmental Planning*. San Francisco: W. H. Freeman, 1978. 818 pp.

Geikie, A. *Text-Book of Geology*. New York: P. F. Collier and Son, 1901.

Gilbert, G. K. "Report on the Geology of the Henry Mountains." Washington, DC: U.S. Geographical and Geological Survey of the Rocky Mountain Region, 1877.

Gilbert, G. K. *Lake Bonneville*. Washington, DC: U.S. Geological Survey Monograph, Vol. 1, 1890.

Gilbert, G. K. "Transportation of debris by running water." U.S. Geological Survey Professional Paper 86, 1914.

Hack, J. T. "Studies of longitudinal profiles in Virginia and Maryland." U.S. Geological Survey Professional Paper, 294-B, 1957.

Horton, R. E. Erosional development of streams and their drainage basins, hydrophysical approach to quantitative morphology. Geological Society of America Bulletin, 56: 275–370, 1945.

Jenny, H. *Factors of Soil Formation*. New York: McGraw-Hill, 1941. 281 pp.

LeConte, J. H. *Elements of Geology*. New York: D. Appleton & Co., 1878.

Leopold, L. B. *Geomorphology: a sliver off the corpus of science*. Annual Reviews of Earth and Planetary Sciences, 32: 1–12, 2004.

Leopold, L. B., and T. Maddock. "The hydraulic geometry of stream channels and some physiographic implications." U.S. Geological Survey Professional Paper 252, 1953.

Leopold, L. B., and M. G. Wolman. *River channel patterns, braided, meandering, and straight.* U.S. Geological Survey Professional Paper 282, 1957.

Mackin, J. H. "Concept of the graded river." Geological Society of America Bulletin 59: 463–512, 1948.

Rubey, W. W. 1938. "The force required to move particles on a stream bed." U.S. Geological Survey Professional Paper 189-E, 1938.

Russell, I. C. *Rivers of North America.* New York: G.P. Putnam's Sons, 1898.

Russell, I. C. *River Development.* London: John Murray, 1909.

Scatena, F. N, and R. D. Varrin. *Fluvial processes in geomorphology and environmental management:* The 2006 Benjamin Franklin Medal in Earth and Environmental Science awarded to Luna B. Leopold and M. Gordon Wolman. Journal of the Franklin Institute 347: 688–697, 2010.

Scheidegger, A. E. *Theoretical Geomorphology.* Berlin: Springer-Verlag, 1961. 320 pp.

Schumm, S. A. "Evolution of drainage systems and slopes in badlands at Perth Amboy, New Jersey." Geological Society of America Bulletin 67: 597–646, 1956.

Thornbury, W. D. *Principles of Geomorphology.* New York: John Wiley & Sons, 1954. 618 pp.

Wolman, M. G. "A cycle of sedimentation and erosion in urban river channels." Geografiska Annaler 49A: 385–395, 1967.

Wolman, M. G., and L. B. Leopold. *River flood plains: Some observations on their formation.* U.S. Geological Survey Water Supply Paper 282 C, 1957.

Preface

THIS BOOK deals primarily with landform development under processes associated with running water. The bias of the book is dictated by our experience and interest and also by our belief that there is a great need at the present time for a review of geomorphic processes.

We have emphasized those things which are best known to us (and about which we feel most is known). Many subjects we have included are by no means treated completely, for they are discussed only from one viewpoint. Others that are treated lightly, such as the evolution of slopes, are ones for which little comprehensive quantitative data are available.

Rather than present a mere rehash of published material which we could not adequately discuss, we decided to omit entirely subjects we have not studied ourselves in the field or laboratory. Some summary monographs are available for wind, shore, and glacial processes, and we have not attempted to cover those subjects here. Combining process and stratigraphy for wind, shore, and glacial morphology would only have enlarged this book to unmanageable bulk; and, as Penck argued many years ago, a case can be made for the thesis that river and hillslope processes provide the central theme of geomorphology.

Our emphasis on process is not intended to minimize the importance of the historical aspects of geomorphology. Unfortunately, because of the limited understanding of geomorphic processes and their associated landforms, we ourselves are unable at present to make a truly satisfactory translation from the dynamics of process to historical interpretation. Better future understanding of the relation of process and form will hopefully contribute to, not detract from, historical geomorphology.

Despite its omissions, we hope that our treatment of geomorphology in this book will provide a logical framework for the subject as a whole, within which students and other readers can integrate material appropriate to their own interests or local physiographic environments.

We have sorely missed our compadre and co-author, John Preston Miller, during the last two years when this book was actively being constructed. Those portions which he prepared were perforce revised during that time. We hope that his principal ideas have been retained and that we have not allowed either divergent viewpoints or errors to creep into his work. Though we can put a book together without him, we can not view the high mountains nor can we pitch a camp in just the same spirit as when he was along.

We are indebted to colleagues and friends too numerous to name who helped in a variety of ways—in technical review of portions of the manuscript, in furnishing data and information, in preparation of copy and illustrations, and in our field work. But some should be noted specifically.

First, Mae E. Thiesen, although this is not the first manuscript which she has prepared for us. It is a pleasure to be able here to acknowledge her thoughtful and untiring help in all aspects of manuscript preparation, without which this book would not have been brought to completion.

We are particularly indebted to A. O. Woodford and James Gilluly for their overall review, and to Ralph A. Bagnold, Ivan K. Barnes, John T. Hack, Meyer Rubin, and Estella B. Leopold for their suggestions on portions of the work.

To the other river boys, William W. Emmett and Robert M. Myrick, our thanks not only for help in the field and in preparing the manuscript, but for their company at many delightful campfires beside many distant rivers.

Particular thanks we owe to William H. Freeman, who encouraged us when we most needed such encouragement.

And finally, we wish to mention two men who long have been close friends, admired colleagues, and friendly advisors, Walter B. Langbein and Thomas Maddock, Jr., whose influence on this work has been perhaps deeper and more significant than that of any others.

LUNA B. LEOPOLD

M. GORDON WOLMAN

May 1963

Contents

PART III. THE EFFECTS OF TIME

Chapter 9. Geochronology   389

Chapter 10. Drainage Pattern Evolution   411

Chapter 11. Channel Changes with Time   433

Part I

THE EVOLVING
LANDSCAPE

Chapter 1   The Changing Scene

When a man makes a pilgrimage to the fields and woods of his boy-
hood, he does not expect to find the hills and mountains dissolved, or the
valleys moved. If other men have not torn up the land to build factories
and towns, he expects his children to see the hills and swales as his
forefathers saw them. And he is almost right. Probably neither he nor
the children will ever notice that in fifty years the surface of the ground
has been lowered perhaps a fraction of an inch. Why should they? But
they might not be surprised to find that the old mill pond behind the
dam is now more mud than water.

Under the action of the force of gravity the land surface is sculptured
by water, wind, and ice. This sculpturing produces the landforms with
which geomorphology is concerned. Some of these forms owe their origins purely to denudational processes; other forms may be depositional;
still others owe their existence to combinations of both processes.

A picture of the dynamics of the earth's surface is by no means complete, however, if only gradation or leveling is considered. Clearly, if
there were no counteracting forces we should expect that the land surface,
given sufficient time, would be continuously reduced. Eventually, little
or no relief would remain. Geologic history demonstrates, however, that
the degradational forces acting on the earth's surface are opposed by
constructional forces. These internal, or endogenous, forces cause the land
to rise, and as they do so it is subjected to attack by the external, or
exogenous, agents. Geomorphology is primarily concerned with the exogenous processes as they mold the surface of the earth, but the internal
forces cannot be disregarded when one considers fundamental concepts
of the origin and development of landforms.

Ideally, the basic principles underlying the development of landforms
can be considered in simple terms. A given land area is composed of a
particular set of rocks, which have particular chemical and mineralogic
compositions and specific physical properties. Because these rocks were

formed at different temperatures and pressures within the earth, when they are exposed at the surface they are no longer in equilibrium with their environment and thus begin to decompose. Where a gradient is created by gravity, the moving water, earth, air, and ice help in the attack upon the rock and remove the products of weathering. In the process, landforms of various aspects are created. In a given environment the physical and chemical constitution of the rocks determines the way in which they will break down and, in turn, the size and quantity of debris made available to the denudational agencies.

Each denudational agent, depending upon its density, gradient, and mass at a particular place, is capable of applying a given stress on the materials available. A certain amount of work may be performed by the application of this stress, and the results of this work are the landforms that we see developed in various parts of the world. In a given climatic and vegetational environment the shape or form of the landscape will vary, depending upon the character of the rock and the type and available stress of the erosional agents. But as the land surface is reduced—so long as the products of weathering and the applied stress remain constant—the form of the land should remain the same.

If one were able to evaluate properly the properties of the rocks and the present and past capabilities of the denudational agencies, he should have no trouble in developing a rational, even mathematical, equation capable of describing the development history and equilibrium form of any landscape. William Morris Davis said essentially the same thing in 1902 when he observed that any landform is a function of the structure of the rocks (including their composition and structural attitude), the processes acting upon them, and the time over which these processes have been active. Only as we study the interrelations of these three factors are we able to discern which combinations produce which particular landforms and how they do so.

Some landforms, such as volcanoes, which may have been unaffected by denudational processes, may be considered purely constructional forms. As soon, however, as they are modified by external agencies, their form begins to represent the resultant of an interaction between the constructional forces, the rock substrate, and the applied stress.

The application of such an ideal concept to any actual landform at the present time is fraught with problems. The natural world is highly variable and the mechanics of uplift, weathering, and erosion are for the most part poorly understood. As will be seen, climate itself is a complex factor, and in most regions of the world inorganic processes are inseparable from the complex organic processes carried on by plants and animals. Although it is frequently convenient and helpful to construct a simplified synthetic picture of the natural environment, we should not lose sight of the fact that a given landscape must be the result of a complex set of factors which encompass the behavior of materials and processes over varying periods of time.

It is important to note that whether one refers to the effect on landforms of different rock types, or to the effect of different rates of uplift, such differences or changes must manifest themselves in the environment of the landform in simple physical terms. A normal fault whose strike is perpendicular to the direction of flow of a river, with downthrown block in the downstream direction, constitutes to the river a merely local increase in gradient. A similar increase in gradient might be effected by local changes in lithology, an abrupt shortening in channel length, or by an abrupt change in discharge downstream. The same physical principles determine the river's subsequent response in each case. The permanence or impermanence of the change, as well as its possible propagation either upstream or down, will depend upon the type and amount of material available and the distribution and quantity of flow. Any true principle enunciated to explain one of the cases must be applicable to the others as well.

Thus, although the application of the principle to any one example may be fraught with difficulty, an understanding of the principle at least reduces the burden of innumerable "unique" cases. Geomorphologists have always sought such unifying concepts, and for a proper view of the field as a whole one must turn initially to the classical concepts of landform evolution.

The influence of William Morris Davis on geomorphology was without doubt greater and longer-lasting than that of any other individual. His major contribution was a genetic system of landform description. Beginning in 1899, Davis developed the concept that during erosion of a highland the landscape evolves systematically through distinctive stages, to which he gave the names, youth, maturity, and old age. This entire sequence of stages he called an erosion cycle (or geomorphic cycle), and the end product was supposed to be a surface of low relief, or peneplain. He elaborated the effects of interruptions in the cycle and argued that the principal factors controlling the character of landforms are geologic structure, geomorphic processes, and the stage of development. Davis' genetic concept of landform development was a brilliant synthesis, which grew directly out of the work by Powell, Gilbert, and Dutton and also from the controversial ideas on organic evolution which were prevalent at the time.

The concept of the erosion cycle was never accepted in Europe to the same degree as in North America. The most serious challenge came during the 1920's from Walther Penck, who attempted to show a direct causal relation between tectonics and the properties of landforms. Many of his conclusions about the trends and ultimate results of tectonics and erosion processes differed only slightly from those of Davis. Penck, however, emphasized slope development, and his theory of slope development is a major contribution that is still being tested and debated.

The principal alternative to the Davisian conception differs mainly in the view of the effect of time, the third of the three fundamental elements, on landforms. Restating and extending the work of Gilbert, Hack (1960) emphasizes the concept of a dynamic equilibrium in the landscape which is quickly established and which responds to changes that occur during the passage of time. This view postulates that there is at all times an approximate balance between work done and imposed load and that as the landscape is lowered by erosion and solution, or is uplifted, or as processes alter with changing climate, adjustments occur that maintain this approximate balance.

More will be said about these different views in subsequent chapters, as various aspects of the landscape are considered in greater detail.

Paralleling developments in other phases of geology, the past decade has witnessed a remarkable increase in the application of analytical and experimental techniques to geomorphic problems. These investigations have taken two principal directions: (1) efforts to describe landforms more precisely through the use of statistics and other analytical techniques, (2) application of physical and chemical principles to field and laboratory studies of geomorphic processes. Although a few geologists—G. K. Gilbert, and later W. W. Rubey—helped to pave the way for this current trend, developments in other fields of science, especially in engineering and physics, were more directly responsible for it. One outstanding example is the field and experimental work on sand transport by R. A. Bagnold during the 1930's. Another is the contribution of fundamental ideas on the development of stream networks by R. E. Horton. Recently many developments in hydraulics and in the application of soil mechanics have attracted the attention of geomorphologists. At present there is greatly increased interest in the use of more precise tools for studying landforms. The pace of research seems to be quickening and there is reason to hope that a new era of discovery is under way.

Geomorphology in North America has gone through a phase during which extensive description of the landscape in terms of the erosion cycle has been carried out. It was apparently believed that the processes were known or could be inferred, and that form could be assessed by eye.

Similarly, one current earth-history view of geomorphology assumes that enough is now known to interpret landforms and deposits in terms of processes that operated in times past. In the most qualitative way this is probably true. However, we believe that the genetic system breaks down when it is subjected to close scrutiny involving quantitative data. At present deductions are subject to considerable doubt, for the detailed properties of landform have not been studied carefully enough and the fundamental aspects of most geomorphic processes are still poorly understood. So long as this is true, the interpretation of geomorphic history rests on an exceedingly unstable base.

Accordingly, we plan to concentrate on geomorphic processes. The emphasis is primarily upon river and slope processes; river processes will receive greatest attention, since the greatest volume of information available is on rivers. Our objective is to synthesize the material on these subjects in an attempt to assess the current status of knowledge and at the same time to draw attention to its shortcomings.

Process implies mechanics—that is, the explanation of the inner workings of a process through the application of physical and chemical principles. We realize that some readers may be more interested in descriptions of landforms than in the detailed analysis of the processes that formed them. So far as possible, we attempt to relate the processes discussed to specific types of landforms. Unfortunately, the gap between our understanding of specific processes in microcosm and the explanation of major large-scale landforms is still wide. It is interesting to note that geomorphologists seem to have a better understanding of depositional than of erosional forms. This may be because the formation of depositional features such as sand dunes, deltas, and flood plains is more easily seen in the field, or because many erosional features retain less clear evidence of their mode of formation.

Detailed understanding of geomorphic processes is not a substitute for the application of basic geologic and stratigraphic principles. Rather, such understanding should help to narrow the range of possible hypotheses applicable to the explanation of different geomorphic forms and surficial earth processes and deposits.

Our approach involves some use of mathematics. We are aware that the feelings of professional geomorphologists about numbers, graphs, and formulas range from acceptance and enthusiasm to bewilderment and forthright hostility. We have not gone out of our way to be mathematical, but wherever we felt that mathematics contributed either clarity or brevity to the discussion, it has been used. Some fundamental principles of mechanics and statistics are introduced in the text where they are appropriate and necessary to an understanding of the subject at hand. Because fundamental principles of geomorphology are drawn from both mechanics and geology, some readers—depending on their backgrounds—will find specific explanations oversimplified to suit their taste, while others will find the same material wanting in simplicity. Although we have attempted to achieve balance in this regard, the wide spectrum of readers' interests and background in the subject suggests that a perfectly happy medium is not likely to be attained at this time.

With those readers who have a conditioned reflex against "quantitative geomorphology" we agree that numerical descriptions can be used to give misleading and even erroneous impressions of erudition. However, the fact remains that one's senses, especially sight, when coupled with a conscious or unconscious bias, sometimes play strange tricks. Thus, a property which seems perfectly apparent, or an "obvious" relation of cause and effect, may upon careful measurement and analysis prove to be exactly the reverse of the "apparent" or the "obvious." Some examples will be cited in the text. From a scientific standpoint, most students agree that numerical data are superior to subjective adjectives—such as big, little, high, low, steep, and gentle—in objective analyses and comparisons.

We recognize that the decision to concern ourselves primarily with the dynamics of processes has some serious pitfalls. The most critical is the fact that field investigations of modern process cannot be segregated completely from historical aspects of landform development. The same statement applies to geologic structure. Each element of the landscape has evolved through a long period to its present configuration, and this heritage doubtless influences the processes now acting upon it. Sequential observations, comparative studies with statistical controls, and perhaps scale models do, however, help to mitigate these problems.

Disclaimers to the contrary, a glance at the chapter headings will show the reader that the book as a whole is arranged according to classical geomorphic principles. Chapters 3 through 7 deal essentially with process, structure, and morphology. The evolutionary or developmental aspect of landforms is treated in Chapters 8 through 11, after the introduction of the concept of time and geochronology in Chapter 8. We hope that this separation will make clearer both the extent and limits of our understanding of surficial processes and landforms.

REFERENCES

Bagnold, R. A., 1941 (reprinted 1954), The physics of blown sand and desert dunes: Methuen Press, London, 256 pp.

Davis, W. M., 1909, Geographical essays: Ginn, New York, 777 pp.

Gilbert, G. K., 1877, Report on the geology of the Henry Mountains: *U. S. Geol. Survey,* Rocky Mtn. Region, Report, 160 pp.

Hack, J. T., 1960, Interpretation of erosional topography in humid-temperate regions: *Am. Journ. Sci.,* v. 258A, pp. 80–97.

Horton, R. E., 1945, Erosional development of streams and their drainage basins: hydrophysical approach to quantitative morphology: *Geol. Soc. Am. Bull.,* v. 56, no. 3, pp. 275–370.

Penck, W., 1922, Morphological analysis of land forms, ed. H. Czech and K. C. Boswell, 1953, Macmillan, London, 429 pp.

Powell, J. W., 1875, Exploration of the Colorado River of the West and its tributaries: Smithsonian Institution, Washington, D. C., 285 pp.

Rubey, W. W., 1933, Equilibrium conditions in debris-laden streams: 14th Annual Meeting, *Am. Geophys. Union Trans.,* pp. 479–505.

Chapter 2 Geomorphology and the Field Problem

Introduction

As in many of the natural sciences, it is difficult to assess the problems without an initial appreciation of the available field evidence. What does one see in nature? What requires explanation? Even with a good grasp of the tools of mathematics, chemistry, physics, or botany, it is not easy to frame fundamental problems in an understandable context unless one begins with a feeling for relations as observed in the natural setting. So we begin here, not with the tools nor even with the processes, but rather with some field observations.

A Mountain Block in a Semiarid Climate

On Highway 66 one may drive eastward in nearly a straight line from Albuquerque, New Mexico, across the Rio Grande Valley, into Tijeras Canyon of the Sandia Range, and up over the mountain. From the bridge over the Rio Grande a panoramic view reveals a flat trough sloping upward to the Sandia Mountains in the east. In the eight miles between the river and the mountain front is an extensive, treeless, sloping plain nearly smooth in appearance, which abuts sharply against the steep face of the mountain block, as seen in the photograph of Fig. 2-1. On the other side of the river, facing west, buff-colored treeless hills seem to rise in a series of stairsteps, toward a horizon not far distant.

In such an environment and in an arid climate, the landscape is open to view. Geological relations are but little obscured by vegetation, and the arid zones are areas of tension where differential climatic and geologic effects are likely to be prominent rather than subdued.

Figure 2-1.

West front of the Sandia Mountains, New Mexico, looking north across the broad, slightly dissected pediment separating the mountains from the Rio Grande, which is off the picture to the left. [Photo courtesy John Whiteside.]

On closer scrutiny the mountains appear to have a steep, nearly vertical face on the river side, but there is a gentle slope backward toward the east. They resemble somewhat an ordinary brick, half-buried in the dirt but tipped slightly from the horizontal (Fig. 2-2). The gently sloping principal face of the half-buried brick is covered with green—a forest consisting principally of pine, but near the top or northeast corner composed in part of aspen and spruce.

The long and narrow face of the tilted brick is steep, but not vertical, and from our vantage point the corresponding mountain face appears to be quite devoid of trees.

This general view alone is enough to pose several questions. Why does the steep face join so abruptly with a broad sloping plain? Why is the upper part of the steep mountain front devoid of trees, whereas at the same elevation the gentle back-slope is covered by forest?

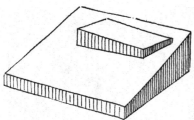

Figure 2-2.

An ordinary brick, half buried in the dirt but tipped slightly from the horizontal, has the general appearance of the fault block of the Sandia Mountains, New Mexico.

A closer look at the river channel and the adjoining valley floor raises additional questions. The channel is wide, sandy, and shallow. Water is confined to a mere trickle of reddish-brown fluid resembling a chocolate malted milk, not really thick, but giving the impression of being less fluid than water.

The banks of the river channel stand vertically, to a height of 3 to 5 feet. Whereas the channel bed is a mixture of fine sand and silt, with surface ripples and bars indicative of past flows, the banks are dark brown in color and composed principally of silt with some admixture of clay. Why this difference in material of bed and bank?

Stretching out on both sides of the nearly dry channel is a broad flat, covered with cottonwoods and geometrically partitioned into rectangular patches of irrigated farmland. From a more distant view the green of the valley floor is but a ribbon of verdure stretching carelessly down the center of the broad trough between mountains. The general picture is sketched in Fig. 2-3. Details could be supplied if one drove eastward toward and into the mountains.

Leaving the river channel, we can see the dark brown soil which supports the natural cottonwoods and is the basis for the irrigated farming. The flatness of this area makes it easy to till, but shallow depressions are white with efflorescence of salts, or what locally is called alkali. What is the origin and significance of this white powder on the surface?

Through the farmland, unlined irrigation ditches are paralleled by drainage ditches. In some parts of the valley floor slight depressions have an outline which suggests that the river channel formerly flowed there. Some of the depressions are closed lakes and appear to be loops of a former river channel.

Proceeding to the edge of the valley flat, we reach a nearly continuous escarpment over a hundred feet high (*E* in Fig. 2-3). The face of the escarpment exposes sand containing lenses of rounded gravel, and the same materials are repeated throughout the full height of the cliff except for the very top layer. There a white zone occurs where the cobbles are covered with a cementlike deposit, to a depth of nearly 30 feet. What is the relation of this coarse material bordering the valley and the fine grained material in the vicinity of the river?

The escarpment is carved by rills and V-shaped valleys of varying size that have notched the face a short distance back. All of these rills and valleys are ephemeral; that is, they are dry except during occasional heavy rainstorms. The larger of these minor rills are seen to be continuous with flat or dish-shaped swales on the sloping plain (*D* in Fig. 2-3). From the top of the escarpment we look down on the green ribbon of the valley floor, and then mountainward up a nearly featureless and treeless plain rising uniformly to the base of the main mass. Even the dish-shaped swales opening into the V notches in the valley-margin escarpment seem to disappear as we look toward the mountains.

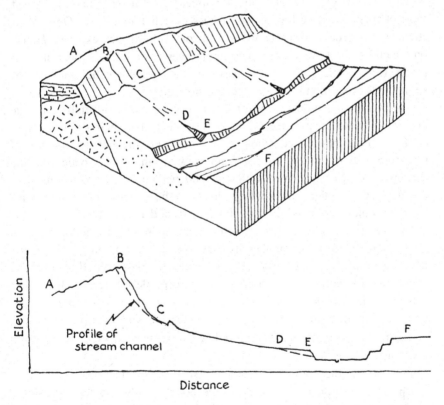

Figure 2-3.   *Block diagram and cross section, showing the main topographic features of the Sandia Mountains, New Mexico, and the Rio Grande.*

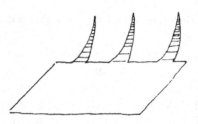

Figure 2-4.

Topographic relation of pediment and the tributary valleys in the adjacent mountain block.

We move upslope on this plain and approach the mountain front, which at close range thrusts itself skyward out of the plain. Immediately at the base of the bold cliff are a few hillocks standing isolated from the mountain mass, like icebergs broken off the main front of a glacier and surrounded by the ocean in which they float.

The mountain front turns out to be broken by deep notches separated by spurs with knifelike crests. At its base the mountain is granite. Each boulder appears weathered and soft. Indeed, most of them could be broken into pieces by a hammer blow. Only where streams have eroded away the weathered zone is hard rock exposed. Higher on the cliff face, however, are layer upon layer of dolomite, which in many places form vertical cliffs. The rocks in the valley notches and on the streambeds are a mixture of granite and dolomite.

The face of the mountain front generally tends to be a jumble of great and small boulders, perched on a slope so steep that their stability seems doubtful. Boulder-covered spurs look down into the narrow notches of the steep valleys which, hidden from view of the broad plain, contain rivulets of clear water lined with oak and walnut trees, shaded most of the day by the high surrounding slopes. Where does this water come from, in valleys bounded on all sides by dry and rocky cliffs?

Observe the striking difference in the manner in which the sloping plain at the base of the range makes its junction with the mountain spur and with the mountain valley. By foot or by car we pass smoothly from the plain into the valley notch without any perceptible change in slope. The valley gradient smoothly and gradually increases as it penetrates deeper into the mountain mass. It is as if the mountain valleys were extensions of the plain, each extension shaped like an arrowhead bent skyward toward the tip (Fig. 2-4).

In contrast, the spurs and the intervalley scarps meet the plain at a sharp angle, nearly as a wall meets the floor. So also do the side slopes of the isolated, often conical hills, which stand like sentinel towers protecting the approach to the main mass of a castle. Why does the mountain mass appear to thrust boldly out of the plain while the valleys seem to be a smooth headward extension of the nearly flat plain?

As higher elevations are attained, the granite is left behind, the road hugs the dolomite cliffs, and pine is replaced by spruce, with occasional patches of aspen. Where a turn of the valley leads the road in a direction truly perpendicular to the main long axis of the mountain block, rather than horizontal, bedding planes of the dolomite can be seen to be dipping eastward.

The dip of the dolomite strengthens the analogy that the mountain is like a partly buried brick whose broadest face is slightly tipping rather than horizontal. The canyon which the road follows finally tops out on this dipping face of the brick, on a broad sloping land covered with pine forest.

The flat floor of the mountain valley draws attention to still another feature, characteristic also of some of the tributary valleys. The flat floors are trenched by rectangular and chutelike arroyos which, though vertical walled, meander in serpentine curves from side to side down the valley (Fig. 2-5). The 20-foot walls of these trenches expose a reddish silt with lenses of gravel, and this type of material changes but little along the length of the trough. Downstream each of these arroyos decreases imperceptibly in depth. Walls 20 feet in height are characteristic of reaches upstream, but why are they only 2 or 3 feet high near the place where the valley opens out to the promountain plain.

Another characteristic of these arroyo trenches is that where they are deep, well within the mountain mass, the floor of the trench is often composed of bed rock. Thus the arroyos, where developed to maximum depth, apparently have eroded out the full depth of the fine-grained material making up the flat valley floor.

Supplementing the observed field evidence with data from wells drilled for water, let us consider briefly some possible interpretations of the evidence, as well as a few questions that arise from it.

Outcrops on the mountain front indicate that in this area the valley of the Rio Grande must be part of a block downthrown with respect to the mountain mass which was uplifted and tilted slightly to the east. The dolomite beds comprising the upper half of the moun-

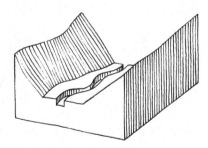

Figure 2-5.

A mountain valley in which the valley alluvium has been trenched by an arroyo.

tain appear to rest unconformably on the Precambrian granite and metamorphosed basement rocks.

Slickensides seen on some surfaces of the mountain front and on the
triangular faces of the intervalley spurs suggest that the mountain front
is part of an eroded fault plane, much of which may be below the plain
adjoining the front. The presence of the fault is also inferred from wells
drilled in the promountain plain at various distances from the mountain
front. Within a distance of about a half mile of the mountain front, wells
penetrate 30 to 100 feet of unconsolidated sand and gravel, below which
is rock, usually granite (*D* to *C*, Fig. 2-6). Water is usually obtained
from the basal gravel lying immediately on top of the rock.

Wells drilled only slightly farther from the mountain are known to
have penetrated hundreds of feet of sand and gravel and were, nearly
without exception, dry holes (*C* to *F*, Fig. 2-6). These relations suggest
a major fault line, *ACB,* roughly parallel to the mountain front and on
the average about half a mile from it. That the bedrock at shallow depth
is granite implies that the mountain front once stood half a mile more
toward the valley than its present position.

Since the time of upthrow of the mountain block, the steep face must
have progressively been worn back the half mile which now separates
it from the major fault. Furthermore, in the back-wearing or retreat of

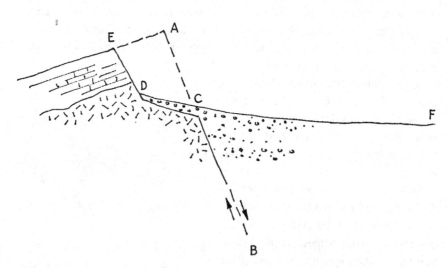

Figure 2-6.   *Diagrammatic cross section through the Sandia Mountains,*
                   New Mexico, showing fault plane and erosional scarp retreat.

the scarp, although less steep than the dip of the fault the present erosional scarp is still very steep.

Does the back-wearing slope attain a particular declivity and retain this angle of slope as it retreats? If so, would this also occur if the environment were a humid rather than an arid climate? Even to begin to answer this kind of question requires some knowledge of the processes involved—weathering, stream action, mass movement, and probably others.

In the retreat of the mountain scarp the thorough weathering of rocks and boulders at the surface prepares them for rapid breakup into small fragments when transported by gravity or water. The steepness of the scarp appears to be associated with the large boulders that weathering produces, and such steep zones have been referred to as "boulder-controlled slopes." When these large weathered boulders are sapped or undercut and begin to roll or wash down the steep slope, they smash into small fragments which can be transported even on a relatively flat slope. It has been suggested that this is a principal reason for the marked break in gradient between the mountain front and the alluvial plain.

Even if this were true, there are many other aspects to the process of slope retreat. What are the relative relations of rill development, valley cutting, and unrilled slopes to the problem of mountain-front retreat? What are the roles of mass movement, overland or sheet flow, gully erosion, and solution?

The scarp bordering the valley of the Rio Grande at E in Fig. 2-3 appears to be the result of a lowering of the channel of the river. Presumably at an earlier time the plain joined the river valley in a smooth unbroken curve when the river flowed approximately at the level of E (Fig. 2-3). Subsequently the regimen of the river was so changed that the river incised itself to its present level. This in turn caused tributaries to notch the resulting scarp, as at E. What were the relative roles of land warping, uplifting, and climatic change? What kind of a climatic change is required to alter the relation between rainfall, runoff, and sediment transport?

These effects are known only in a qualitative, not a quantitative sense. For example, it is not known how much the rainfall must change to alter significantly the rainfall-runoff relations. Nor is it clear what factors in climate would have to be changed. Is a certain change in the amount of summer rainfall more, or less, effective than an equal change in spring rainfall? Would a change in rainfall intensity without a change in the seasonal amount have the same effect?

This relatively detailed discussion of a single physiographic environment serves here only as an illustration of a large number of basic problems in geomorphology. Partial explanations of these problems can be offered, but more complete explanations require much more knowledge of processes than is presently available. Restated, a few of the important questions would be these:

1. What factors govern the longitudinal profile of an alluvial plain? What is the relative importance of discharge, particle size, frequency and magnitude of floods, vegetation, and infiltration losses?

2. What are the mechanisms by which diastrophic movements result in alteration of river slope and channel elevation? How do the adjustments of channel caused by diastrophism differ from those caused by changes in climatic factors? How will the evidence differ in either case? What climatic factors are controlling, and where and how are their effects felt?

3. When degradation of a channel occurs, what factors govern the way in which the tributary valleys react? How does the regrading of a rejuvenated tributary progress in time, and what controls the process of regrading? When is a local scarp or knickpoint preserved, and under what conditions is it eliminated?

4. What factors control the declivity and form of a scarp or hillslope? With the passage of time, through what evolutionary sequence will such a slope pass?

A Meandering River Cut into Bedrock

Any river system or terrain will pose similar problems. A briefer look at a classic physiographic region in the eastern United States will perhaps emphasize this point, while exposing some additional fundamental questions. In the present discussion small details are magnified to emphasize fundamental questions. A road map of Pennsylvania would serve as the basis for these physiographic observations; the basic features of a portion of the area are shown on the topographic map in Fig. 2-7.

The North Branch of the Susquehanna River rises in New York State north of Binghamton. Both it and a tributary to the west, the Chemung River, pass through areas of irregular hummocky topography and through valleys strewn with deposits of sand and gravel. The two join near the New York–Pennsylvania line and, after flowing south for about 20 miles to Towanda, Pennsylvania, the river flows in a southeasterly course to Scranton. In this stretch the course of the river consists of series of huge meandering loops that show clearly on a map. Each meander has a "wavelength" of about 3 to 4 miles and an "amplitude" of perhaps 3 miles.

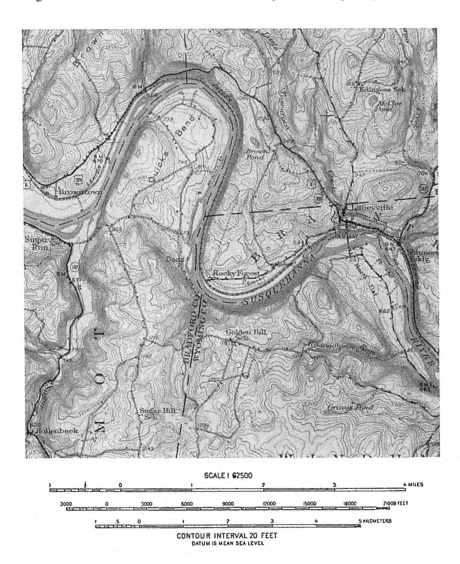

SCALE 1 62500

CONTOUR INTERVAL 20 FEET
DATUM IS MEAN SEA LEVEL

Figure 2-7.

Portion of Meshoppen Quadrangle, Pennsylvania, showing large meanders of the Susquehanna River, bordered by steep bedrock cliffs.

A map would not necessarily show that each meander is cut in bedrock. The regular meander pattern of the river traverses the grain or strike of the topography—here formed primarily of sandstone and shale formations. Although gravel deposits occur here and there along the highway, primarily on the insides of the meandering bends, the river throughout much of its course is actually flowing on bedrock.

Comparison with other streams of similar size flowing in alluvium reveals that the meander loops of the North Branch of the Susquehanna are much larger than those for a river of comparable size flowing in alluvium. Some of the gravel deposits contain cobbles and boulders indigenous to northern New England, a region now separated by a drainage divide from the Susquehanna drainage. These deposits, coupled with the lakes shown on the road map and with other deposits found in the area, provide clear evidence of glaciation.

The presence of the large meanders, however, poses several problems. First, how and when were such meanders initiated? Second, why do they appear to be unusually large? Third, how are they maintained as the stream crosses the bedrock? Was the meander form inherited from an earlier time when the river flowed at a higher level, perhaps on a depositional or erosional surface no longer evident in the present topography? Or do the large meanders suggest that at one time much larger volumes of water were carried by the Susquehanna? If so, did these larger rivers provide greater energy with which to mold the large bends in bedrock? Despite our increased understanding of meandering rivers, we can not give unequivocal answers.

To the west, in the vicinity of Williamsport, Pennsylvania, the road map shows the West Branch of the Susquehanna as it passes the city, flowing parallel to the strike of Bald Eagle Mountain, passing east around the mountain in a nose or loop and thence regaining a generally southward course. Elsewhere, however, the river does not appear to pass around the plunging mountain masses, but passes instead directly through the ridges in a series of impressive gorges. Farther south the Susquehanna flows south and then southeast and in the vicinity of Harrisburg, Pennsylvania, passes successively through one resistant bedrock ridge after another. These ridges trend roughly northeast-southwest, or at right angles to the course of the river, and most are capped by resistant quartzite or sandstone. Even the road map shows names classic in the geomorphic literature—Kittatinny Mountain, Blue Mountain, and Tuscarora Mountain. Between the ridge crests, in limestone and shale valleys, the streams tributary to the Susquehanna flow in valleys parallel to the ridge crests.

Field observations and profiles indicate that many of the ridge crests are of relatively uniform elevation. Occasionally the crests of the ridges are broken by shallow notches or depressions, which in some cases appear to be aligned with each other from ridge to ridge and with portions of the courses of the rivers that now flow far below them in the valley. The notches themselves and their alignment suggest the courses of former rivers. Some investigators have reasoned that the aligned notches and the even crests of the ridges represent remnants of a much higher surface over which the ancestral Susquehanna may once have passed. Thus it has been postulated that this ancestral river gradually eroded a course athwart the resistant mountains.

But evidence of this former surface exists only in the resistant ridge crests. A reasonable alternative hypothesis is that the apparent evenness of the crests, as well as their elevation, occurs because they are composed of resistant sandstones and quartzites. Structural weaknesses and progressive downcutting and adjustment to stratigraphic and lithologic controls must then be called upon to explain the unexpected course of the river. Neither hypothesis is as yet fully explained or validated.

Throughout the river system, flat bench lands are evident at irregular intervals along the valleys. In some places these benches or surfaces are underlain by bedrock thinly veneered with gravel. At others the surfaces are quite dissected by tributary streams, and some display characteristic soil profiles. The number of bench levels also varies from place to place. Although it is evident that these are alluvial terraces, neither their correlation with climatic and glacial history nor their relation to the bedrock and to the large meanders is clear.

Massive piles of rubble and debris flank the resistant ridges. Some of these, as on the north face of Bald Eagle Mountain, contain boulders 5 to 10 feet in diameter, resting roughly at the angle of repose against the mountain front. Occasionally large boulders are found in tonguelike pattern extending outward from the mountain into the valley floor below. The boulders in these tongues do not appear to be moving today. But the slopes of debris at steeper angles appear to break or slide down the surface occasionally, spilling onto highways and railroads. What then is the origin of these diverse deposits? Are the processes which created them still active today or are these perhaps relics of an earlier age?

One would logically associate a different climate with the evidence of glaciation noted in the Susquehanna drainage basin. Are the deposits and rubble features attributable to colder and perhaps wetter climatic conditions, or to frost action which might have produced more debris and a more favorable environment for the mass movement of materials on slopes? How different from today was such a climate?

Here then are a host of basic geomorphic questions readily discernible in an entirely different environment from our previous example of a mountain block in a semiarid climate.

Benches along a Sea Coast

Along the coast of California yet another set of physiographic features are evident to the traveler. In places, mountain ranges along the coast plunge directly into the sea. Elsewhere, 30 to 40 feet above sea level, flat benches can be seen stretching inland for hundreds of yards, ending abruptly against steeply rising cliffs. In places these benches or platforms are covered with a thin veneer of gravel. In others the gravel is virtually absent, and it can be seen that the surfaces are not flat but are gently beveled or concave. Because they are along the coast, one would readily assume that these platforms are in some way related to the sea. They are, however, above the range of present waves—even storm waves. What level of sea do they represent?

In some locations rivers from the interior passing through the mountain ranges debouch into the ocean in the vicinity of these platforms. At the elevation of the ocean platforms, remnants of benches paralleling the course of the river inland may be observed. What is the relation of these terraces or benches associated with the river to those facing the sea? Are they of simultaneous origin? If so, has the river been affected by the change in sea level? Or has the drainage basin of the river been altered perhaps by climatic conditions associated with the changing sea level? If no glacier existed in the drainage basin, under what conditions would the climate have caused the river to deposit its load at one time and at a later time to cut through the deposited material, leaving depositional remnants along the margin of the valley?

There is of course a difference between understanding the mechanisms of erosional processes and recognizing the evidence of such processes in the field. Our three somewhat rhetorical examples do no more than call attention to some of the problems of geomorphology and to the way in which they manifest themselves in the field. Recognition of such problems is at least a first step toward their solution.

In subsequent chapters we deal with various aspects of comparable problems in more detail. In some instances even informative and detailed analyses do not yet provide the critical translation from analytical and laboratory studies to field evidence. This may be because the field evidence is often subject to several possible interpretations. Plural processes can produce similar evidence. It is therefore usually impossible to explain all the details of the formation of many features of the landscape, at least to the satisfaction of those who desire mathematical certainty. But this level of uncertainty should not blind us to the fact that the mechanics of erosion and deposition and the correlation of existing landforms with the processes which appear to be forming them are currently both inadequately described and poorly understood. Before it is possible to unravel the large and admittedly complex geomorphic questions, further theoretical and descriptive studies of the processes themselves are necessary. They should be coupled with the study of the stratigraphic and erosional evidence which such processes leave on the landscape.

REFERENCE

Reiche. P.. 1949, Geology of the Manzanita and North Manzano Mountains, New Mexico: *Geol. Soc. Am. Bull.,* v. 60, pp. 1183–1212.

Part II

PROCESS
AND FORM

Chapter 3 Climate and Denudational Processes

Introduction: Processes of Upbuilding and of Downwasting

In discussing the origin of the earth's surface features, Griggs (1954) computed that if the forces of erosion and hydrostatic buoyance alone were operative, the earth as a whole would be eroded to a featureless plain near sea level in a period of less than 100 million years. The buoyant force, it will be recalled, results from the fact that the continental land masses float upon the denser mantle below. The average density of the granitic crust is approximately 2.84 g/cc, while that of the basaltic mantle is approximately 4.93 g/cc. Griggs' calculation was based upon the assumption that over the surface of the earth as a whole the time constant of erosion is about 10 million years, while that of the buoyant rise is closer to 1 million years. The time constant is inversely proportional to the rate of denudation. Thus, if we assume that the average worldwide rate of erosion is on the order of, say 2.7 cm or about 1 inch per thousand years (see Table 3-1), the average land elevation of the continents (840 m) could be reduced to sea level in about 31 million years. If, however, for each centimeter removed, buoyant forces cause the land surface to rise 0.42 cm, then the total time required for reduction of the land to a featureless plain is closer to 1.42 times 31 million, or 44 million years.

This computation, of course, is only approximate, but it gives some idea of the magnitude of the forces of erosion involved in denudation. At the same time, one is aware that rates of erosion are highly variable from place to place and—from the available evidence—may vary in time as

Table 3-1.   Past and present rates of denudation. [After Menard (1961).]

1	2	3	4	5	6	7	8
Region	Area Denuded (10⁶ km²)	Time[a] (10⁶ yr.)	Volume Deposits (10⁶ km³)	Denuda-tion (km)	Past Rate (cm/10³ yr)	Present Rate (cm/10³ yr)	Ratio: col. 6/col. 7
Appalachian	1.0	125	7.8	7.8	6.2	1.1[d]	5.6
Mississippi	{ 3.2 (1.6)[b]	150	11.1	6.9	4.6	4.6[c]	1.0
Himalaya	1.0	40	8.5	8.5	21	100	0.2
Rocky Mt. (L. Cret.)[e]	0.8	25	0.6	0.7	3	—	—
Rocky Mt. (U. Cret.)[e]	0.4	40	2.2	4.8	12–20	—	—

[a] Geological Names Committee, 1958, U. S. Geol. Survey.
[b] Area denuded in past.   [c] Gilluly (1949).
[d] Suspended load of rivers + 33%.   [e] Suspended load of rivers + 10%.

well (Table 3-1). Spatial variations even over large areas, however, are great; as we would expect, regions of high relief, steep slopes, and abundant moisture shed sediment much more rapidly than lower-lying plains and hills (Table 3-2). These differences will be reflected not only in the character of the sediments brought down from the upland and deposited in basins, but also in the landforms themselves in the source area, the region of transport, and the subareal margins of the depositional basins. It is this association that provides the tie between geomorphology and sedimentation.

Reconstruction of the environment of the source areas of sediments constitutes an attempt to reconstruct the landforms and processes that produce the sediments. To date, such reconstructions—based on broad principles such as tectonic uplift, subsidence, available slope or energy, and distance of transport—state that where uplift and/or subsidence is rapid, sediments are large and poorly sorted. With increasing distance of transport, sorting is likely to improve. Climate is assumed to be a function of uplift or relief. Obviously, the precise forms of the landscape can only be conjectured.

It is clear, however, that different climatic conditions, even where relief is the same, may produce different weathering products. These in turn may be transported by different agents and physically altered or

Table 3-2.   Relative rates of denudation in uplands and lowlands and in different climates. Figures are not of uniform quality due to limited sample and variable estimates of contribution of suspended load, bed load, and dissolved load o rivers. [After Corbel (1959).]

Physiographic Environment	Estimated Rate of Denudation (cm/1000 yrs)
LOWLANDS: Slope ≤ 0.001	
Climate with cold winter	2.9
Intermediate maritime climate (Lower Rhine, Seine)	2.7
Hot-dry climate (Mediterranean, New Mexico)	1.2
Hot-moist climate with dry season	3.2
Equatorial climate (dense rain forest)	2.2
MOUNTAINS: Slope ≥ 0.01	
Semihumid periglacial climate	60
Extreme nival climate (Southeastern Alaska)	80
Climate of Mediterranean high mountain chains	45
Hot-dry climate (Southwestern United States, Tunisia)	18
Hot-moist climate (Usumacinta)	92

redistributed by these agents to form aggregates or deposits having different properties, including textures and structures. One distinguishes between such deposits as alluvial, glacial, eolian, or turbidites, of course, and the distinguishing characteristics of some of them are discussed in the geologic literature. But association of process and product is still meager and much interpretive argument of the geologic record is based upon a relatively limited sampling of the modern environment in which process and deposit are related to one another. The record is even less satisfactory when it comes to relating process, landforms, and erosional products. In this chapter, we will attempt to classify the erosional processes, their possible relation to climatic controls and other governing factors, and their relative magnitudes or rates.

Characteristics and Classification of Processes

Denudation as a whole includes three separate parts: weathering or breakdown of rock materials, entrainment of weathered debris, and transportation and deposition of debris. The word erosion is often used synonymously with denudation, although erosion classically applies to entrainment and transportation and not to weathering.

Erosion and weathering may be thought of as the result of forces or stresses applied at the surface of the earth. At the point of application such stresses are opposed by internal resisting forces. Ideally, by considering the properties of the materials involved and their behavior under different stresses, one might derive the resultant landforms. In nature, however, the geologist is rarely concerned with only one process, and both materials and processes are complex. Nevertheless, following the outline proposed by Strahler, we can begin at least by classifying processes and materials according to their basic physical properties. In succeeding chapters, many of these processes are considered in more detail.

Table 3-3 divides the stresses into two kinds, gravitational and molecular. Chemical processes are grouped separately in the table, although these too produce stresses, as the discussion of weathering in Chapter 4 indicates. We will be concerned initially with large-scale geomorphic processes.

All erosional processes are driven by gravity. In some, such as streamflow, downhill creep, or downslope movement of glacial ice, the medium usually moves down a gradient from higher to lower elevation. In the case of the wind, solar heating has created differences in densities of masses of air, which tend to be equalized by air flow from regions of higher to lower pressure. Shore currents, if due to wave action, may be traced back to gravitational stresses through the winds responsible for their formation; if due to tides, the currents may be traced to gravity flow responding to differences in water level. These wind and shore processes produced indirectly by gravitational stresses are separated in Table 3-3 from those produced directly by gravitational stresses.

The materials of the earth can be classified in terms of their response to an applied shear stress. The behavior of each of these substances under an applied stress is given in Fig. 3-1: (1) elastic solids, such as crystalline rocks, in which the strain is directly proportional to shear stress (shown at **A**); (2) plastic solids, such as some soil materials, which deform by flowing when a limit or yield value is exceeded (**C**); (3) viscous or Newtonian fluids, such as water, which deform or flow upon application of a shear stress **B**).

If an elastic solid is deformed within the elastic limit, and the deforming stress is then released, the material returns to its original form. At the yield stress, it ceases to be elastic. Where the yield stress is exceeded, the elastic solid fails and shear or tension fractures are formed. The most appropriate geomorphic examples are rockfalls and landslides. In most

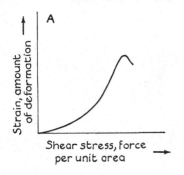

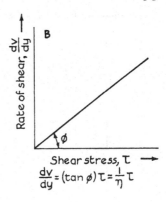

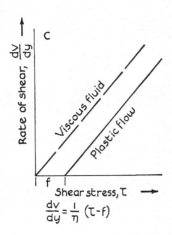

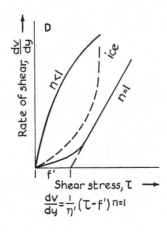

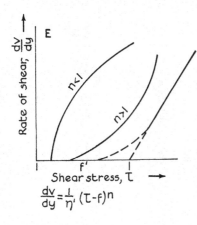

Figure 3-1.

Generalized relation of strain or rate of strain to applied stress. **A.** *Elastic solid.* **B.** *Viscous fluid.* **C.** *Plastic flow.* **D.** *Quasi-viscous flow.* **E.** *Quasi-plastic flow.*

Table 3-3.   The dynamic basis of geomorphology. [From Strahler (1952).]

A. GRAVITATIONAL STRESSES

MATERIAL INVOLVED IN MOTION	PROPERTIES OF MATERIAL	TYPE OF FAILURE (STRAIN)	GEOMORPHIC PROCESSES AND FORMS
1. Crystalline rocks, arenaceous rocks, limestones; dry soils	Rigid, elastic solid or "elastic continuum." Obeys Hooke's law: strain ∝ stress	Sudden rupture along shear surfaces or tension fractures	Landslides: slump, slide (compressional stress); rock-fall (tensional stress)
2. Glacial ice, near surface	Plastic solid in region of elastic behavior	Sudden rupture along shear surfaces or tension fractures	Crevassing, overthrusting, calving of glaciers
3. Argillaceous rocks and soils	Plastic solid in region of slow creep	Continuous, slow laminar flow (distributed shear)	Large-scale creep phenomena. Superficial deformation of clays
4. Unconsolidated rock, soil + water	Plastic solid in region of flowage. Obeys Binghams law: Rate of shear proportional to stress, above a yield limit	Plastic flow (shear between grains) when stress exceeds yield limit. Movement ceases below yield value. Flow laminar or turbulent	Solifluction, earth flow, mudflow, highly turbid streams, turbidity currents
5. Glacial ice under heavy load	Plastic or quasi-plastic solid. Nonlinear increase of shear rate with increase of stress	Plastic or quasi-viscous flow above "yield" limit	Continental and alpine glaciers. Erosion forms due to ice abrasion. Depositional forms molded by ice flow

6.	Water film on sloping surface	Newtonian fluid. No yield value. Linear increase of shear rate with stress. Subject to capillary influences	Laminar flow, ceasing when water thins below capillary control limit	Sheet runoff on slopes and rock surfaces. Slope reduction by removal of ions, colloids, clays. Fluting, grooving of limestone
7.	Water in permeable rock, soil. (No surface slope)	Newtonian fluid, subject to capillary influences	a) Silts: laminar flow following Darcy's law b) Sands: mixed laminar and turbulent flow c) Gravels: turbulent flow	Infiltration of precipitation, carrying down of ions, colloids, clays, silts (illuviation). General slope reduction. Karstic forms in highly soluble rocks
8.	Water layer on sloping surface	Newtonian fluid	Sheet runoff in turbulent or mixed turbulent-laminar flow	Slope erosion, transportation. Slope forms of fluvial drainage basins
9.	Water in sloping linear channel	Newtonian fluid	Stream flow, turbulent except in bed layer	Stream erosion, transportation, deposition. Drainage systems. All fluvial landforms

INDIRECT RESPONSES TO GRAVITATIONAL STRESSES

10.	Standing water bodies. Oceans, lakes	Newtonian fluid	Turbulent flow as a) Pulsating or oscillating currents caused by waves b) Tide-induced currents	Shoreline processes of erosion, transportation, deposition. Shoreline landforms: cliffs, benches, beaches, bars, spits
11.	Air	A gas: compressible fluid of extremely low viscosity	Turbulent flow induced by pressure gradients (gravitational stress on air masses)	Wind erosion, transportation, deposition. Deflational and abrasional forms. Dunes, loess

Table continues

Table 3-3 continued

B. MOLECULAR STRESSES

MATERIALS INVOLVED	PROPERTIES OF MATERIAL	STRESS AND CAUSE	KIND OF FAILURE	WEATHERING PROCESS AND FORM
1. Rock: strong, hard crystalline, glassy, or crystal aggregate	a) Elastic solid, nonhomogeneous	Shear stress due to nonuniform expansion-contraction in cyclic temperature changes	Rupture by shear or tension fractures between grains, along cleavages, joints, bedding planes	Granular or blocky disintegration of rocks, esp. coarse-grained crystalline rocks
	b) Elastic solid, homogeneous	Shear stress set up by thermal gradient from surface heating	Rupture between layers paralleling rock surface	Exfoliation of rock by fire, lightning; solar or atmospheric heating-cooling
2. a) Permeable rock + water	a) Elastic solid	a) Shear stress set up by interstitial ice crystal growth	a) Rupture between grains, cleavage pieces, joint blocks, beds	a) Frost disintegration of rocks. Felsenmeer
b) Clay soils	b) Plastic solid	b) Stress from growth of ice lenses, wedges	b) Plastic deformation of clays adjacent to ice	b) Heaving of clay soils, frost mounds, polygons
3. Permeable rock or soil + water and salts	Elastic solid or elastic continuum	Shear stress set up by interstitial growth of salt crystals	Rupture between grains, cleavage pieces, joint blocks, or beds	Efflorescence, granular disintegration in dry climates. Caliche heaving

	MATERIALS	PROCESS		FORMS PRODUCED
4.	Rock or soil + colloids and water	Shear stress set up by dilatation accompanying water adsorption and drying	a) Rupture between grains b) Plastic deformation of clays during swelling	Exfoliation of basaltic, granitic rock upon alteration of silicates. Slaking of shales, argillaceous ss
5.	Rock or soil + capillary water	Shear stress set up by dilatation accompanying changes in capillary film tension	Rupture between grains or masses of soil	Disintegration of granular permeable rocks. Heaving or subsidence of clays, silts
6.	Rock or soil + plant roots	Shear stress set up by swelling of rootlets under osmotic pressure	Rupture between grains, cleavage pieces, joint blocks, or beds	Disintegration of rock by prying of roots. Deformation of soils
7.	Strong, hard monolithic bedrock	Shear stresses of tectonic origin stored as elastic strain at depth	Rupture of rock on planes paralleling surfaces after release of confining pressure	Exfoliation of domes, slabs, shells. Quarry rupture, rock-burst

C. CHEMICAL PROCESSES

	MATERIALS	PROCESS	FORMS PRODUCED
1.	Soil, rock + acids, water	Reaction between acid ions and mineral surfaces. Removal of products in solution	Lowering of rock and soil surfaces. Pitting, cavitation of rocks, esp. carbonates. Cavern and karst forms. Weakening of bonds between mineral grains
2.	Soil, rock + water	Simple solution (ionization) of susceptible minerals	Cavitation of soluble salt formations. Slow attrition of exposed mineral surfaces

such failures the fracture is followed by rapid motion of the detached segment. Slumps, falls, and landslides are usually associated with particular rock types, and failures often occur along joint planes or planes of sedimentary discontinuities.

At the opposite end of the scale, a viscous fluid deforms continuously under an applied shear stress and the *rate* of shearing is directly proportional to the stress. In Fig. 3-1, **B,** the rate of shear is given by the expression dv/dy. The proportionality between shearing rate and shearing force or stress is the viscosity, v. Flowing water is of course the most ubiquitous geomorphic viscous fluid.

Between the elastic solid and the viscous fluid there is a complex range of materials, including many soils and glacial ice, which deform either as plastic solids, as pseudoplastic solids, or as pseudoviscous fluids. By definition, in plastic flow the rate of shear is proportional to applied stress after a yield value is exceeded. This relation (Fig. 3-1, **C**) is given by the equation

$$\frac{dv}{dy} = \frac{1}{v}(\tau - f),$$

where f is the yield limit. If the yield limit is not zero, then the flow is considered plastic. The curves of Fig. 3-1, **D** and **E,** show the behavior of a variety of possible materials. The curve for glacier ice is only approximate, but its form shows that the stress (rate of strain relation) is nonlinear. At low stresses ice appears to flow as a viscous material (Meier, 1960). Many surficial materials, such as soils containing silts and clays, deform as plastic solids; others undoubtedly have more complicated flow laws. In nature these materials that deform by various plastic or viscous flow laws often deform by creep, which usually involves slow movement of masses of material rather than rapid movement of detached segments. The latter is typically associated with the rupture of elastic solids. Mass movement on slopes, debris floods, avalanches, etc., take place in materials having highly variable properties subject to many influences.

The properties of the materials considered under gravitational stresses pass progressively from rigid rocks to flowing fluids, in descending order in the left column of Table 3-3. This order is one of descending "consistency," so to speak, or descending resistance to shear stress. Table 3-4 shows these "consistency" or viscosity differences for several types of geologic processes. By comparing, for example, landslides (1), mudflows (4), and sheetflows (8) in Table 3-3, one can see that the simple addition of water may alter the mode of failure and of movement.

Table 3-4.   Approximate viscosities of several geomorphic agents.

Geomorphic Agent	Viscosity (poises)[a]	Source
Glaciers	10^{13}–10^{14}	Sharp (1954)
Rock glaciers	2×10^{14}–8×10^{14}	Wahrhaftig and Cox (1959)
Mudflow	2×10^{3}–6×10^{3}	Sharp and Nobles (1953)
Solifluction	10^{3}	Hopkins and Sigafoos (1952)
Rivers (70°F)	10^{-2}	King (1954)

[a] 1 poise = 1 gram mass per centimeter second. 1 slug per foot second = 478.8 poises.

Because the processes behave according to different flow laws, a comparison based on consistency or viscosity is qualitative rather than quantitative, and is useful only as a guide. Nevertheless, the approximate viscosities of various geomorphic agents in Table 3-4 give at least the range of these materials.

It will be noted that the molecular stresses listed in Table 3-3 primarily concern what we have called the weathering process. On the other hand, the various flow phenomena may also involve molecular alterations—as, for example, in the flow of glaciers involving deformation of ice crystals along specific planes. These molecular and chemical phenomena are considered in the chapter on weathering.

Erodibility

We have described briefly the flow and yield properties of various geomorphic agents. Some of these agents, such as running water or glacial ice, are in a sense separate from or act upon the surface of the earth, whereas others, such as mass movements of the soil mantle, both act upon and form in themselves part of the landscape. In both cases particular forms of the landscape arise as a result of the interaction of the geomorphic agents and the available earth materials. The ease with which these earth materials give way is called erodibility. Erodibility is indirectly dependent upon susceptibility to weathering as well as susceptibility to removal and transportation. Given a certain condition of applied stress, erodibility may be measured in units of volume or mass per unit area per unit time. It often is used to note relative susceptibility to gullying or rill formation but in an ill-defined way. Although one readily refers to one type of rock as more erodible or "weaker" than another, these terms are vague because relatively little is actually known about how differences in rock material lead to distinctive landforms in different environments.

To date no adequate standard of erodibility has been developed, but a number of studies since 1930 have shown that permeability and the relative aggregating or binding properties of soil materials are factors of importance. André and Anderson (1961) suggested that erodibility would depend upon the ratio between the portion of the soil surfaces requiring binding—that is, the portion comprised of particles of fine sand and larger—and the quantity and quality of the binding silts and clays.

Such a ratio was found to be correlated with sediment yields from watersheds in Oregon where soils developed on acid igneous rocks were roughly 2.3 times as erodible as soils developed on basalt. Serpentine soils proved least erodible, only about three-fourths as erodible as basalt.

The amount of calcium and magnesium absorbed on the clays was also highly correlated with the surface-aggregation ratio—suggesting, as one would expect, that chemical as well as physical properties affect erodibility. Reinforcing the idea that the ratio of binding to nonbinding particles is significant, Woodburn and Kozachyn (1956) have also shown that the amount of erosion due to the splash of artificial rain increases as the percentage of sand increases, and decreases as the percentage of water-stable aggregates increases. The latter is defined as the percentage of a sample of the material which remains on a sieve of a particular mesh during a process of wet sieving.

Laboratory studies of the resistance to erosion of soils subject to the shear stress of flowing water have also shown that as the percentage of clay or aggregating materials increases, soils are less erodible (Smerdon and Beasley, 1959). A common engineering measure of soil properties, shear strength, also appears to be at least indirectly related to erodibility.

The shear strength of a soil is defined experimentally as

$$s = c + \sigma \tan \phi,$$

where s is the shear strength in units of force per unit area, c is the cohesion, σ the stress normal to the plane of shear, and ϕ is defined as the angle of shearing resistance. These quantities are shown on the diagram in Fig. 3-2. Presumably, factors tending to reduce cohesion or the resistance due to forces tending to hold the particles together, or the internal friction due to interlocking of the particles, should increase erodibility. Many factors in nature, such as the quantity of organic material and the clay mineral composition, influence the shear strength in a complex fashion (Grim, 1962, p. 233). But the effect of two simple parameters—particle-size distribution and water content—on shear strength is of some interest here. The effect of the particle-size distribution is felt in two ways. First, even a small admixture of clay size particles may markedly increase the cohesion of a given soil. Second, experiments have also shown that the internal frictional resistance increases as the range of particle size increases. In more familiar terms, the poorer the sorting, the greater the shear

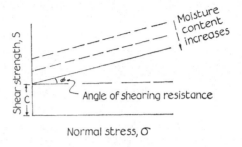

Normal stress, σ

Figure 3-2.

Empirical and generalized relation of shear strength to normal applied stress in a natural cohesive soil, as a function of moisture content.

strength. In addition, as Fig. 3-2 shows, shear strength of a given soil material decreases with increasing water content.

The above discussion has considered properties of surficial materials which appear to affect erodibility. In practice, it is the correlation of sediment yield with an index of erodibility that indicates the relation of erodibility to land sculpture. Erodibility describes a rate of action which must, in turn, be related to the development of specific landforms.

It is the weathered mantle that is subjected to what we have been calling erosive stresses. If, by virtue of its chemical and physical constitution, the weathered mantle has a low resistance to shear, a low value of cohesion, or a high surface aggregation, one should expect topography developed upon it to be relatively subdued compared to that on less erodible materials. Chorley (1959) has indeed shown a relationship between the structural properties of the weathered mantle and the gross topographic form of an area in England. These results, stated in terms of the properties of earth materials, are shown in Table 3-5.

Greater relief exists in areas underlain by sandstone formations than by the clays. The former has higher shear strength, as shown in the table. This high shear strength is correlated with poor sorting and relatively high density of the material when dry.

Although engineers in the field of soil mechanics have developed methods of analyzing the mechanical behavior of earth materials under some conditions, how these mechanisms produce particular landforms is less clear. Save in the case of some types of landslides, neither the mechanics of the processes nor the properties of the earth materials are as yet well

Table 3-5.   Topographic relief and shear strength of underlying soil
materials. [After Chorley (1959).]

Geologic Formation	Mean Shear Strength (lb/sq in)	Coefficient of Permeability (cm/sec)	Relief
Upper greensand	795	0.00084	Moderate
Shotover sand	685	.0043	Greatest
Kimmeridge clay	426	.0013	Lowland
Oxford clay	344	.00080	Lowland

enough understood to allow many useful applications to the task of explaining specific landforms.

Morphogenetic Regions

The preceding classification has emphasized primarily the constitution
and behavior of earth materials that are acted upon or themselves act
as agents in geomorphic processes. Theoretically, any or all of these processes could be active in any region at any time, but in reality earth processes are closely related to climate, the distribution of precipitation, wind,
and temperature. The existence of a glacier demands both precipitation
and freezing temperatures. Sand dunes would be impossible without wind,
even in the presence of discrete particles of sand of low resistance to shear
stress. Lakes require precipitation and storage, rivers require runoff, and
creep phenomena may require moisture.

It is clear that weathering as well as erosion may be vastly different
as climate differs. Frost action in cold moist climates may be important
in fracturing rocks; elsewhere, as in the humid tropics, increasing moisture
and higher temperature hasten chemical reactions and the relative importance of mechanical fracturing is reduced. The effect of a cold-moist
climate on a rock usually altered mostly by solution is dramatically illustrated by the photograph in Fig. 3-3.

By separating weathering from erosion and further dividing the weathering process into two categories, mechanical and chemical, probable regional distributions of geomorphic processes have been constructed by
several authors. The resulting regions in which particular geomorphic
processes are supposed to predominate are called morphogenetic regions.
Peltier, following Budel, has attempted such a classification, using mean
annual temperature and mean annual precipitation as indicators of cli-

mate. The absolute values developed in such a scheme are subject to great error. On the other hand, the rationale of such a classification provides a useful framework. Such a morphogenetic pattern is developed here, considering separately the two facets of weathering and three mechanisms of transport—wind, gravity movement or mass wasting, and running water.

The mean annual temperature and precipitation at a number or representative localities throughout the world, as well as a descriptive climatic classification according to Koppen, are shown in Fig. 3-4, **A** and **B.** These data provide some reference points for the hypothetical morphogenetic regions.

There appear to be two major requirements for intensive mechanical disintegration of rocks—alternating freeze and thaw, and the presence of moisture. A second factor of possible importance under some conditions is alternate heating and cooling, although the significance of this phenomenon has often been exaggerated.

A qualitative indication of the relative intensity of operation of these factors in different regions can be defined in terms of various combinations of mean annual temperature and precipitation. On a graph of annual temperature as a function of annual precipitation (Fig. 3-4, **C**),

Figure 3-3.

Weathering of limestone in cold moist climate. Upper section shows effects of solution, lower part is frost shattered; Somerset Island, Northwest Territories. [Photograph by J. Brian Bird.]

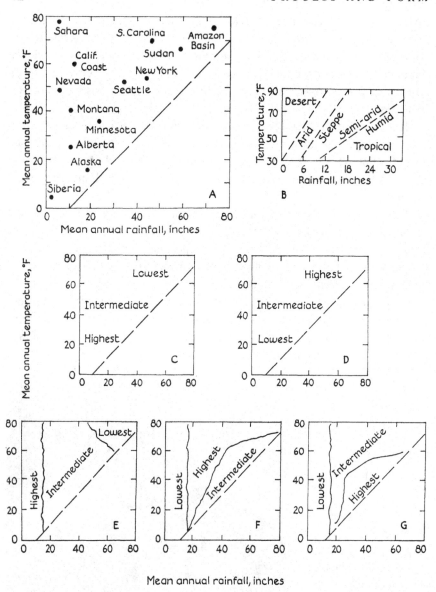

Figure 3-4.   *Relations of various factors in development of a generalized pattern of morphogenetic regions.* **A.** *Reference locations for climatic conditions.* **B.** *Köppen classification of climates.* **C–G.** *Hypothesized variations in intensity of different geomorphic processes in different climates.*

it is then possible to indicate areas on the graph (combinations of temperature and precipitation), other things being equal, where mechanical weathering is relatively enhanced or suppressed.

To emphasize the purely relative character of these designations the words "highest," "lowest," and "intermediate" are used to characterize each process; these words are placed appropriately on the graph.

In a similar fashion the principal effects of precipitation and temperature on chemical weathering can be outlined. It is known that the rate of chemical reactions increases with increasing temperatures, everything else remaining constant. On the diagram of Fig. 3-4, **D,** a gradient of increasing significance of chemical weathering should parallel the ordinate.

Many weathering reactions also involve hydrolysis or the combination of constituents of rocks with water. Water is the solvent within which all forms of ionic dissociation occur and is thus essential to the process of chemical weathering. In addition, increasing availability of water provides for progressive leaching or removal in solution of the products of weathering. One should expect, then, that chemical weathering would be strengthened by both high temperature and high precipitation and thus would be greatest in humid tropical regions and least in arid regions with progressive variation between. Thus Fig. 3-4, **D,** indicates a gradient of chemical weathering in the direction of increasing rainfall as well as increasing temperature.

Consider now the climatic factors governing the distribution of the wind as a geomorphic agent. Inasmuch as wind must blow to be effective, it might be reasoned that in the region of the doldrums on either side of the equator, corresponding for the most part to a region of relatively high temperature, wind action as a geomorphic agent would be at a minimum. Elsewhere the factors determining the effectiveness of the wind are essentially external to the wind itself. Assuming a high wind velocity, the wind's effectiveness should increase as the protective cover or resistance of the surface to erosion decreases. If then, the absence of vegetation promotes wind effectiveness, it should be expected that the regions of highest wind activity would lie in the arid lands of the poles and deserts. The trade wind belts play a major part in determining the location of major deserts. Grossly considered, they occupy a belt between latitudes 20 and 30 degrees in the northern and southern hemispheres. Putting these observations together, two regions of highest wind activity can be designated (Fig. 3-4, **E**) both of low precipitation but one a region of high temperature, the other a region of low temperature. Where both precipitation and temperature are high, vegetation should be great and wind action correspondingly inhibited. Such a locus is shown as a region of least wind activity in Fig. 3-4, **E.**

As seen in the brief review of earth materials and their behavior, the behavior of "viscous" fluids, which flow or creep under gravitational stresses, is influenced by moisture. Dry soil will not deform plastically and hence will not creep. Thus, mass movement related to climatic controls, as distinct from geologic, should be of most importance in wet regions. At the same time, the moisture must be available to the soil; hence, assuming a given quantity of precipitation, a relatively even distribution throughout the year rather than a concentration of heavy precipitation should favor mass wasting processes over those of surface runoff. Thus there might be expected some dominance of mass wasting as a geomorphic agent in wet regions having relatively even distributions of precipitation.

If freezing takes place first at the surface, the thickness of the frozen zone increases as moisture is drawn toward the surface. If the entire mass of soil freezes and remains frozen, it will have become an elastic solid or perhaps a plastic one, but in either case it will move only upon application of a rather high shear stress. In contrast, if the climate is one in which freeze and thaw alternate repeatedly, there should be a more favorable environment for effective mass movement. Such cold moist regions are often called periglacial. Strictly speaking, these are regions peripheral to glaciers, characterized as a rule by the presence of permanently frozen ground (permafrost). The term periglacial, however, is often applied to similar cold regions not necessarily peripheral to a glacier. Although melting of the surface may occur in summer, the soil at depth remains frozen. Both melting and the presence of the frozen substrate promote entrainment and transfer by mass movements. On the basis of such reasoning we may tentatively designate regions of high precipitation as ones within which wasting should be effective. Within this broad designation, areas of frequent freeze and thaw rather than those of extreme cold should favor mass movement; this combination is approximated in Fig. 3-4, **F.**

It has always been assumed, and probably with good reason, that running water is the most important of the earth's erosional agents. With the gradual abandonment of the notion that most desert topography is the result of wind action and the concomitant realization of the significance of running water in the denudation of arid regions, the areas of the earth in which running water can be asserted to be of little or no significance are few. Its relative effectiveness, however, can be analyzed. For runoff to occur at all, precipitation must exceed the sum of evaporation and transpiration by plants—the combination called evapotranspiration. Evaporation increases with temperature and wind velocity while transpiration increases with the amount of vegetation. Erosion by running water then should be at a minimum in dry regions. With increasing runoff in the absence of vegetation, erosion intensity should increase, but with increasing vegetation, transpiration will also increase. To the extent that increasing temperature furthers the growth of vegetation, it can be expected that above some threshold value of precipitation, runoff will decrease as temperature increases.

Considering these hypothetical gradients of relative runoff effectiveness as a function of precipitation and temperature, on the diagram in Fig. 3-4, **G** a region of relatively low precipitation and runoff might be considered to be the area of minimum effectiveness of running water, whereas the region of maximum effectiveness should coincide with an area of maximum runoff. At either extreme, however, vegetation itself directly influences erosion. The effectiveness of runoff is increased by the sparsity of vegetation in arid regions, as was the effectiveness of wind action. Similarly, where vegetation is dense, as every conservationist knows, erosion is inhibited. Thus the region of highest erosion intensity by running water presumably lies above zero runoff, below maximum precipitation, and displaced somewhat away from the hottest regions where evaporation is greatest.

To synthesize these major weathering and erosional processes into groups representing climatically controlled morphogenetic regions, the parts of Fig. 3-4 are superimposed one upon the other to develop Fig. 3-5. This

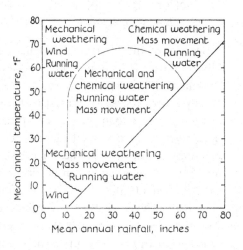

Figure 3-5.

Hypothetical morphogenetic regions drawn from reasoning expressed in Fig. 3-4. The possible relative importance of specific weathering and erosional processes is indicated by the order in which they are listed.

composite is a description of the probable coincidence of several geomorphic elements. From it one can say, if the foregoing reasoning coincides
with reality, that in midtemperate, humid regions the action of running
water is probably dominant in the sculpture of landforms. In cold-arid and
warm-arid regions, running water competes with wind action; in cold but
slightly moister regions mass movement may compete with both wind and
running water in molding the land surface.

So far, we are in a position to make only the most general statements
regarding *the relative magnitude of a given process in a given region*. Of
even greater significance, however, is the fact that in the absence of quantitative information, statements regarding *the relative importance of different processes in a given region* must be extremely guarded. The qualitative appraisal provides a useful framework. On the other hand, a truly
adequate definition of morphogenetic regions awaits the collection of adequate measurements. There are only a few processes which have been
studied in any detail and over broad areas. Several of these form the
basis for the following section.

Interaction of Vegetation, Runoff, and Sediment
Yield: An Example

Because of the obvious economic interest in rivers for drinking water,
industrial use, irrigation, and navigation, records of flow and sediment
movement in rivers and reservoirs are far more extensive than they are
for any other process of interest to the geomorphologist. Although records
are rarely longer than 50 years, and areal coverage is by no means complete, river discharge is sufficiently well documented to permit a tentative
quantitative statement of the relationships discussed in the preceding section. Such an illustration can be drawn from the recent work of Langbein
and Schumm (1958).

In the derivation of the hypothetical morphogenetic regions, the obvious fact was noted that if there is no runoff there will be no sediment
removed by runoff from the surface area drained by a river. In the absence of vegetation it should be expected further that, moving to regions
of successively higher precipitation and runoff, the sediment yield from
drainage basins in these regions will also increase progressively.

Comparison of maps of the distribution of vegetation with those showing the distribution of precipitation also suggests that the type of vegetation, and perhaps the quantity, varies with the quantity of precipitation.

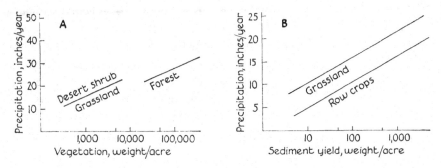

Figure 3-6.   *Quantitative relations between sediment yield and vegetative growth, respectively, and annual precipitation. [From Langbein and Schumm, 1958.]*

There are few data in which different types of plant cover are compared on a common base in terms of bulk per unit area. Nevertheless, Langbein and Schumm were able to construct the quantitative relation between annual precipitation and weight of vegetation shown in Fig. 3-6, **A.**

Vegetation bulk increases as some power of the annual precipitation as expressed by the equation

$$V \propto P^x,$$

where V is the weight of vegetation in pounds per acre, P is the annual precipitation in inches, and x is a number greater than 1. It will be noted that with increasing precipitation the vegetation changes from desert shrub to grassland to forest. Between grassland and forest there is a break in the curve suggesting that for a given annual precipitation, the forest vegetation produces a greater bulk per acre than does grass.

A large number of agricultural studies have shown that an increase in vegetation inhibits erosion. The data from Musgrave (1947), plotted in Fig. 3-6, **B,** show that for a given precipitation the quantity of sediment eroded from fallow land or land in row crops is roughly 80 times the amount eroded from grassland. If one considers the open desert shrub as comparable, at least in kind, to the fallow or row cropland, and the pasture and forest comparable to the natural grassland and forest, it should be expected that at some point the hypothetical relation between increasing sediment yield and increasing precipitation should reach a maximum and begin to decline as the inhibiting effect of vegetation makes itself felt.

That this is what happens in nature on a regional scale is shown by the

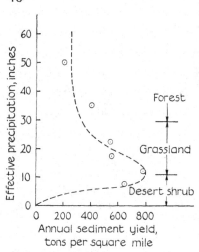

Figure 3-7.

Climatic variation of yield of sediment as determined from records at sediment-measuring stations. [From Langbein and Schumm, 1958.]

graph in Fig. 3-7. The actual sediment yield per square mile reaches a maximum where the precipitation is approximately 15 inches per year, There is scatter of points on such a curve and the maximum point should not be thought of as precise or universally applicable. As a guide, it might be pointed out that in the United States a mean annual precipitation of about 15 inches corresponds roughly to an annual runoff of 0.5 inch, a value characteristic of central Kansas and Nebraska.

In terms of morphogenetic regions, if one were interested in locating more precisely the theoretical point or region of maximum effectiveness of running water, on the basis of the foregoing quantitative analysis he would be forced to locate it well into the "arid" region on the graph in Fig. 3-5, rather than toward the region of higher rainfall—which initially seems most plausible. Since the Dust Bowl of the 1930's people are acutely aware of the fact that the region thus defined here as a maximum for pluvial erosion is also on the margin of a region similarly noted for intense wind action. To evaluate the relative importance of wind and running water in this environment, it would be necessary to construct a curve similar to Fig. 3-7, showing the relation of climate to pounds of material per square mile eroded by wind. Adequate data with which to construct such a curve are not now available. It is probable, however, that the total quantity of material removed from an area by wind will be small relative to that transported by water. Only in regions of virtually no precipitation can one expect that the wind will be the dominant erosive agent.

The quantitative example illustrates how climate and geomorphic processes interact. The parameters discussed thus far, however, are quite general, and to understand better how landforms are related to climate and geomorphic processes requires still closer examination of where, when, and how these elements interact.

Effective Climate in Geomorphology

In considering hypothetical morphogenetic regions two simple and very general parameters—mean annual temperature and mean annual precipitation—were used to characterize climate. An equally general expression —mean annual runoff—can be used to describe the relation between regional precipitation and evapotranspiration. The use of average parameters does allow one to describe rather large areas on the basis of climatic data most often available. Such gross descriptive terms, however, may obscure the true relations between climate and geomorphic processes. Thus, although average runoff correlates with sediment yield, it may well be that sediment is actually eroded from drainage basins by intermittent rains of high intensity that bear only an indirect relation to mean rainfall. To the extent that such intermittency correlates with mean annual runoff, sediment yield will be adequately expressed by correlation with mean annual runoff, but the timing and direct control of the sediment production will not be revealed by such a correlation. The following discussion introduces some facets of hydrology and effective climate that appear to be of particular importance to geomorphic processes.

The Annual Water Budget

The circulation of water from ocean to atmosphere back to land and ocean is known as the hydrologic cycle. Heat from the sun provides the energy by which water is evaporated from the surface of land and ocean and transpired from vegetation. In vapor form this water is carried in the atmosphere, later to be precipitated as rain or snow. The soil takes the first increment out of the total water made available by precipitation. What the soil can imbibe depends not only on its porosity and permeability but also on the amount of water already taking up space in the interstices between the soil particles. What water is not taken in by the soil runs off the surface and enters directly into the network of surface channels, which progressively join downstream into rills, creeks, and thence rivers. Some water which infiltrates into the ground surface moves only a short distance as ground water, to reappear as streamflow. Infiltrated water may be held absorbed onto the surface of soil particles until drawn out of the ground by evaporation or by the roots of plants and returned to the atmosphere. Some infiltrated water, however, moves into the saturated zone and may be stored as ground water for long periods. Some indeed is infiltrated into the ground close enough to the ocean that it moves directly to the ocean through the earth mantle rather than as surface water in channels.

The hydrologic cycle is sufficiently well known that it needs no further elaboration here, except to note that there are many subcycles or short-cuts shunting water from one part of the main cycle to another. The main cycle, the flow of water from earth and ocean to the atmosphere and reprecipitation back to earth and ocean, involves each year a large volume of water, but still only a small part of the total water on the earth. This aspect of the cycle, the hydrologic budget, is not so generally known; in fact, it can be discussed quantitatively only for a few portions of the globe, owing to dearth of information.

It is easy to imagine that until the advent of modern physics, the flow of water in rivers and from springs and the occurrence of rain were viewed as things magical. In the Middle Ages it was believed that water in rivers flowed from the center of the earth. Late in the 17th century Halley, the famous English astronomer whose name is still associated with a comet he discovered, made a computation of the water budget of the Mediter-ranean Sea. He made measurements of the rate of evaporation. By adding up the amount of water flowing in rivers to the Mediterranean he found that their flow is about equal to the water falling as rain and snow on the area drained by the rivers.

About the same time two Frenchmen, Perrault and Marriotte, measured the flow of the Seine River and concluded that there was more than enough precipitation falling on the catchment area to account for the flow of the river.

These are the earliest known instances of anyone having correctly reasoned that precipitation feeds lakes, rivers, and springs, and having presented observational evidence to support the generalization.

Enough data are available for the United States to set up an annual water budget with considerable accuracy. More is known, however, of the quantities of water moving through the cycle than of the total water in storage in various phases of the cycle.

Though any of several units might be used, it is perhaps easiest to visualize the budget items expressed in terms of inches of water over the United States. One inch of water would be equivalent to that required to cover the country 1 inch deep. One inch over the United States would be equivalent to 161 million acre-feet or 5½ times the storage capacity of Lake Mead behind Hoover Dam.

The annual budget may be thought of in terms of credit and debit. The credit or annual increment consists of rainfall and snowfall. This averages 30 inches for the whole country. The debit items include the removal of water from the land by the flow of streams, deep seepage of ground water to the oceans, the transpiration from plants, and evaporation from lakes, ponds, swamps, rivers, and from the moist soil.

Of the 30 inches received annually as precipitation over the United States, about 21 inches are returned to the atmosphere by evaporation and transpiration. The remaining 9 inches contribute to the flow of rivers, replenish ground storage, and a small part flows directly as ground water seepage to the ocean. Although there are 30 inches delivered as rain and snow, the atmosphere receives back only 21 inches from the land as evapotranspiration. There is, therefore, an item of 9 inches that represents transport of water from the oceans to the atmosphere, which moves from ocean to continent. This balances the 9 inches delivered by the rivers to the sea.

Only in recent years have measurements become available by which it is possible to estimate the gross transport of moisture by air masses, which flow from ocean to continent and back to ocean. Each year the atmosphere brings over the continental United States about 150 inches of water from the oceans, but 141 inches are carried back to the oceans. Thus it is seen that precipitation and evaporation are only a part of the whole atmospheric movement of moisture across the continent.

Distribution of Total Water Supply

Most measurements have been directed toward finding how the stock of water in different places changes from season to season and year to year. Therefore, much less is known about the total capital stock. A sample distribution sheet for the United States is presented in Table 3-6. A similar rough accounting for the world was prepared from published sources by Nace (1960) and is shown in Table 3-7. These tables summarize present knowledge of the amount of water in storage and in transit in various parts of the hydrologic cycle.

This compilation reveals some unexpected facts. It may not be surprising that the oceans of the world contain 97% of all the water in the

Table 3-6.   Approximate distribution of water in the United States
(48 Contiguous States). [From Langbein et al. (1962).]

Water	Area (sq mi)	Volume (cu mi)	Annual Circulation (million acre-ft per year)	Detention Period (yr)
FROZEN WATER				
Glaciers	200	16	1.3	40
Ground ice (seasonal only)				
LIQUID WATER				
Fresh-water lakes[a]	61,000	4,500	150	100
Salt lakes	2,600	14	4.6	10
Average in stream channels	—	12	1,500	.03
Ground water				
Shallow	3,000,000	15,000	250	200
Deep	3,000,000	15,000	5	10,000
Soil moisture				
(3-ft root zone)	3,000,000	150	2,500	.2
GASEOUS WATER				
Atmosphere	3,000,000	45	5,000	.03

[a] United States part of Great Lakes only.

world. But the amount in the atmosphere is surprisingly small, about $\frac{1}{1000}$ of 1%. As indicated in the discussions of the hydrologic cycle, the transfer of water from earth to atmosphere and back is the mechanism by which precipitation provides the water that sustains the vegetation and flows in the rivers of the world. Yet the atmospheric moisture constitutes only a minor part of the total water of the world.

Of the water on and in the continents, 78% is locked up in icecaps and glaciers. All water in the ground in all continents constitutes the bulk of the remainder. All surface water other than that in icecaps constitutes only $\frac{1}{2}$ of 1% of the water on the continent.

The large lakes on the North American Continent contain about 8,000 cubic miles, or about $\frac{1}{4}$ of all the fresh liquid surface water on the globe.

Frequency Distribution of Climatic Events

The statement that the mean annual rainfall of a given region is 30 inches gives only a limited amount of information about the characteristics of the precipitation. It might be useful to know if this total falls bit by bit at a rate of 0.12 inch per day, whether it is seasonably distributed.

or whether half of the annual amount falls regularly in cloudbursts of a few hours' duration. In a word, one is immediately interested in the frequency distribution of the precipitation.

Because this book emphasizes fluvial processes, a more detailed discussion of the frequency characteristics of streamflow is used here to show both the nature of the variability of this parameter and the techniques of analysis which may be used in analyzing frequency distributions of other climatic factors. Variability is one of the most important characteristics of climatic parameters, and its meaning and the methods for its use are important in understanding many hydrologic and other phenomena.

The characteristics of variability are of practical and theoretical value in estimating the probability of given values of total runoff—that is, of volume of stream discharge. To introduce the principle, annual values of runoff will be dealt with first. As will be shown, the same technique is applicable to the runoff in a 10-year period, 20-year period, and so on.

Data on annual discharge of a river are available for 7,000 stream-gaging stations now operating in the United States, and similar data are

Table 3-7. Distribution of the world's estimated supply of water (All quantities rounded). [From Nace (1960).]

Location	Surface Area (thousands of sq mi)	Volume of Water (thousands of cu mi)	Percentage of Total Water
World (total area)	197,000	—	—
Land area	57,500	—	—
Surface water on the continents			
Polar icecaps and glaciers	6,900	7,300	2.24
Fresh-water lakes	330	30	.009
Saline lakes and inland seas	270	25	.008
Average in stream channels	—	.28	.0001
Total surface water	7,500	7,360	2.26
Subsurface water on the continents			
Root zone of the soil	50,000	6	0.0018
Ground water above depth of 2,640 ft	—	1,000	.306
Ground water, depth of 2,640 to 13,200 ft	—	1,000	.306
Total subsurface water	50,000	2,000	.61
World's oceans	139,500	317,000	97.1
Total water on land	—	9,360	2.87
Atmospheric moisture	—	3.1	0.001
Total, world supply of water		326,000	100

available for an additional 3,000 sites. These data are published in the water-supply papers of the United States Geological Survey.

For the present example the annual discharge in acre-feet for the Colorado River at Lees Ferry, Arizona, will be used. Including figures for early years obtained by correlation with other stations in the area, a 61-year record widely used for engineering purposes has been compiled for the period 1896–1956. The arithmetic mean flow during this period was about 15 million acre-feet (Fig. 3-8). When for each year of record the deviation

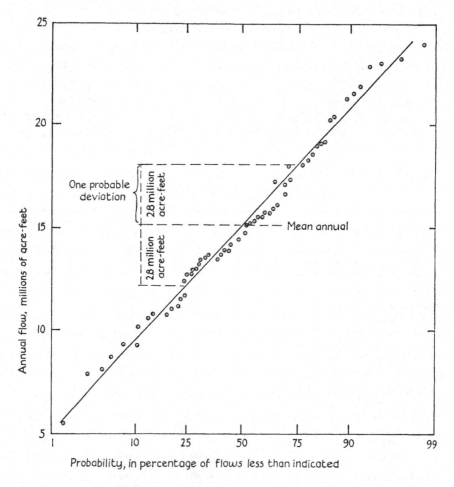

Figure 3-8.   *Cumulative distribution curve of annual runoff values, Colorado River at Lees Ferry, 1896–1956.*

from the mean is determined and the sums of the squares of deviation totaled, then

$$\text{Standard deviation} = \sqrt{\frac{\sum x^2}{n-1}},$$

where x is the difference between the value of an individual measurement and the mean of all the measurements in the sample, and n is the number of measurements in the sample.

It must be kept in mind that any record of streamflow is a sample in time. It therefore has the characteristics of any sample of a large population. When the distribution of sizes or quantities in a population follows the so-called normal law, the histogram (frequency distribution of data among size categories) will be a bell-shaped curve. In streamflow this implies that the annual discharge values include a few exceptionally large ones, a few small ones, and a preponderance of discharges centered around some central value.

Normal distributions have certain mathematical properties. One is that the arithmetic mean of the values is identical with the mode, the category having the largest number of cases. Another characteristic is that the range, which includes one standard deviation on either side of the mean, includes two-thirds of all the values in the sample.

For the series representing the annual flow of the Colorado River at Lees Ferry in the period 1896–1956, the standard deviation comes out to be 4.22 million acre-feet. The data closely approximate a normal distribution, and two-thirds of the values fall in the range between 10.96 and 19.40 million acre-feet. It can be said, then, that in any given time period in the future, provided the climate remains the same, two-thirds of the annual flow values probably will fall in this same range.

Another correlative property of normal frequency distributions is that 50% of the values fall in a range described by one probable deviation on either side of the mean (Fig. 3-8). The probable deviation is the product of the factor 0.6745 times the standard deviation.

To test whether a series of data is normally distributed, the values may be arranged in order of magnitude and plotted on probability paper, as in Fig. 3-8. Normally distributed data plot as a straight line on such a graph. The abscissa position of any individual item in the series is obtained by using the formula

$$\text{Plotting position} = \frac{M}{N+1},$$

where M is the rank of the individual number in the array and N is the total number of cases in the sample. It can be seen that the flow values in the 61-year record of the Colorado River at Lees Ferry are normally distributed, for the points align themselves reasonably close to a straight line.

As indicated earlier, not all hydrologic data are normally distributed. Data which do not fit the normal distribution pattern must be transformed to provide a series that will be normal before the usual statistical treatments discussed here can be applied.

The use of cumulative distribution curves plotted on probability paper to determine the chances of the recurrence of an event of particular magnitude is a simple and useful technique. It must be emphasized that a statement of probability is not a forecast. The past record of hydrologic events should be used as an indication only of the probability that certain events will occur in the future, not as a forecast.

Where individual values in a sample are normally distributed, the means of groups of data in the sample will also be normally distributed. For example, in a streamflow record the means of 10-year periods may be computed, and the values of these 10-year means treated as items in another sample. In a 61-year record only 6 independent 10-year means are available—a relatively small sample. Nevertheless, within determinable confidence limits the technique may be used to estimate the probability of a given 10-year mean occurring in the future (for details see Leopold, 1959).

But this requires consideration of one of the important discoveries in hydrology made in recent years—what may be called a persistence effect, an effect first described by Hurst (1950). Experience in our daily lives verifies the fact that rainy days occur together and dry days occur together. Probably for analogous meteorological reasons, wet years tend to occur in groups and dry years similarly tend to follow one another. This tendency for grouping is called persistence.

If the means of groups that included a nonrandom assortment of individuals were computed, the spread or deviation among the means would be greater than if the groups consisted of a random selection of individuals. As an example, imagine measurements of the mean height of 10-men groups on a college campus. Suppose, first, that the groups were made up by a random process of selection so that each 10-man group tended to include some tall men and some short ones.

In contrast, suppose that one of the 10-man groups was drawn from the basketball team, another from the coxwains of the crew, and so forth. The mean height of the basketball players would be greater than the

mean of 10 randomly selected individuals, and the mean height of
the coxwains would be smaller.
Thus, the variation of the means
would be larger in nonrandomly
selected groups than in randomly
selected ones. Likewise, the
spread of groups consisting of
wet years and those consisting of
dry years would be greater than
if the groups consisted of randomly ordered individual years.

H. E. Hurst (1950), working
with the longest record of river
stage in the world—nearly 1,000
years of recorded stage on the
Nile—showed that the tendency
for wet years to occur together
and dry years together increases
the variability of means of vari-

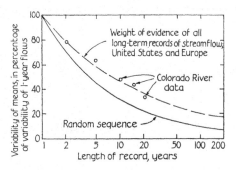

Figure 3-9.

*Variability of mean values of streamflow
for records of various lengths. The solid
curve shows the variability expected in
randomly selected data. The dashed curve
is the average relation of variability to
record length in streamflow data. The circles are values from the Colorado River
data and fall close to the average curve
for river data.*

ous periods. The idea has been confirmed by other scientists working with
independent data.

Using a list of 13 long-term records of streamflow of rivers in various
parts of the United States and Europe, means of 5-, 10-, 15-, and 20-year
periods were computed, and variability of the means was measured. The
results are shown in Fig. 3-9. Variability of the means is plotted on the
ordinate scale as percentage of the variability of 1-year flows. This percentage is plotted against length of record, which appears on the abscissa
scale, the relation being shown as the dashed line.

The full line shows variability as a function of length of record for data
randomly ordered. With regard to the solid line, it is known from statistical
considerations that variability of means decreases inversely as the square
root of the number of items making up the mean. Thus the means of 100-
year groups would be $1\sqrt{100}$ or $\frac{1}{10}$ as variable as 1-year values. Thus the
solid line goes through the ordinate value of 10% for an abscissa value of
100 years.

The difference in variability of naturally occurring groups of streamflow
data and the same data randomly ordered is defined by the difference
between the dashed and solid lines. To explain the importance of this

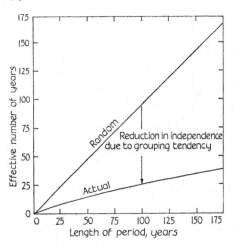

Figure 3-10.

The effect of grouping tendency in streamflow data. This graph shows that the variability of 100-year means of streamflow would be the same as that of 25-year means were the flows to occur in random sequence.

difference, consider a horizontal line through the ordinate value of 20%. This line passes through the solid curve at an abscissa value of about 25 years and the dashed curve at an ordinate value of 100 years. This means that 100 years of actual streamflow record has about the same variability as means randomly ordered in a 25-year record. A record of 100 years of streamflow is required to estimate the mean flow with the same confidence as could be estimated from 25 years of record were the flows to occur in random order. A 100-year record, then, has an effective length of a 25-year record of randomly distributed data. This concept of effective length of record is shown in a different way in Fig. 3-10, which shows the reduction in independence due to the nonrandom character of actual streamflow data.

To summarize, a tendency toward persistence in meteorologic and climatic events leads to a larger variability of mean values than if the events occurred in random sequence. It is perhaps also valuable to note here that one is accustomed to classify climate in terms of mean values of various parameters. Thus, a change in climate must be defined in terms of some degree of change in these mean values, and the greater the variability the more difficult it becomes to specify a "climatic change."

Streamflow Fluctuation

Streamflow is what is left over after precipitation has supplied the demands of vegetation and the process of evaporation. Leftovers or differences tend to vary greatly with time; that is to say that the variability of residues is greater than the variations in the original quantities. For example, suppose the rainfall in one year is 40 inches, which would be typical for a place such as Washington, D. C. Evaporation and transpiration

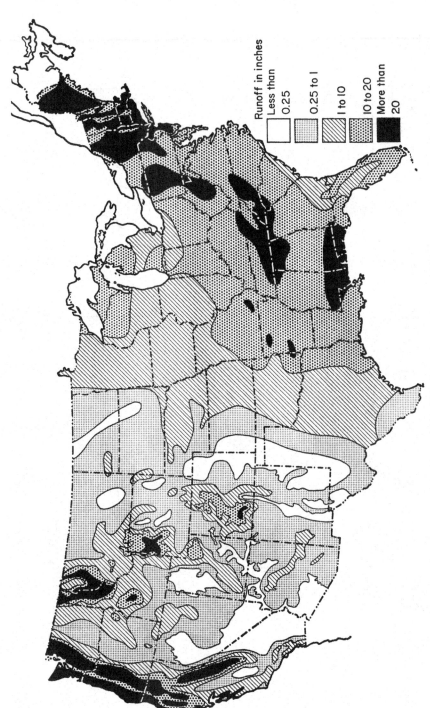

Figure 3-11.   Average annual runoff in the United States, 1921–1945. Categories are expressed in inches of runoff. [After Langbein et al., 1949.]

Runoff in inches

Less than
0.25

0.25 to 1

1 to 10

10 to 20

More than
20

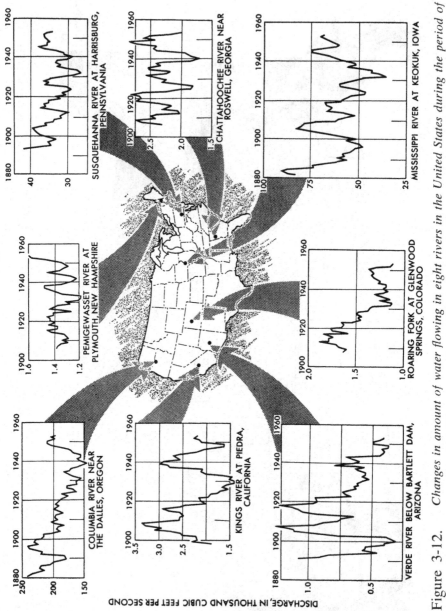

Figure 3-12.   Changes in amount of water flowing in eight rivers in the United States during the period of measurement. These examples include some of the longest records in the United States.

might be 20 inches in a given year. This leaves 20 inches to be carried off by streamflow. Suppose that in the following year the precipitation is 30 inches, 25% less than in the year previous. If evaporation and transpiration were the same, which is quite possible, streamflow would be only 10 inches. This is only 50% of that which occurred in the year previous. Thus a 25% change in precipitation became a 50% change in runoff— a demonstration of the sensitivity of streamflow to changes in rainfall.

Streamflow or runoff decreases generally from east to west over the United States in response to climatic characteristics, but is as large on the mountains of many Western States as in the more humid East (see Fig. 3-11). Variability of runoff tends to be greater with more arid conditions. Climatic variability is a characteristic inherent in the arid zones of the world.

To exemplify the variations in streamflow which have been observed during the period of record, Fig. 3-12 shows the flow in eight rivers in different parts of the United States. Rivers have been chosen for which records are relatively long; in each instance the record began between 1880 and 1910. The longest continuous record of stream discharge by the U. S. Geological Survey began in 1889, on the Rio Grande at Embudo, New Mexico. Some partial records were collected as early as 1860.

The graphs in Fig. 3-12 show the flow expressed as a 5-year moving average, and so some of the variability has already been eliminated. Yet the characteristics most apparent in the graphs are a large variability at each of the stations shown, and a decided lack of simultaneity in the ups and downs of flow among the rivers in different parts of the country. There appears, however, in the majority of these curves a tendency for high flows in the early part of the record and a discernible downward trend during the period. A large number, but by no means all, river records show this same trend.

The overall picture for runoff in the United States can be seen in Fig. 3-13. There was a general decrease in streamflow during the first three decades of the present century. The lowest streamflow of record occurred in the decade 1930–1940. This trend was reversed during the next decade, which included the war years, but the records in 1950–1960 seemed to show a return toward lower runoff.

These observed trends in runoff have been similar in direction to records of other climatologic parameters. There is, during the period of record through most of the northern hemisphere, a tendency toward warmer and drier conditions, at least since about 1870. About 1870 most glaciers had

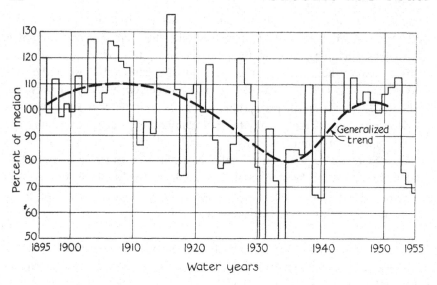

Figure 3-13.   *Variation in annual runoff in the United States as a whole.*

advanced to points farther downvalley than they had attained for some centuries previously. Since that date glacier retreat has been rapid and marked, but by no means universal.

Temperatures at long-term stations have also shown a trend toward increasing values. These changes toward drier and warmer conditions have been more marked and consistent at high than at low altitudes and have been generally more marked in the northern than in the southern hemisphere. The differences between hemispheres are, however, not established clearly because there are relatively fewer observation stations in the southern hemisphere.

It can be said unequivocally that the observed changes in streamflow are not sufficiently cyclic to allow any meaningful forecast of future streamflow by extrapolation of apparently repetitive cyclic behavior of flow. It is a debatable point as to whether or not one should refer to the variations in runoff displayed in Fig. 3-12 as evidence of climatic change. Most hydrologists and meteorologists would probably not so consider them. Streamflow has varied from 10 to 20% of the median in a period of 25 to 30 years, however—certainly in itself a significant variation. The extent to which such variations will affect the landscape requires close analysis of climatic change if it is to have meaning in geomorphology. The values so far treated are central ones—means or medians. The problem is made more difficult to the extent that the climatic elements of geomorphic significance are the extreme or unusual events, or consist of complex interrelations between climatic factors having different frequency distributions.

Flood Frequency

Floods, for example, are usually considered climatically controlled events of geomorphic significance. The occurrence of floods is studied as a probability problem, and knowledge of the probability of flood occurrence is needed for a variety of engineering and economic reasons. These needs and potential uses have led to a considerable effort over many years devoted to developing procedures and to the production of systematic studies of flood probability. Because of the geomorphic significance of such events, probability studies of flood occurrence are important tools in geomorphology.

In contrast with records of annual or monthly values of mean runoff, flood occurrences may be treated as random events, for the meteorologic and hydrologic factors affecting flood production do vary with time sufficiently that the combinations have many characteristics of chance events. The underlying premise is that the floods occurring during a period of time constitute a sample of an indefinitely large population in time. Specifically, if in a period of 30 years of record the largest flood recorded was of a certain size, it is probable that the next 30 years will also contain a flood of equal magnitude.

Several procedures, all relatively simple, are used for the computation of flood frequency. One of the simplest and most practical begins with the tabulation of the highest discharge in each year of record at the station. Momentary peak discharges are used for this array if they are available. The sample then includes only one event in each year. The mean of this series is called the mean annual flood.

The plotting position for individual items in the array is determined by the formula

$$\text{Recurrence interval} = \frac{N+1}{M},$$

where N equals the number of years of record and M is the rank of the individual item in the array. Such data for an individual station are usually plotted on logarithmic probability paper (see Dalrymple, 1960). Mathe-

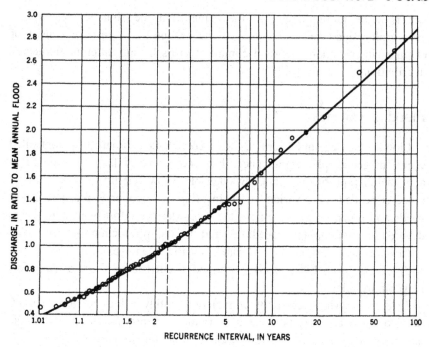

Figure 3-14.   *Regional flood-frequency curve for Youghiogheny and Kis-*
kiminetas river basins, Pennsylvania. To determine the re-
currence interval of a particular flood flow at point, the mean
annual flood at that point is measured or estimated; the
ratio of the flood flow to the annual flood is then entered on
the ordinate and the recurrence interval of the flood is read
on the abscissa.

matical analyses have shown that the mean annual flood has a recurrence
interval of 2.33 years; that is, once every 2.33 years, on the average, the
highest flow of the year will equal or exceed the mean annual flood.

A similar procedure utilizes the data for all high flows, even when several occur in the same year. The data array includes all flood peaks having
a value greater than some chosen base. The mean recurrence interval for
events in such an array, called a partial-duration series, is computed by use
of the same formula given above. Because it is likely that very large floods
will occur only once in a given year, the recurrence interval of these large
floods will coincide in the annual and partial-duration flood series. Langbein (1949) has shown that the recurrence intervals approach one another
and are nearly identical for records of 10 years or longer.

Because there are differences in the flood experience even in neighboring streams, and because of differences in flood potential that exist among lithologic and topographic types, it is necessary to generalize flood experience over geographic areas. This is usually done by plotting graphs similar to that shown in Fig. 3-14, in which the ordinate scale is the ratio of any given flood to the mean annual flood. By using this dimensionless factor, the flood experience of areas of various sizes and length of record may be grouped. On the basis of similarity in such curves, geographic areas are delineated that have comparable flood potential. Details of the procedure for developing regional flood frequency curves have been published by Dalrymple (1960).

Data on flood frequency have been compiled for many states and for river basins in the United States and elsewhere. For about two-thirds of the United States one can estimate the probability of various flood discharges from published data. Publications on flood frequency are listed in the pamphlet, "Publications of the Geological Survey."

It should be kept in mind that flood-frequency compilations and analyses of the type described yield values of a parameter—recurrence interval. This is the average interval of time within which a flood of a given magni-

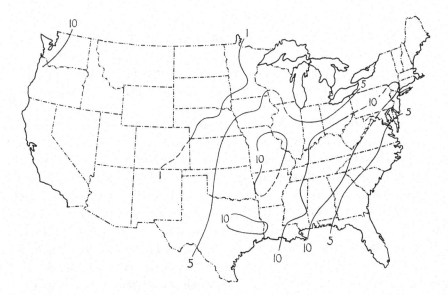

Figure 3-15.   *Flood potential: the mean annual flood in thousands of cubic feet per second from drainage basins of 300 sq. mi. area.*

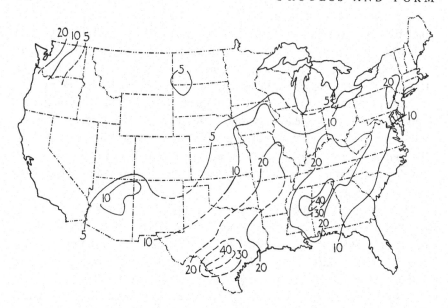

Figure 3-16.   *Flood potential: the 10-year flood in thousands of cubic feet per second from drainage basins of 300 sq. mi. area.*

tude will be equaled or exceeded but once. It is also a statement of probability. Thus, a flood having a recurrence interval of 10 years is one that has a 10% chance of recurring in any single year. Such a flood, or even one so large that it would not be expected more than once in a hundred years, might occur next year. Thus the recurrence interval, as stated earlier, is not a forecast.

The flood potential of different regions can also be compared by using drainage basins of equal size as units of comparison. For the United States we have compiled two maps, Figs. 3-15 and 3-16, showing respectively the flood "potential" of basins of 300 square miles, for the mean annual flood and for floods having a recurrence interval of 10 years. On drainage areas of 300 square miles the greatest potential for floods of moderate size—say, a 10-year recurrence interval—occurs in the Southern Appalachians and in south-central Texas (Fig. 3-16). It is perhaps in these areas that we should expect floods to have particular significance in landform development, for it will be shown in the next section that frequent events of moderate effectiveness are of great geomorphic importance. A 10-year flood (one having a recurrence interval of 10 years) is not a great or catastrophic flood but one of moderate size.

The Frequency Concept and Geomorphic Processes

The significance of different frequency distributions of factors affecting geomorphic processes is not easily evaluated because of the paucity of data. We are familiar, however, with the tremendous "geomorphic activity" associated with events classified on the human scale as catastrophes. Hurricane destruction of beaches, flood alteration of alluvial channels, and gullying of hillsides by cloudburst rains are recognized examples. Less easily reckoned is the cumulative effect of high waves, floods, or rains which recur once a year or perhaps once every 3 to 5 years.

One simple example of the possible cumulative effect of events of only moderate intensity is provided by an analysis of the energy of a typical rainstorm. Figure 3-17 is a histogram showing rainfall intensity measured in inches per hour as a function of time for a moderate rain of 2.62 inches. As the variation of raindrop size with rainfall intensity is known, the kinetic energy produced by the storm can be computed as well as the proportion of the kinetic energy contained in each increment of rainfall.

The percentage of the total kinetic energy of the storm contributed during each period is given by the figures written at the top of each vertical bar in Fig. 3-17. The greatest intensity, 1.2 inches per hour, was experienced in the half hour between 4 and 4:30 P.M., during which 26% of the total energy was contributed. Wischmeier and Smith (1958) have shown that the yield of sediment is proportional to the product of the maximum 30-minute intensity experienced during the storm and the total kinetic energy of the rainfall; thus, in this example a large proportion of the sediment produced may be attributed to the precipitation with intensities less than 0.9 inch per hour.

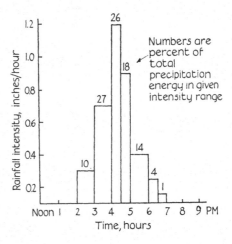

Figure 3-17.

Histogram of rainfall intensity as a function of time during a rainstorm. The percentage of the total kinetic energy produced during the storm by each portion of the storm is shown as the figure written near the top of each bar.

There are several ways of considering the geomorphic work performed by events of varying frequency. First, one may consider the quantity of material actually transported by events of increasing magnitude which occur with decreasing frequency. This, when summed, will provide an estimate of the total amount of work performed in transport of material, which, in turn, can be apportioned among the events of different frequencies. It does not follow, however, simply because the volume of material moved by a great number of events of moderate magnitude is large, that these events in turn are responsible for the formation of specific features of the landscape. Such aspects of the landscape such as drainage density, slope of hillside, and river pattern may, indeed, be related to specific events that occur at specific recurrence intervals. These recurrence intervals, however, need not coincide with those for the events that are the largest transporters of the mass of material from a given region.

Consider first the debris load carried past a particular point by a river. It is to be expected, of course, that the debris load of a flood is larger by far than that of a moderate flow, which, in turn, is larger than the load carried at low flow. To obtain the total load, multiply the load for the flow of each given size class by its relative frequency of occurrence, and sum for all size classes. This also permits a comparison of the relative importance of each size class in the transport of the total debris load.

The data needed to describe the frequency of each size class are given by the duration curve. The duration curve is a frequency distribution of the mean daily flows at a particular location on a stream. For the two rivers in Fig. 3-18, having about the same drainage areas, the cumulative curves of

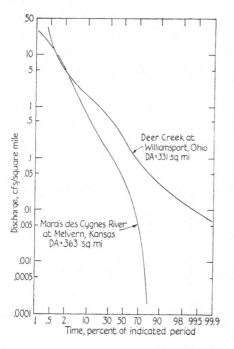

Figure 3-18.

Flow-duration curve or frequency distribution curve of mean daily flows for each of two rivers—Deer Creek at Williamsport, Ohio, and Marais des Cygnes River, at Melvern, Kansas.

daily flows are nearly straight lines on the log-normal graph and hence approximate log-normal distributions. Note that Deer Creek in Ohio has a flow equal to or more than 0.03 cfs per square mile or 9.9 cfs (0.03 × 331 square miles) 90% of the time. The Marais des Cygnes in drier Kansas is dry about 20% of the time. The position of the duration curve gives the magnitude of the flow, and the slope of the curve is a measure of flow variability.

Flow-frequency distribution curves are useful for the analysis of the flow values associated with principal transport of sediment and dissolved load. Data on discharge, suspended sediment load, and dissolved load for the Bighorn River at Thermopolis, Wyoming, are plotted in Fig. 3-19. The data represent conditions before the closure of Boysen Reservoir, which has subsequently changed the flow duration curve and traps a large amount of the sediment. The river at the measuring point drains 8,080 square miles and has a mean flow of 1,900 cfs. A large part of the basin is semiarid but the headwaters are in the high mountains, which receive considerable snow. The period of highest sustained runoff occurs in early summer from snowmelt.

Looking first at the duration curve and histogram of streamflow, Fig. 3-19, **A,** it can be seen that discharges in the range from 1,000 to 2,000 cfs occur about 33% of the time—33% of the days in the year, on the average.

The mean flow, 1,900 cfs, occurs at the 40% position on the duration curve, or the river flows at a rate less than the mean flow during 60% of the time. This figure lies between 60 and 75% for many rivers. Similarly, many rivers flow at a rate less than half the mean flow about 25% of the time.

Now consider the contribution of different ranges of discharge to the total volume of runoff, the solid line in Fig. 3-19, **B.** This is computed by multiplying the number of days having a given range of discharge by the mean discharge of that range. For example, in the Bighorn River discharges between 400 and 600 cfs were observed on 74 days out of 2,026 days of record, or 3.6% of the time. The product of 74 days times a mean rate of 500 cfs gives 37,000 cfs days, which is 0.9% of the total volume of flow during the period of record.

The discharge range from 1,000–2,000 cfs occurs 33% of the time, as indicated on Fig. 3-19, **A,** and contributes 24% of the total amount of runoff (Fig. 3-19, **B**). Flows of 10,000 cfs and above contribute only about 6% of the runoff. All flows greater than 5,000 cfs contribute only

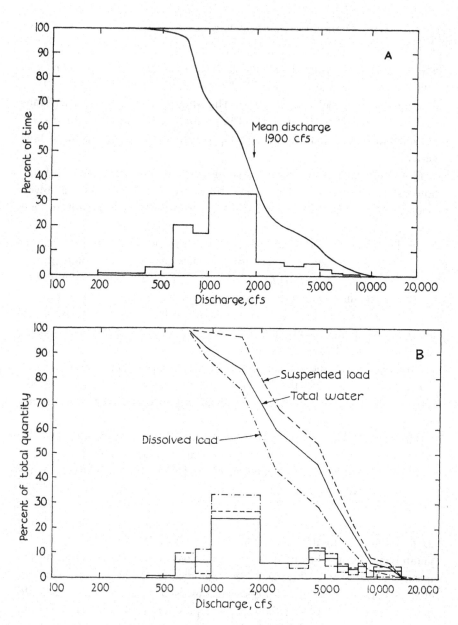

Figure 3-19. *Histograms and cumulative graphs of water and load, Bighorn River at Thermopolis, Wyoming.* **A.** *Duration curve of water, showing percentage of time various discharge rates are equaled or exceeded.* **B.** *Cumulative graphs and histograms showing relative contribution of various discharge rates to total flow of water, suspended load, and dissolved load.*

40% of the total runoff. Floods provide large discharges but occur on so few days that they do not contribute as much water as lower flows, which occur much more often.

Next observe the contribution of various ranges of discharge toward the total dissolved load of the river. High flows contribute a smaller percentage of the total yield of dissolved salts than they represent in percentage of contribution to total runoff. This results from the fact that low flows have larger concentration of dissolved salts than do flood flows.

Figure 3-19, **B,** also shows the contribution of various ranges of discharge to the transportation of suspended sediment load. The method of computation is similar to that already explained for dissolved load. Concentration of sediment tends generally to increase with discharge except at the highest discharges.

The increase in concentration with discharge might lead one to expect that high flows carry the largest percentage of the total load. But the data in Fig. 3-19, **B,** show that the extremely high flows, though effective in geomorphic work of erosion and transportation, are so infrequent that their contribution to the total work is less than that of more modest discharges which occur more often. Flows of 10,000 cfs or greater transport only about 9% of the total measured load. A large percentage (more than one quarter) of the total measured load is carried by flows of 1,000 to 2,000 cfs. Flows occurring 9 out of 10 days (flows less than 5,500 cfs) carry 57% of the total suspended load. Thus, in terms of the transportation of sediment, the major work is accomplished during the more modest but relatively frequent floods rather than during the larger but rarer catastrophic floods.

The results of similar computations for Brandywine Creek in Pennsylvania and for the Rio Puerco in New Mexico are shown in Table 3-8. Despite climatic and physiographic differences, 50% of the total suspended load of both of these streams was transported by flows which occur on the average one day or more per year (Table 3-8, column 4). Although many fewer flows were required to transport the remaining 50% of the sediment, the data in the table indicate that, as in the Bighorn example, at least half of the suspended sediment is removed from these drainage basins by low and moderate flows.

The previous data were derived from computations based upon the quantity of material and the frequency of daily flows; but rarer or catastrophic floods may be considered by analyzing the maximum floods of each year of record. Such a record of discrete events is discontinuous and

Table 3-8.   Percentage of suspended sediment transported by flows of different magnitudes.

1	2	3	4	5	6	7
River and Station	Distribution Measure of Flow Used in Analysis	Magnitude of Flow Below Which 50% of Total Sediment Is Transported	Frequency of Occurrence of Flow in Column 3	Magnitude of Flow Below Which 90% of Total Sediment Is Transported	Frequency of Occurrence of Flow in Column 5	Remarks
Rio Puerco near Bernardo, New Mexico	Daily discharge duration curve	950 cfs	Equaled or exceeded 6 days/yr	3,400 cfs	.7 days per year[a]	Zero flow approximately 70% of year
Brandywine Creek at Wilmington, Delaware	Daily discharge duration curve	1,900 cfs	Equaled or exceeded 11 days/yr	8,200 cfs	.2 days per year[b]	
Bighorn River at Thermopolis, Wyoming	Daily discharge duration curve	4,800 cfs	Equaled or exceeded 50 days/yr	9,500 cfs	4.1 days per year	

[a] .7 days per year is 7 days in 10 years or about one day in 1.5 years.
[b] 1 day in 5 years.

hence less satisfactory than the continuous record of daily discharges used above. The data in Table 3-8 (column 6) show that 90% of the sediment from the Rio Puerco and Brandywine basins is transported by storm runoff or discharges which recur at least once every 5 years. Although the proportions vary markedly for the two river basins, the results seem to indicate that a large part of the work, considered in terms of the bulk of sediment transported from the river basin, is performed by what may be considered moderate events that recur relatively often. A catastrophic flood may by itself transport an immense volume of sediment. On the other hand, it recurs so infrequently that its cumulative effect is less than that of the more moderate but much more frequent annual and biannual floods.

Let us consider now the different importance of frequent and infrequent events when viewed in terms of the sediment load directly, apart from the water discharge. This difference derives from the fact that in many instances the sediment is not carried during the times of maximum discharge. In many snow-fed streams, for example, the maximum discharge occurs during the period of snowmelt. However, the maximum sediment load per unit time is frequently transported during times of thunderstorm summer rainfall. Table 3-9 shows the percentage of the total suspended load carried during various intervals of time. It can be seen that the Colorado River at the Grand Canyon transports approximately 4% of its total load during the maximum 10-day period. But the Rio Puerco in New Mexico, a stream noted for its high sediment load, transports 31% of its total suspended load during the maximum 10 days. Similarly, the Cheyenne, a river which experiences considerable fluctuations in discharge as well as in sediment content derived from the tributary badland topography of its drainage basin, carries 28% of its total suspended load during the 10 maximum days. The Niobrara River, on the other hand, a stream whose flow is quite steady due to the regulation of inflow provided by the Sand Hills of Nebraska, which it drains, transports only 7% of its total load during the 10 days of maximum transport. These relations are essentially reversed when we consider the last column in Table 3-9. These show that the Colorado and the Niobrara require 31 and 95 days respectively to transport 50% of their load, whereas the Rio Puerco and Cheyenne require only 4 days per year to transport 50% of their load. For all four of the streams, 98–99% of the total load is carried during events which occur more frequently than once in 10 years. The more variable the flow of the streams, the larger the percentage of load that is carried in a relatively few isolated events or in a relatively few days.

Table 3-9.   Time required to transport various percentages of total suspended load.

River and Station	Drainage Area (sq mi)	Maximum Day	10 Maximum Days	Events Which Recur 1 Day/Yr	Days/Year Required to Transport 50% of Load
		Percentage of Total Suspended Load Carried During:			
Colorado River at Grand Canyon, Arizona	137,800	0.5	4	92	31
Rio Puerco at Rio Puerco, New Mexico	5,160	5	31	82	4
Cheyenne River near Hot Springs, South Dakota	8,710	5	28	78	4
Niobrara River near Cody, Nebraska	3,000[a]	2	7	95	95

[a] Approximate.

The more even the flow, the more evenly distributed in time is the transport of sediment. It should be noted that in these comparisons only suspended load has been considered. In the examples cited, with the exception of the Niobrara, the suspended load probably accounts for 90% of the transported clastic load. For the Niobrara perhaps one-half of the total sediment discharge is carried in suspension.

Each of these examples suggests that events of moderate frequency rather than catastrophic events account for a large percentage of the total sediment removed from a drainage basin. Thus the work in terms of transport done by moderate events of relatively frequent occurrence is considerable. It has also been found that, in addition to the variability of flow, or perhaps as a correlative of it, the smaller the drainage basin, the larger the percentage of the load contributed by relatively infrequent events is likely to be. Thus, those factors that contribute to variability of flow, such as limited drainage area as well as intermittent rainfall, also contribute to the irregularity of the period of sediment transport.

In the examples given it must be emphasized that in considering transport we are including only flows above a given threshold value. Thus, there is no transport if the flow does not exceed the competent or threshold value required for transport of the available material.

Another approach to the problem of analyzing the relative geomorphic significance of events of different frequencies is provided in the relation between the dissolved and the suspended load transported by a river.

In the United States, dissolved load increases with increasing annual runoff up to about 10 inches. For higher annual runoff dissolved load is about 125 tons per square mile per year. In these wetter regions the dissolved load is a function of the availability of salts in the rocks—that is, it depends on rock composition and weathering rate and not on the available runoff. Inasmuch as the dissolved load is less dependent than sediment load upon the quantity of flow, it follows that the larger the percentage of the load carried in solution, the larger the percentage of material which is being removed from the basin by more frequent flows of smaller magnitude. The concentration of dissolved material in a river channel decreases with increasing magnitude of flow; it does not increase as does the suspended load with increasing magnitude of flow. Although the process of solution may be aided by floods and streamflow of high velocity, it is more dependent upon the presence of soluble, permeable rocks and abundant total precipitation to percolate through them.

Figure 3-20 shows for a number of streams the relation between the dissolved load in tons per day and the discharge in cubic feet per second. As the slope of each curve on the graph is less than one, in all cases concentration decreases with increasing discharge. By combining this information with the frequency distribution of large and small discharges,

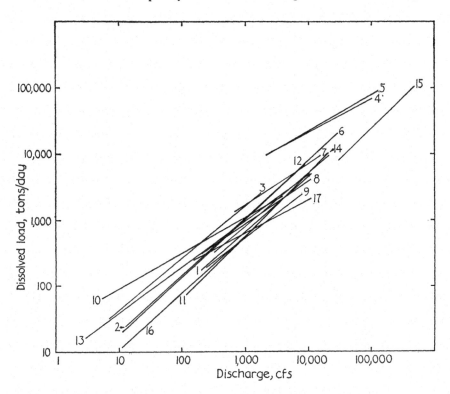

Figure 3-20.

Relation of dissolved load to discharge at measuring locations in selected river basins in the United States.

1. *Rio Grande at Otowi, N. M.*
2. *Rio Grande at San Acacia, N. M.*
3. *Rio Grande at San Marcial, N. M.*
4. *Colorado River at Lees Ferry, Ariz.*
5. *Colorado River at Grand Canyon, Ariz.*
6. *San Juan River at Bluff, Utah.*
7. *Bighorn River at Thermopolis, Wyo.*
8. *Shoshone River at Byron, Wyo.*
9. *Wind River at Riverton, Wyo.*
10. *Saline River at Russell, Kan.*
11. *Iowa River at Iowa City, Iowa.*
12. *White River near Kadoka, S. D.*
13. *Moreau River at Bixby, S. D.*
14. *Cedar River at Cedar Rapids, Iowa.*
15. *Columbia River at Int'l. Boundary.*
16. *Ute Creek near Bueyeros, N. M.*
17. *Allegheny River at Red House, N. Y.*

Table 3-10.   Dissolved and suspended load in selected rivers in different climatic regions of the United States.

River and Location	Elevation (ft)	Drainage Area (sq mi)	Average Discharge, Q (cfs)	Discharge ÷ Drainage Area (cfs/sq mi)	Years of Record in Sample[a]	Average Suspended Load	Average Dissolved Load	Total Average Suspended and Dissolved Load	Total Average Load ÷ Drainage Area (tons/sq mi/yr)	Dissolved Load as Percent of Total Load (%)
						(millions of tons/yr)				
Little Colorado at Woodruff, Arizona	5,129	8,100	63.3	.0078	6	1.6	.02	1.62	199	1.2
Canadian River near Amarillo, Texas	2,989	19,445	621	.032	1	6.41	.124	6.53	336	1.9
Colorado River near San Saba, Texas	1,096	30,600	1,449	.047	5	3.02	.208	3.23	105	6.4
Bighorn River at Kane, Wyoming	3,609	15,900	2,391	.150	1	1.60	.217	1.82	114	12
Green River at Green River, Utah	4,040	40,600	6,737	.166	26–20	19	2.5	21.5	530	12
Colorado River near Cisco, Utah	4,090	24,100	8,457	.351	25–20	15	4.4	19.4	808	23
Iowa River at Iowa City, Iowa	627	3,271	1,517	.464	3	1.184	.485	1.67	510	29
Mississippi River at Red River Landing, Louisiana		1,144,500[b]	569,500[b]	.497	3	284	101.8	385.8	337	26
Sacramento River at Sacramento, California	0	27,000[c]	25,000[c]	.926	3	2.85	2.29	5.14	190	44
Flint River near Montezuma, Georgia	256	2,900	3,528	1.22	1	.400	.132	.53	183	25
Juniata River near New Port, Pennsylvania	364	3,354	4,329	1.29	7	.322	.566	.89	265	64
Delaware River at Trenton, New Jersey	8	6,780	11,730	1.73	9–4	1.003	.830	1.83	270	45

[a] Computation of load, dissolved or suspended, depends on discharge for same period. Years of record pertain to number of years used for related values of discharge and of suspended and dissolved load. Where two figures are shown, the first is for suspended load and the second is for dissolved load.
[b] From USGS records for Vicksburg, Mississippi station.
[c] Estimated.

we may compute the percentage of material transported as dissolved load by flows of varying magnitude which occur with varying frequency. As in the example of the Bighorn River, the largest portion of the dissolved load is carried by relatively frequent flows.

The data indicate that the greater the percentage of the total load carried in solution, the more significant from an erosional standpoint will be the flows of smaller magnitude. It is interesting to note that in a sample of about 70 rivers in different regions of the United States approximately 20% of the total measured load is carried in solution. A representative selection of these rivers is given in Table 3-10. From the table it can be seen that this percentage of load in solution varies from 1% in the Little Colorado River in Arizona to 64% in the Juniata River in Pennsylvania.

As can be seen in Table 3-10, this regional variation appears to be systematic and not random. In arid and semiarid regions where vegetation is sparse and sediment production high, dissolved load is a relatively small part of the total. Where precipitation is greater, runoff higher, and vegetation more abundant, the percentage of the total load which is dissolved increases. Dissolved load constitutes only 9% in areas of low runoff and increases progressively to 37% for areas where discharge is equal to or greater than 0.7 cfs per square mile (Table 3-11).

Table 3-11.   Variation with climate of ratio of dissolved load to total load.

Climatic Indicator, Discharge per Square Mile (cfs/sq mi)	Number of Streams in Sample	Dissolved Load as a Percentage of Total Load
0–0.1	22	9
0.1–0.3	19	16
0.3–0.7	7	26
>–0.7	22	37

The same question of percentage of dissolved load was approached in a similar way by Langbein and Dawdy,[*] who used the published records of 170 stations where chemical quality data are obtained. The measurements were reduced to tons per square mile per year and were grouped by mean annual runoff. The data were also expressed in terms of concentration.

[*] Langbein, W. B., and Dawdy, D. R., 1963, Some general comments on the occurrence of dissolved solids in surface waters of the United States: *U. S. Geol. Survey* (unpublished).

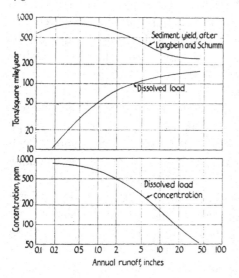

Figure 3-21.

Dissolved and suspended load in the United States, as a function of annual runoff; data for dissolved load represent 170 station locations. [From Langbein and Dawdy, unpublished.]

Langbein and Dawdy's conclusions are shown in Fig. 3-21. With no flow there is no load, and as runoff increases the dissolved load does also. When runoff exceeds 3 inches, additional runoff does not produce much additional dissolved load. At about 10 inches of runoff the load approaches 150 tons per square mile per year, a maximum value. At about this value, and thus when the annual precipitation exceeds 25 inches annually, weathering proceeds at about the maximum rate and additional precipitation does not produce any greater quantity of soluble products.

Concentration, on the other hand, decreases from an average of 800 ppm in an arid climate, and—as Langbein and Dawdy point out—where runoff exceeds about 10 inches, concentration decreases with runoff as a straight dilution effect.

These investigators compare dissolved with suspended load where both data are collected, and their comparison can be seen in Fig. 3-21. In dry climates only a small part of the total load is dissolved, but the percentage in solution increases with annual runoff, as shown in Table 3-11, and here it approaches equality with the load in suspension where runoff exceeds 25 inches annually.

Data presented by Durum, Heidel, and Tison (1960) and by Corbel (1959) for streams in diverse regions of the world also suggest that the percentage of the dissolved load is less where land slope is high. The data are far from adequate, but since gradient or slope is a principal determinant of clastic load it is to be expected that for comparable climatic conditions the relative importance of solution should decline as gradient increases.

The qualification regarding constant climate is of considerable importance, however. Figure 3-22 shows the relationship between denudation rate and altitude in the Colorado River System, including the San Juan and Green. In this river system the major part of the flows are derived from the headwaters, but the maximum sediment yield comes from different climatic zones of lower precipitation, runoff, and gradient. Thus geologic evidence of rapid deposition in a sedimentary basin does not necessarily imply the presence of an adjacent or nearby high-gradient source area. Climatic factors may, in fact, override topographic factors in controlling rates of sediment production in the source area.

From the standpoint of work done in erosion and transport, then, the evidence from measurements of the dissolved and suspended load carried by rivers suggests that in many drainage basins a large proportion of the work is performed by relatively frequent events of moderate magnitude. The results of the examples presented can be generalized as follows. To the extent that the movement of sediment by water or air is dependent upon shear stress, rate of movement can be described by the equation

$$q = k(\tau - \tau_c)^n,$$

where q is the rate of transport, k a constant related to the characteristics of the material transported, τ the shear stress per unit area, τ_c the critical or threshold shear required to move the material, and n an exponent. If the frequency distribution of such stresses, determined by climatic and meteorological events, follows a log-normal distribution as we have sug-

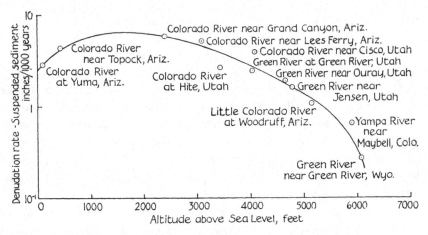

Figure 3-22.   *Relation of denudation rate to elevation, Colorado River System. Denudation rate is here derived from suspended load data expressed as inches derived from the drainage basin per unit of time.*

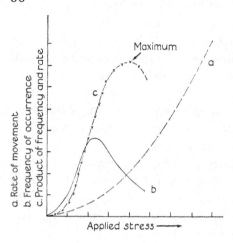

Figure 3-23.

Generalized relation of effective work done by events of different magnitude.

gested, then the product of the weight of material moved times the frequency will always reach a maximum such that the greatest quantity of material is transported by frequent events rather than by the extremes. The recurrence interval or frequency at which this maximum occurs is controlled by the rates of change of the rate of movement with the stress and of the stress with time (Fig. 3-23). It is essentially this maximum which we have defined as the interval or frequency interval during which the maximum amount of erosional work is done upon the landscape. Inasmuch as the laws governing transport by wind and by water are very similar, what has been said regarding the events responsible for the transport of sediment in streams can be readily extended to the transport of sand and silt by wind (Wolman and Miller, 1960).

Landforms in Relation to Frequency of Climatic Events

Catastrophic events of rare frequency may have a great effect on the characteristics of particular landforms, even though rare occurrences may perform but a moderate proportion of the total work of moving materials out of the basin. The effect on surface form, however, is not easily studied. Rivers in flood are known to exercise tremendous force upon objects in their paths, but inhospitable and dangerous working conditions in the field have limited the number of direct observations made during such events. Most of the knowledge of the geomorphic effects of floods is built on observations made ex post facto—a definite disadvantage.

It is known that the size, shape, and pattern of river channels are closely related to the flows which they transmit. Although the mechanics involved in the adjustments of channel form, pattern, and discharge are not completely understood, there appears to be a close correlation between discharge and specific aspects of channel form. The wavelength of a meandering river, for example, is roughly proportional to some measure of discharge. It is reasonable to suppose that the pattern of the channel itself is formed by flows which apply sufficient force to mold the channel but are also retained within the channel, rather than by those which occupy the entire cross section of a valley during periods of flood. In many rivers the bankfull stage recurs on the average once each year or perhaps once every two years. Although a channel may be scoured during individual high flows, flows much above the bankfull stage and out of the channel cannot be responsible for the regular meandering pattern. Such flows often move downvalley directly across the pattern and not coincident with it.

Additional observations indicate that the cross section of a straight reach of channel is adjusted to a range of discharges which provide a shear stress balanced by the resistance of the banks. A meandering channel migrates both across and down its alluvial plain. The stability of the meandering channel is a function, then, of the shear stress on the outer bank and of the associated deposition on the inner or convex bank, a subject covered in some detail in Chapter 6. In either case, the channel is formed and re-formed during a range of flows lying between the lower limit of competence and an upper limit at which the flow is no longer confined within the channel. The range so defined consists of flows which recur more often than once in a year or once in two years. Thus it appears that the channel- and pattern-forming discharge is one which recurs frequently. Although great erosion does occur during exceedingly large flows, in streams with developed meander patterns and floodplains the channel-forming discharge does not seem to be that associated with the infrequent or catastrophic events.

The process of channel formation and its relation to competence and to the time intervals during which the major portion of the work is done can be tied together by considering a particular river section. Seneca Creek near Dawsonville, Maryland, about 20 miles north of Washington, D. C., drains an area of 100 square miles. It lies within rolling farm country in the Piedmont province, in an area underlain by deeply weathered schist and sandstone lying outside the boundary of glaciation. A photograph of the reach is shown in Fig. 3-24.

At the measuring point the channel is 75 feet wide, the bed is gravelly, and banks of silt stand about 5 feet high. The mean flow is 100 cfs. Flow is less than the average value (less than 100 cfs) 66% of the time or 240 days per year. At the 50% point of the duration curve the discharge is 72% of the mean flow, and at that stage the average depth is 0.7 feet.

Figure 3-24.

A. *Seneca Creek near Dawsonville, Maryland, looking downstream at low flow; bar in middle ground is composed of gravel.*

Thus half the time the channel is less than 0.18 of its bankfull depth. The discharge at bankfull stage is approximately 1,330 cfs, and the recurrence interval of this discharge is 1.1 years.

Over a period of 7 years of study of the reach of Seneca Creek it has been our observation that even sand grains do not move to any great extent at mean flow and below, probably because the cobbles on the bed protect the smaller grains from exposure to the flow. Cobble gravel bars which are out of water at mean flow are stripped of their surface rocks

Figure 3-24.

B. *Seneca Creek, looking downstream at flow about half bankfull; note that riffle is drowned out so that no visible evidence of it appears on the water surface.*

at less than bankfull stage and rocks of the same size replace those moved, for the gravel bars have not changed in position, form, or elevation after a bankfull flow has receded.

Such observations indicate that the work of such perennial streams in scour and fill and in transport of debris is accomplished principally by flows near or above bankfull stage—flows which occur less than 0.4% of the time or roughly once a year. Such events are not normally considered catastrophic.

In steep narrow valleys the likelihood is greater that infrequent events of major magnitude will so devastate the channel and valley that long intervening periods of more moderate flows have a relatively small effect on the landscape. At best these intervening flows slowly "repair" the hillslopes and valley floor. Hack and Goodlett (1960, p. 42) have reported upon the effects of a single devastating storm in 1949 in the Appalachian mountains of Virginia. This storm created major slides on the hillslopes of many first-order valleys, and in the channels themselves large masses of debris were eroded from some localities and deposited in others. Although 13 years later vegetation covered many of the erosional and depositional features created by the storm, the field evidence clearly shows that similar devastating events have produced similar results in the past. As the subsequent discussion of hillslopes shows (Chapter 7), however, the vegetation and topography are also related to more common or frequent moisture regimes, even though the rare events may be responsible for some of the most significant forms.

Tricart and collaborators (1961, p. 276) describe very nearly the same devastating effects of a catastrophic flood in the high Alps of France. Their observations suggest that a flood of 1957 created fresh gullies and slides, moved quantities of debris, and altered the valley floors which had not been similarly affected perhaps since the Pleistocene. Jäckli (1957) also compares a single prehistoric rockfall in a region of the Swiss Alps with recent processes resulting from a catastrophic storm and with average rates of movement of materials in the region. That single event moved over 1,500 times as much material as the average annual vertical movement of all current processes measured in the area. (Some of Jäckli's data are included in Table 3-13 and are discussed later.)

Obviously mountain streams which contain large boulders in their beds will not be adjusted to flows that are incapable of moving the sizes of material found within the channel. The threshold and channel-forming discharge will be one competent to move the sizes available. The larger the material, the larger will be the discharge necessary to provide the stress required for moving the material. In the preceding illustrations showing the relative importance of frequent events of moderate magnitude, it was pointed out that the frequency distributions included only events beyond the threshold level. If the required competence increases upstream, then the effective frequency decreases upstream.

Additional illustrations of the effectiveness of individual climatic catastrophic events are not difficult to find, even if the descriptions lack desirable precision. Major avulsions or changes in channel direction and form occur regularly, particularly in semiarid regions and even in humid regions, during catastrophic, rare floods. A single rainstorm of 3 days' duration in California in 1938 produced as much as 150 cubic yards of sediment per acre, primarily from cultivated lands, and initiated about 700 new gullies on an area of 40,000 acres (Bamesburger, 1939, pp. 11 and 14). Boulders as large as houses have been moved by floods and mudflows.

Gullies and tributaries have been begun or enlarged by similar floods in virtually every climatic or physiographic environment. Whole beaches along coasts have been destroyed or built up during individual storms that lasted for no more than 6 to 8 hours. Mudflows are often produced in high mountain areas under short but intense storms, and these flows have produced—at the foot of or on the flanks of the mountains—deposits which are removed only through long periods of wash or erosion. During an earthquake in August 1959 a whole mountainside moved downhill, completely blocking the valley of the Madison River in Montana. Large alluvial fans in many cases have been constructed by the transport of material during intermittent events of considerable intensity. Because these events are so striking, they perhaps need less emphasis than do the less apparent but more common phenomena. Most important, however, is a better understanding of the way frequent and infrequent events combine in denuding the landscape and in molding its forms.

Ephemeral streams—that is, streams which are dry part of the year—have quite different characteristics. In the headwaters of streams near Washington, D. C., for example, hillside rills and channels 1 to 2 feet wide, and not fed by springs, carry water roughly 11 days each year, and usually only for a few hours at a time during heavy rainstorms. Such rills may also carry a trickle of water for a few days as a result of melting snow. The rill we have been observing (Fig. 3-25) not far from Seneca Creek drains a forested watershed of second-growth tulip, hickory, and beech trees. The drainage area is about 2.3 acres and the rill, with a mean gradient of 0.17 foot per foot, has a width of 1.5 feet near the mouth. Both the forest floor and the channel are carpeted with fallen leaves. In 1961, during which there was 37.6 inches of precipitation, there were 11 events during which runoff occurred in the rill, and this runoff totaled about 0.21 inch, or less than 0.6% of the precipitation.

There are many such headwater rills in well-vegetated areas. What is the effective force responsible for degrading the land when flow is so rare? Observations lead us to believe that water flows in the rill channel without any of the water flowing to it over the land surface. All the rainfall ap-

Figure 3-25.

Headwater rill at Sisters Creek, a tributary to Cabin John Creek, near Washing-ton, D.C. The drainage area is 2.3 acres. The vegetation is primarily tulip and hickory.

parently infiltrates into the soil, and what appears in the rill has moved laterally through the soil. Under these circumstances the water carries prac-tically no sediment. Whatever mineral material moves down the rill must be excavated from the rill. Yet the history of the vegetation, and thus the land, during at least 150 years as read from the tree rings and from com-position of the vegetation, indicates that the rill must have deepened, but very slowly. Another possibility is that mass movement tends to narrow the rill, and water-flow keeps the channel in equilibrium with the soil creep.

Considering the large amounts of material carried in solution in this climatic region, it is quite possible that reduction of the rill proceeds pri-marily by weathering and the export of weathering products in solution. In limestone terrain the role of solution is perhaps more easily accepted. Nevertheless, in both cases solution plays a significant role in evolution of features of the landscape, and individual catastrophic events are prob-ably of relatively little importance.

Most processes, of course, are far more complex than the example pro-

Figure 3-26.

Watts Branch, near Rockville, Maryland, a perennial stream having a drainage area of 4.2 sq. mi. **Top.** *View showing meander pattern, typical silt in cut bank, and in foreground new deposition on point bar.* **Bottom.** *Bank sloughing during early spring as a result of sapping by ice action and slump when bank is thoroughly wetted.*

vided by the transport of discrete particles on the bed of a stream. Erosion
of cohesive materials, for example, is associated with temperature and precipitation in a complex manner. The threshold of erosion is controlled by
the degree of preparation of the substrate or geologic material upon which
the shear stress must act. Freeze and thaw and moisture in the soil greatly
lower the shear stress required for erosion, and thus in many instances
erosion of a cohesive material is greatest at times when a moderate flow
attacks a land surface or channel boundary previously prepared by precipitation and frost action. This coincidence of events is illustrated by erosion of a cohesive river bank of a small stream near Washington, D. C.,
shown in the photograph of Fig. 3-26. Erosion is most rapid from Novem-
ber through March. This winter period experiences moderate flows, occasionally to the bankfull stage. Peak floods occur from summer thunderstorms. At this time, however, the stream banks are dry and resistant to
erosion. There was practically no erosion during the maximum of flood
of record experienced in July 1956 (see Fig. 3-27).

The importance to geomorphic processes of a coincidence in time of
meteorologic conditions is also illustrated by the relation between temperature, precipitation, and frequency of rockfalls in Norway above 60° north
latitude. Measurements along railroads over a period of years show that
earth slides and rockfalls are most frequent during the months when thawing of ice in the ground is at a maximum. Precipitation in the locality is
rather evenly distributed throughout the year (see graph at top of Fig.
3-28). The largest number of rockfalls occur in June and September with
an intervening decline when temperatures are higher in July and August.

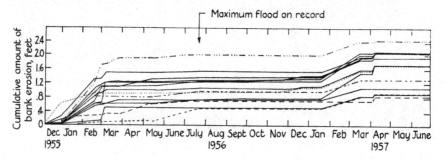

Figure 3-27.   *Cumulative bank erosion from December 1955 through June
1957 at Watts Branch near Rockville, Maryland, having a
drainage area of 4.2 sq. mi. Each line on the graph represents erosion at a single cross section. Maximum flood on
record occurred on date shown, in July 1956.*

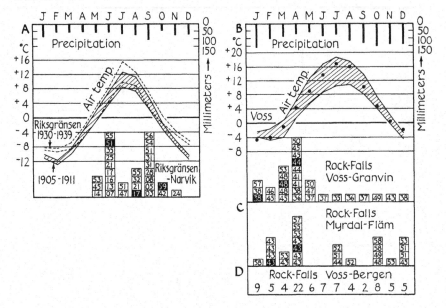

Figure 3-28.   *Diagrams showing annual periodicity of mass-movements on some railway lines in Scandinavia. **A.** Rockfalls and earth-slides recorded on the Riksgränsen—Narvik railway in 1902–1959 (68 N. lat.). **B. C. D.** Rockfalls recorded on three railway lines in western Norway (61 N. lat.). Black square means more than one fall etc. in one day. The year of each case is marked out within the square. The curves show the average max. and min. temperatures. Most of the falls are probably smaller than 10m³. Falls in tunnels are excluded. **B.** Voss—Granvin railway, 1935–1958. **C.** Myrdal— Flåm railway, 1943–1958. **D.** Voss—Bergen railway, 1920– 1958. [From Anders Rapp, 1960.]*

Rockfalls are at a minimum in winter, when the soil is frozen. On the Voss-Granvin railway rockfalls are most frequent when precipitation is least but temperatures are near the freezing point and the spring thaw has begun. This annual periodicity, dependent upon thawing of frost in the Arctic, is comparable to the periodicity observed in the latitude of Washington, D. C., where freeze-thaw cycles are perhaps more numerous but persist over roughly the same time period each year.

An additional example from this same region is shown in Fig. 3-29. Here avalanches occur following thawing after periods of snowfall. Rainfall apparently contributes to dirty avalanches (mixture of snow and debris). The relation of active movement to thawing is striking.

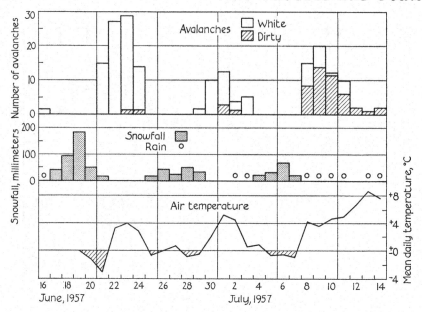

Figure 3-29.   *Correlation between frequency of avalanches from E.-facing walls, snowfall and air temperature respectively at the Mikka glacier, Sarek, Lappland. Avalanches occur mainly during thaw periods after snowfall. The number of dirty avalanches increase at the end of the snow-melting, which in this year was at least two weeks later than usual. [From Anders Rapp, after Josefsson, 1958.]*

The Blackhawk rockslide in the Lucerne Valley, California, is a feature 5 miles long, 2 miles wide, and 30 to 100 feet thick, derived from a resistant mountainside of marble and deposited on an alluvial apron at the foot of the mountain. Geological evidence, including the details of form of the deposit, indicates that the slide traversed the gentle alluvial slope as a non-deforming sheet of breccia moving at more than 50 miles per hour, riding on a layer of compressed air rather than on water or mud. From his theoretical analysis, Shreve (1960) concludes that this mode of lubrication is possible. This large slide is nearly unique in the area. In contrast, nearby desert areas have colluvial deposits flanking granitic hills, and these small fans or colluvial footslopes show ample evidence that small mudflows constitute an important source of debris.

In a different semiarid physiographic region, but also in Southern California, the familiar sequence of drought, forest fires, floods, erosion, and landslides illustrates the significance of particular combinations of climatic events in geomorphic processes. To date, there is little or no information on the frequency of such sequential events or on their relative significance in determining the morphology of the landscape.

Sand dunes and beaches provide much more transitory illustrations of the adjustment of geomorphic forms to both magnitude and frequency. Each of these forms is associated with a particular stress applied in a particular direction over a given or specified period of time. An equilibrium form of a beach or sand dune would presumably be one that is maintained over a reasonable period of time. It is a form which the geomorphologist is likely to measure in the field as the mean of a series of measurements of slope, altitude, wavelength, or whatever parameter is chosen to characterize a beach or set of sand dunes. These landforms will be associated with particular wind and wave patterns, characterized by frequency, magnitude, and direction. The orientation of a dune field is a function of the force of the wind, its direction, and the time over which the wind has blown. Although a catastrophic or hurricane wind may deform an entire sand region, during the long interval between such catastrophic winds the dune field will become reoriented and will assume a series of forms appropriate to the force and direction of the prevailing sand-moving winds. In some cases a balance may be achieved between short-lived high-intensity winds from one direction and winds of longer duration but less magnitude from another direction. The resultant prevailing sand-moving winds may be larger than the minimum required for movement, but considerably less than those of catastrophic or hurricane velocity. The form and orientation of the sand dune then can be a complex function of events of different magnitudes occurring with varying frequency. Small as well as large events play a significant role.

The conclusion must be drawn that both magnitude and frequency strongly control the significance of different geomorphic processes and of different climate controls of geomorphic processes. In some cases the extremes or the variability itself of individual climatic events are closely correlated with the mean values, a condition which would perhaps lend support to such simplifications as our use of means in the derivation of hypothetical morphogenetic regions. Fraser (1959), for example, has demonstrated for parts of central Canada that frequency of freeze and thaw is closely associated with the mean annual range in temperature. Similarly, it is known that areas of moderate precipitation or moderate runoff often have highly variable precipitation regimes. Thus many semiarid regions which have low rainfall often have rains of high intensity. It should also be noted, however, that the rains of highest intensity are associated with areas of high average precipitation.

It is not surprising to any natural scientist that detailed investigations of specific geomorphic processes reveal that the combination of biological and physical processes in nature is immensely complex. The relative importance of various geomorphic processes in most regions is known only qualitatively. Interestingly enough, this is particularly true for the mid-latitude, temperate regions where the large concentrations of population and earth scientists are to be found. Perhaps because the rates of such processes are believed to be particularly rapid in high latitudes and periglacial regions, more measurements have been made in these areas. Rapp (1960) compiled a table showing both the rates of movement of various

Table 3-12.   Annual movement of material by different processes in Kärkevagge, Northern Scandinavia, and in the Upper Rhine valley, Switzerland. [After Rapp (1960) and Jäckli (1957).]

Process[a]	Volume[b]		Tons per Square Kilometer		Vertical Movement[b]	
	Kärke-vagge (m³)	Upper Rhine (10⁶ m³)	Kärke-vagge	Upper Rhine	Kärke-vagge (ton meters)	Upper Rhine (10⁶ ton meters)
Rockfalls	50	0.016	3	—	19,545	14.3
Avalanches	88	0.25	14	9,000	21,850	45
Earth slides (rapid)						
Bowl slides	170	—	20	—	75	—
Sheet slides	190	—	23	—	20,000	—
Sheet slides continuing as mudflows	150	—	18	—	70,000	—
Other mudflows	70	—	8.4	—	6,300	—
Schist slides	—	14,000	—	100,000,000	—	224
Rapid slides	—	15	—	5,400,000	—	27
Total slides as given	580[c]	14,000+			96,375	251
Creep						
Talus creep	300,000	—	—	—	2,700	—
Solifluction	550,000	41.6	—	115,000	5,300	0.75
Creep	—	1,200	—	19,000,000	—	23
Total slow mass movement	850,000	1,200+			8,000	23.75
Running water						
Dissolved salts	150	0.224	26	135	136,500	1,196
Fluviatile erosion	—	4	—	1,740	—	—

[a] Processes are matched as closely as possible for the two areas but designations are not precisely the same in each case. Period of record: Kärkevagge 1952–1960, Rhine region 1949–1957, plus historical records dating back 350 years.
[b] Note variation in units due to differences in the size of the areas covered.
[c] May be too large due to unusually heavy rainstorm in 1959.

Table 3-13. Amount of displacement in Upper Rhine area: Estimate of yearly average. [From Jäckli (1957).]

	Mean Thickness (m)	Areas Concerned (10⁶ m³)	Areas Concerned (%)	Volume Moved (10⁶ m³)	γ (t/m³)[e]	Weight Moved (10⁶ t)[e]	Amount Moved During the Year (10⁶ mt vertically)[e]
Rockfalls	—	—	—	0.016[a]	2.0	0.032	14.3
Avalanches	0.065[a]	650[a]	15	41.6[b]	1.8	74.9	0.75
Soil flow	14.1[b]	1.26[b]	0.03	17.8[b]	1.9	33.8	2.5
Rockflows							
Earth slides							
Rapid	3[a,e]	5[a,c]	0.12	15[a]	1.8	27	27
Creep	10[a]	120[a]	2.8	1,200[a]	1.9	2,280	23
Schist slides	50[a]	280[a]	6.5	14,000[b]	2.0	28,000	224
Total fluvial effect:	—	—	—	4.0015	1.8	7.2	5,933.27
Deposition in Upper Rhine area		178.3[b,d]	4.15	1.7415[a]	1.8	3.13	1,456
Valley bottoms	0.002[b,d]	25[b,d]	0.58	0.05[a]	1.8	0.09	72
Lakes	0.00066[a,d]	2.3[b,d]	0.05	0.0015[a]	1.8	0.0027	0.27
Alluvial fans	0.01[a,d]	145[a,d]	3.38	1.45[a]	1.8	2.61	1,040
Artificial fill	0.04[a,d]	6[b,d]	0.14	0.24[a]	1.8	0.43	344
Deposition in lakes	0.0009[a,e]		—	2.26[a]	1.8	4.07	4,477
Correlated erosion in drainage areas		4,128.7[a,e]	95.85	4[a]	1.8	7.2	—
Glacial transport	0.2[a]	50[b]	1.15	10[a]	2.7	27	27
Avalanche transport	0.005[a,c]	50[a]	1.15	0.25[a]	1.8	0.45	45
Chemicals in solution							
Carbonates	0.00004[b,e]	4,307[b]	100	0.172[b]	2.6	0.449	920
Total dissolved load	0.000052[b,e]	4,307[b]	100	0.224[b]	2.6	0.584	1,196
Year's total				—		—	15,804.09
Comparative total							
Durnagelbach near Linthal, aggradation of Wildbachschutt on Aug. 24, 1944	2.4[b,d]	0.189[b,d]		0.45[b]	1.8	0.81	810
Flimser-rock slide, prehistoric	250[a,e]	40[a,d]		10,000[a]	2.2	22,000	24,200,000

[a] Rough estimate.
[b] Exact measurement.
[c] Corresponding to the area of erosion.
[d] Corresponding to the area of sedimentation.
[e] In the column headings, t is tons and mt is meter tons.

processes and the amounts of material which each has removed from the slopes of a small valley in northern Sweden. Jäckli (1957) made a similar analysis for a somewhat shorter period of record in a periglacial region of the upper Rhine in Switzerland. These studies are major contributions to our meager understanding of the quantitative characteristics of dynamic geomorphology. The data are summarized in Tables 3-12 and 3-13. Comparison of processes within each region shows that in northern Sweden various kinds of earth slides and slow creep move the greatest volume of material from the mountain slopes. Slides are similarly important in the Alpine region but creep is less so. It is also interesting to note that a large quantity of material is removed from the drainage basin in northern Sweden in solution despite the cold climate and limited precipitation.

Although the data from the two areas are not precisely equivalent, having different degrees of accuracy and somewhat different designations of processes, such studies highlight the need for comparable and continuing observations in a multitude of climatic and physiographic regions. In more moderate climates the apparent slowness of geomorphic processes has helped to dampen enthusiasm for such studies. Inaccessibility and danger have similarly limited observations of catastrophic events in progress and have thus limited information to that which can be gotten after the fact. Both kinds of study offer fruitful avenues for future research.

To summarize, information is inadequate to draw any but the most general conclusions about the relative importance of various geomorphic processes, or the relative significance of different frequencies of climatic events controlling such processes. Such information, of course, is necessary to an understanding of the processes of land sculpture. It becomes doubly important in seeking to understand how climatic changes known to have occurred have influenced the development and preservation of landforms. Any concept which stresses the importance of rare or infrequent events in land sculpture must include those rare events in the definition of the "climate" of the period. These may have frequencies of perhaps once in 300 to 500 years. What, then, constitutes climatic change? In contrast, small changes in annual precipitation of very moderate intensities may have a marked effect on the growth of vegetation in an area. The latter, in turn, markedly influences sediment yield and channel development. Thus, a subtle change in climate associated with relatively frequent events or a modest increase in the frequency of rare events may have a pronounced effect on geomorphic events and landforms. Considerably more information is necessary for a fundamental grasp of these influences.

REFERENCES

André, J. E., and Anderson, H. W., 1961, Variation of soil erodibility with geology, geographic zone, elevation, and vegetation type in northern California wildlands: *Journ. of Geophys. Res.*, v. 66, pp. 3351–3358.

Bamesburger, J. G., 1939. Erosion losses from a 3-day California storm: *U. S. Dept. of Agric., Soil Cons. Serv.*, U. S. Govt. Printing Office. Washington, D. C., 21 pp.

Chorley, R. J., 1959, The geomorphic significance of some Oxford soils: *Am. Journ. Sci.*, v. 257, pp. 503–515.

Corbel, J., 1959, Vitesse de l'érosion: *Zeits. für Geomorphologie*, v. 3, pp. 1–28.

Dalrymple, T., 1960, Flood frequency analysis: Manual of Hydrology, part 3, Flood flow techniques. *U. S. Geol. Survey Water Supply Paper* 1543-A, 80 pp.

Durum, W. H., Heidel, S. G., and Tison, L. J., 1960, World-wide runoff of dissolved solids: International Asso. of Scientific Hydrology, General Assembly of Helsinki, Publication No. 51, pp. 618–628.

Fraser, J. K., 1959, Freeze-thaw frequencies and mechanical weathering in Canada: *Arctic*, v. 12, pp. 40–53.

Griggs, D. F., 1954, High pressure phenomena with applications to geophysics; Modern Physics for the Engineer, ed. L. N. Ridenour: McGraw-Hill, New York, pp. 272–305.

Grim, R. E., 1962, Applied clay mineralogy: McGraw-Hill, New York, 422 pp.

Hack, J. T., and Goodlett, J. C., 1960, Geomorphology and forest ecology of a mountain region in the central Appalachians: *U. S. Geol. Survey Prof. Paper* 347, 66 pp.

Hopkins, D. M., and Sigafoos, R. S., 1952, Frost action and vegetation patterns on Seward Peninsula, Alaska: *U. S. Geol. Survey Bull.* 974C, pp. 51–101.

Hurst. H. E., 1950, Long-term storage capacity of reservoirs: *Am. Soc. Civil Engrs. Trans.*, v. 116, pp. 770–808.

Jäckli, H., 1957, Gegenwartsgeologie des bundnerischen Rheingebietes-ein Beitrag zur exogenen Dynamik Alpiner Gebirgslandschaften: *Beiträge zur Geologie der Schweiz*, Geotechnische Serie, Leiferung 36, p. 126.

Josefsson, B., 1958, Studier över temperaturgangen i fast berg och lavinaktivitet vid Mikka glaciaren sommaren 1957. G.I.U. (unpublished).

King. H. W., 1954, Handbook of hydraulics: McGraw-Hill, New York, 563 pp.

Langbein, W. B., et al., 1949, Annual runoff in the United States, *U. S. Geol. Survey Circ.* 52, 14 pp.

———, 1949, Annual floods and the partial-duration flood series: *Am. Geophys. Union Trans.*, v. 30, no. 6, pp. 879–881.

Langbein, W. B., and Schumm, S. A., 1958, Yield of sediment in relation to mean annual precipitation: *Am. Geophys. Union Trans.*, v. 39, pp. 1076–1084.

Langbein, W. B., et al., 1962, Scientific hydrology report of the ad hoc panel on hydrology to the Federal Council for Science and Technology, Washington, D. C., 37 pp.

Leopold, L. B., 1959, Probability analysis applied to a water-supply problem: *U. S. Geol. Survey Circ.* 410, 18 pp.

Meier, M. F., 1960, Mode of flow of Saskatchewan glacier, Alberta, Canada: *U. S. Geol. Survey Prof. Paper* 351, 70 pp., esp. pp. 44–45.

Menard, H. W., 1961, Some rates of regional erosion: *Journ. Geol.*, v. 69, no. 2, pp. 154–161.

Musgrave, G. W., 1947, Quantitative evaluation of factors in water erosion —first approximation: *Journ. Soil and Water Cons.*, v. 2, no. 3, pp. 133–138.

Nace, R. L., 1960, Water management, agriculture and ground-water supplies: *U. S. Geol. Survey Circ.* 415, pp. 1–11.

Rapp, Anders, 1960, Recent development of mountain slopes in Kärkevagge and surroundings: *Geografiska Annaler*, v. 42, p. 185.

Sharp, R. P., and Nobles, L. H., 1953, Mudflow of 1941 at Wrightwood, Southern Calif.: *Geol. Soc. Am. Bull.*, v. 64, pp. 547–560.

Sharp, R. P., 1954, Glacier flow; a review: *Geol. Soc. Am. Bull.*, v. 65, pp. 821–838.

Shreve, R. L., 1960, Geology of the Blackhawk Landslide, Lucerne Valley.

Calif.: *Geol. Soc. Am. Bull.*, v. 71, no. 12, pt. 2, pp. 2075–2076.

Smerdon, E. T., and Beasley, R. P., 1959, The tractive force theory applied to stability of open channels in cohesive soils: *Missouri Univ. Agr. Expt. Sta., Res. Bull.* 715, 36 pp.

Strahler, A. H., 1952, Dynamic basis of geomorphology: *Geol. Soc. Am. Bull.*, v. 63, pp. 923–938.

Tricart, J., et collaborateurs, 1961, Mécanismes normaux et phénomenes catastrophiques dans l'évolution des versants du bassin du Guil (Htes-Alpes, France): *Zeits. für Geomorphologie*, v. 5, pp. 276–301.

Wahrhaftig, C., and Cox, A., 1959, Rock glaciers in the Alaskan Range: *Geol. Soc. Am. Bull.*, v. 70, pp. 383–436.

Wischmeier, W. H., and Smith, D. D., 1958, Rainfall energy and its relationship to soil loss: *Am. Geophys. Union Trans.*, v. 39, pp. 285–291.

Wolman, M. G., and Miller, J. P., 1960, Magnitude and frequency of forces in geomorphic processes: *Journ. Geol.*, v. 68, no. 1, pp. 54–74.

Woodburn, R., and Kozachyn, J., 1956, A study of relative erodibility of a group of Mississippi gully soils: *Am. Geophys. Union Trans.*, v. 37, pp. 749–753.

Chapter 4 Weathering

Land, then, is not merely soil; it is a fountain of energy flowing through a circuit of soils, plants, and animals. Food chains are the living channels which conduct energy upward; death and decay return it to the soil. The circuit is not closed; some energy is dissipated in decay, some is added by absorption from the air, some is stored in soils, peats, and long-lived forests; but it is a sustained circuit, like a slowly augmented revolving fund of life.

ALDO LEOPOLD

A Sand County Almanac
(OXFORD UNIVERSITY PRESS)

Rocks and Water—the Components of Weathering

Valleys, hills, and mountains arrayed across the surface of the earth present almost infinite variety of form and character. Such a complex topographic configuration reflects ultimately the interaction of all the dominant forces and subtle influences that modify the earth's surface. Erosional sculpture of the landscape, our primary concern in this book, involves removal of debris formed by weathering of rocks. Sound, fresh rock is almost immune to attack by the various gradational agencies; broken and decomposed residues of rocks are the materials transported by gravity, flowing water, and wind. As a necessary prelude to erosion and mass wasting, weathering is a fundamental geomorphic process.

Knowledge of weathering processes and products enters into practically all phases of dynamic and historical geomorphology. It has been said that streams are the gutters down which flow the ruins of the continents. Silt, sand, gravel, and solutes carried by the streams are the "ruins" produced largely by weathering. In most places, bedrock is hidden from view by a mantle of soil. We can see in roadcuts and excavations that the soil is weathered rock, with greatest alteration near the surface and grading downward to fresh rock. Thickness and composition of soils are quite variable. The deepest weathering takes place in warm, humid areas with low relief. Differences in elevation of sandstone ridges and limestone valleys, for example in the Appalachian Mountains, must reflect in part different resistances to weathering of the rocks. Buried soils, which are especially common in Pleistocene and Recent deposits, provide important clues to past environments.

Formally, we may define weathering as the spontaneous and essentially irreversible response of rocks to conditions at the earth's surface. Physical disintegration and chemical decomposition transform massive rock into the clastic, colloidal, and soluble states. The visible product of weathering is an unconsolidated mantle, which includes in varying proportions (1) fragments and residues of the original rocks and minerals, (2) newly formed substances produced by regrouping of constituents present in the original rocks, and (3) organic matter in various stages of decomposition. The energy that drives weathering processes is derived partly from the sun (circulated by the atmosphere and stored in organisms), and, in part, from rock minerals themselves. Inorganic chemical reactions of weathering are exothermic (give off heat), and weathering results in a decrease in the energy level of the materials affected.

The importance of weathering in the preparation of land surfaces for action by agents of landscape sculpture is recognized; however, the detailed knowledge of weathering is meager and restricted to certain aspects. Thus with regard to some of the principal problems of geomorphology the pertinent information regarding weathering is qualitative. Keller (1962) gives an exposition of the principles and processes of weathering. Because of the importance of the subject to an understanding of geomorphic processes in general, however, a brief review is warranted here.

There is a large difference in coherence or mechanical competence between the fresh crystalline rock and the unconsolidated materials produced by grain boundary reactions. It should be emphasized that a rock is not a convenient concept for the discussion of the chemical weathering process itself. Rather, a rock should be viewed as an assemblage of phases (minerals) and that each phase has its own intrinsic properties, such as solubility, that are nearly independent of the presence of the other phases. The aqueous solutions resulting from weathering reactions should be regarded as the product of the sum of each individual reaction found in the weathering zone. The weathering reaction of calcite, for example, may be written as

$$H^+ + CaCO_3 = Ca^{2+} + HCO_3^-$$

and so treated without regard to the presence of the other phases. It also should be noted that when weathering is considered in terms of mineral phases rather than rocks, the nature of the granular material supplied to the erosive agencies can be treated more rationally. Many phases or minerals are common to a variety of rock types. Quartz, biotite, muscovite, plagioclase, and orthoclase may be varied in their proportions to produce a wide variety of rock types. In spite of the rock type, many of the weathering reactions would be the same. It is only when the reactions of the particular minerals are considered that order may be brought out of the chaotic literature on weathering. Many weathering studies have been undertaken to demonstrate the effect of "climate" or "rock type" on weathering and have, as a consequence, produced ambiguous results.

In consideration of the loci of weathering action, it should be kept in mind that grain boundaries in crystalline rocks would be expected to be especially reactive because compressibilities and thermal expansions are not uniform for all minerals. Even in carefully annealed crystalline rocks, grain boundaries are more highly stressed than the interiors of the grains, leading to higher solubilities. Furthermore, with regard to monomineralic rocks, the compressibility and thermal expansion, being vector properties, are not the same in adjacent grains with different crystallographic orientations. In consequence, weathering reactions should be especially noticeable along grain boundaries in igneous and metamorphic rocks.

Extensive disaggregation resulting from chemical action on grain boundaries would be expected to be accompanied by but slight changes in average chemical composition. Confusion on this point has led to the overemphasis of "physical weathering" because the importance of chemical action in this disaggregation was not appreciated.

With these considerations in mind, the present chapter is organized as follows. First we discuss briefly earth materials, then weathering processes and products with some discussion of the effects of particular controlling factors. This is followed by a discussion of soils, in which are found the integrated effects of the various factors seen in the field.

Materials Subject to Weathering

The composition and abundance of rocks exposed to weathering at the earth's surface are of importance in viewing weathering processes. Although there are nearly 2,000 minerals, dozens of named varieties of rocks, and 90-odd natural chemical elements, only a few in each of these categories occur at the earth's surface in appreciable amounts.

Almost the entire chemical composition of all the earth's crust is accounted for by only eight elements: oxygen (47% by weight), silicon (28%), aluminum, iron, magnesium, calcium, sodium, and potassium (see Mason, 1958). On a volume basis, the crust is composed almost entirely of oxygen anions, which are bonded to interstitial metal cations. The average composition of the crust is approximately the same as the average for igneous rocks because sedimentary (and metasedimentary) rocks comprise less than 5% of the volume of crust.

The total area of the earth above sea level, and hence subject to subaerial weathering, is about 57 million square miles (149 million square kilometers). Sedimentary rocks of various kinds are exposed over roughly 75% of this area, leaving only 25% for the outcrop area of igneous and metamorphic rocks. Areal distribution of various rock types at the earth's surface is complex and cannot be treated adequately in this summary. However, for our purposes it will suffice to note the relative proportions of the most abundant rock types. Five classes of rocks occupy more than 90% of the continental area, distributed approximately as follows:

Shale	52%
Sandstone	15
Granite (and granodiorite)	15
Limestone (and dolomite)	7
Basalt	3
Others	8

Combining the mineralogic data with the relative abundance of rock types gives the approximate proportions of common minerals exposed to weathering at the earth's surface:

Feldspars	30%
Quartz	28
Clay minerals and micas	18
Calcite and dolomite	9
Iron oxide minerals	4
Pyroxene and amphibole	1
Others	10

In summary, the zone of weathering involves only half a dozen major rock types, composed mostly of 8–10 mineral groups and 8 chemical elements.

Rocks of similar chemical and mineralogical compositions may react very differently to the same weathering environment. For example, coarse-grained granites weather more rapidly than fine-grained granites. Likewise, volcanic ash deposits composed of glass shards may be converted to fertile soils within a few years whereas large masses of similar glasses remain essentially unaltered for thousands of years. Weathering requires the presence of water, and hence rock texture is important because it largely controls penetration of water into rocks. Porosity data for the common rocks are given in Table 4-1. Permeable materials must be porous, but quantita-

Table 4-1. Approximate average porosities for rocks and unconsolidated deposits. [Compiled from several sources, especially Spicer (in Birch et al., 1942) and Kessler and Sligh (1940).]

ROCK TYPE	POROSITY	UNCONSOLIDATED	POROSITY
Granite	1%	Clay	45%
Basalt	1	Silt	40
Shale	18	Sand	35
Sandstone	18	Gravel	25
Limestone	10		

tive relations between these two properties are obscure and variable. Available information suggests the following relative values of permeability:

ROCKS	RELATIVE PERMEABILITY
Igneous (and most metamorphic rocks)	1
Shales	5
Limestones	30
Sandstones	500
UNCONSOLIDATED	
Clays	10
Sands	1,100
Gravels	10,000

Whatever the accuracy of these figures, they emphasize the tremendous range of natural materials in their capacity to transmit water.

Composition of Rain and Snow in Relation to Weathering

Weathering processes of all kinds require the presence of water. Naturally, whatever is dissolved in the water will affect the chemical reactions that occur. Rain and snow are the purest forms of water in nature, but even they are not perfectly pure. Small amounts of atmospheric gases and inorganic impurities of several kinds occur in rain. Apart from the gases, chemical composition of rainwater is highly variable, depending on distance from the sea, terrestrial contamination by natural means, and contamination by man. The major dissolved constituents are sulfate, bicarbonate, chloride, nitrate, calcium, sodium, magnesium, ammonium, potassium, and hydrogen ions.

The proportions of the ionic constituents vary seasonally and areally. The supply of natural terrestrial dust and its composition is important, in addition to sea salt sources, as has been shown by Junge and Werby (1958). Average rain compositions tend to reflect the availability and composition of terrestrial dust, with calcium and sulfate concentrations being high over the American arid southwest. Potassium in average rain has been ascribed to terrestrial sources by the same authors in view of the fact that the potassium-sodium ratio in rain is much higher (1 as compared to .04) than in the sea. The sodium and chloride content of average rain decreases inland, reflecting the oceanic sources of much of the sodium and chloride. Junge and Werby (1958) find the maximum and minimum monthly average composition of rains over the continental United States to be as follows:

	MAXIMUM (ppm)	MINIMUM (ppm)
Ca	6.5	0.27
Na	8.0	.14
K	0.8	.06
SO_4	10.8	.69
Cl	8.85	.09

Although the maximum concentrations in precipitation may approximate values found in natural streams, as the data for streams in the Sangre de Cristo range, New Mexico, suggest (Table 4-2), most of the solutes in the waters are from terrestrial sources, not from precipitation. Gorham (1961, p. 803) cites analyses of Japanese rivers which indicate that the terrestrial contribution of metallic ions is roughly 60 to 90%, although 40% of the calcium in one sample was derived from precipitation. Close correlation of ionic composition of rivers and specific rock types also demonstrates that it is the weathering reactions which give rise to the higher solute concentrations which are observed.

The pH of rain reportedly ranges from 4 to 8, with the average value approximately 6 in places where air is not appreciably contaminated. The pH data from field and laboratory measurements are commonly different for the same water sample. The pH of rain is almost exclusively controlled by the concentration of dissolved CO_2. The solubility of CO_2 is, in turn,

Table 4-2.   Dissolved constituents in streams draining various lithologies in the Sangre de Cristo range, New Mexico. [From Miller (1961).]

	GRANITE		QUARTZITE		SANDSTONE	
	Range	Average	Range	Average	Range	Average
	(11 samples)		(5 samples)		(7 samples)	
Total dissolved (ppm)	20–43	35	7–20	14	87–186	143
Percentage						
SiO$_2$	17–30	26	22–30	26	2–6	3
Fe	0–0.7	0.12	0–0.07	0.03	0–0.01	<.01
Ca	9–14	10	10–17	13	17–23	20
Mg	1.0–2.6	1.5	0–1.5	0.6	0.5–3.0	1.9
Na	4.5–6.3	5.8	2.9–4.4	3.6	0.6–1.6	0.9
K	1.4–2.4	1.8	2.0–2.7	2.2	0.3–0.6	0.4
HCO$_3$	40–50	44	37–47	41	52–67	59
SO$_4$	7–12	9	7–17	12	5–20	13
Cl	0–1.3	0.5	0–1.5	0.4	0–0.3	0.1
NO$_3$	0–1.0	0.3	0–1.5	1.0	0–0.15	0.0ზ
Field *p*H	6.9–7.3	7.1	6.4–6.7	6.6	7.7–8.1	8.0
Drainage area (square miles)	0.52–33.6		1.2–5.1		0.82–16.2	
Altitude (feet)	6,700–10,400		9,600–10,700		8,300–10,100	

controlled by the temperature of the solution and the concentration of CO$_2$ in the coexisting gas. Laboratory atmospheres commonly contain 10 times as much CO$_2$ as the normal earth's atmosphere and are rarely at the same temperature as the sample under field conditions. Only *p*H data from field observations are of significance in formulating weathering reactions.

The principal source of hydrogen ions (H$^+$) is the equilibrium between water and atmospheric carbon dioxide, which yields a *p*H between 5 and 6.5. As Keeling (1960) has shown, the CO$_2$ content of the atmosphere is remarkably uniform at 306 to 318 ppm. Because of temperature and barometric pressure ranges, a single *p*H of rain cannot be predicted. Another source of H$^+$ ions in soil water is the activity of plants. During growth plants produce and exchange H$^+$ for cation nutrients. Decay of organic matter produces organic acids and CO$_2$, which dissolve and lower the *p*H of soil water. When rain reaches the ground, additional carbon dioxide and various cations may be dissolved very quickly. This alters the chemical character of the water and determines what additional material can be dissolved at any particular time. Little is known about the changes in the chemical character of water as it passes from the surface through soil and rock and into the zone below the water table.

Rainwater and snowmelt are not saturated with respect to any mineral. To put it another way, all minerals have a solubility greater than zero in water derived from atmospheric precipitation. All minerals consequently undergo some reaction with water, and the tendency in the reaction is toward a more stable state. The water composition tends toward saturation with the minerals which are present. Some minerals alter to secondary minerals (solid alteration products) having lower solubilities than the primary mineral—another reaction tending toward a more stable state. The solubility of solids is somewhat dependent on the size of the particle. The effect becomes particularly important at sizes less than 2 microns. The pH and oxidizing or reducing conditions are other factors that affect solubility.

Mineral-Water Reactions

Mineral-water reactions may be discussed according to several types. The first, expressed by the equation

$$\text{Mineral} + \text{water} = \text{solution},$$

is one in which the same phases (mineral and water) are present before, during, and after the reaction. Minerals undergoing such a reaction leave no alteration products. If they react completely (dissolve), there may remain no record of their initial presence. The importance of such a reaction may lie in the fact that the parent material may never be reconstructed by an observer who has available only a weathered zone. The inability to reconstruct unambiguously the parent material precludes the possibility of obtaining appropriate chemical reactions to describe the weathering process. Many simple minerals—such as calcite, gypsum, anhydrite, the various forms of silica, hematite, goethite, and dolomite—dissolve in rainwater. The list could be greatly extended.

A reaction involving ion exchange leads to a change in the composition of both phases. Thus,

$$\text{Mineral (composition 1)} + \text{solution (composition A)}$$
$$= \text{mineral (composition 2)} + \text{solution (composition B)}$$

Montmorillonite, for example, readily exchanges sodium and calcium with coexisting aqueous solutions. Silicates with framework or sheet structures show greatest exchange capacity. The most readily exchangeable cations of silicates are those which perform the function of balancing excess charge that results from the substitution of ions within the mineral structure, such as Al for Si in silica tetrahedra. The most common exchangeable ions are Na, Ca, and K.

The exchange capacity of the weathering product varies greatly with content and nature of the clay and organic matter. Rate of exchange is generally rapid, requiring only a few minutes. However, it depends on the exchange site, and is greatest on the surface of mineral particles. Temperature has only a slight effect on rate of ion exchange. Neutralization of acid clays (clays with exchangeable H^+) requires much longer than ordinary ion exchange.

Except for soluble constituents such as $CaCO_3$, clay acting as an ion-exchange material is the most reactive part of a weathering horizon (Kelley, 1948). As shown in Table 4-3, montmorillonite and vermiculite have the highest exchange capacities, and kaolinite the lowest, with illite and chlorite intermediate. But these values of exchange capacity are small relative to those of organic colloids; the result is that only a few percent of

Table 4-3.   Environments of common soil clays, compiled from various sources.

Mineral	Conditions of Formation	Soil Types in Which Clay Is a Dominant Mineral	Exchange Capacity (meq/100 g at $pH = 7$)
Kaolinite	Leaching, either oxidizing or reducing, $pH \geqq 4$ and <7	Laterites; red, yellow, gray and brown podzols	3–15
Illite	Fairly wide range of conditions, but not intense leaching and oxidation; pH not critical	Prairie, rendzina, solonetz, alkali, planosol, desert, chernozem, podzol	10–40
Montmorillonite group	Neutral or alkaline, unstable under acid leaching; $pH \geqq 7$	Chernozem, prairie, chestnut humic-gley, weisenboden, rendzina, tropical black, solonetz planosol, alkali, some latosols	60–150
Vermiculite	Fairly wide range of conditions, pH not critical	Podzol, tundra	100–150
Chlorite	Fairly wide range of conditions	Podzol, tundra, possibly many others	10–40

organic matter may greatly increase the total exchange capacity of a soil. Except for the K^+ of illite, cations of clay minerals are readily exchanged.

Exchange reactions are reversible, and different ions may replace one another. There is no single universal order of the replacing power of either cations or anions. Rather, this relation depends on (1) composition of the exchange material, including both minerals and organic matter, (2) exchange capacity, (3) nature of the ions, including size, charge, etc., and (4) concentration of the solution containing the replacing ion. Capacity for holding cations increases with increasing pH; the reverse is true for anions.

In contrast to the preceding two reactions, in a third,

$$\text{Mineral A} + \text{water} = \text{mineral B} + \text{solution,}$$

a new phase (mineral B) appears during the reaction. Such, for example, is the oxidation reaction,

$$\text{Pyrite} + O_2 + \text{water} = \text{limonite} + \text{solution.}$$

Oxidation in weathering processes commonly involves literal combination with oxygen. This is not a necessary requirement, however, for by definition a substance is *oxidized* when it loses electrons and *reduced* when it takes on electrons. Iron, titanium, manganese, copper, and phosphorus are the principal elements of rocks and soils that undergo oxidation and reduction. For iron, the relation is

$$Fe^{2+} = Fe^{3+} + e^-.$$

Other common constituents that take part in oxidation and reduction reactions are organic materials—nitrogenous and sulfur compounds. By comparison with the tremendous amount of oxidation and reduction associated with growth and decay of organic matter, mineral soil is relatively inert in this respect.

One of the most important weathering reactions is hydrolysis. In this reaction water is a reactant and not merely a solvent and the hydrogen ion, H^+, enters into the silicate structure:

$$\text{Feldspar} + \text{water} = \text{clay} + \text{solution.}$$

The net physical effect of hydrolysis is an exchange of H^+ ion from the water for a cation of the mineral, leading to expansion and decomposition of the silicate structure and increasing the pH of the water.

The increase in volume which accompanies hydrolysis is the most important physical effect of chemical weathering and greatly exceeds in effect the various mechanisms of physical weathering. Granular decay of coarse-grained rocks such as granite may be attributed more to hydrolysis of the feldspars than to any other action.

Even in the desert hydrolysis is of major importance. Both exfoliation, the physical mechanism by which the exterior surfaces of rocks spall or break off, and granular disintegration have been clearly associated with moisture even in deserts (Schattner, 1961, pp. 254–258). In the Sinai desert, where solar insulation is least, granular disintegration and exfoliation are greatest. The latter takes the form of large-scale exfoliation in sheets of several square meters on rock faces, spheroidal weathering in concentric spheres around boulders, and scaling off of small thin pieces primarily on hard, fine-grained rocks.

Products of Chemical Weathering and Some Determining Factors

The materials observed in any weathered zone will depend on the original materials to be weathered and on a variety of other factors, the separate effects of which are not easily distinguished because of the joint interaction of all. The observed products further depend on what has been removed and what has been newly formed during the weathering process. The removal of materials during weathering implies that evidence of both the original conditions and the controlling factors may be lost by that removal, a fact that must be kept in mind when interpreting the origin of the weathering products observed in the field. A major problem common to many processes affecting the earth's surface, including weathering, is that genesis must be inferred from an analysis of the end results with little or no direct knowledge of what actually happened or of the time that it took. The time factor is extremely difficult to appreciate in the field and impossible to duplicate in the laboratory.

Some products resulting primarily from chemical decomposition and associated movement of materials, as well as some of the factors which affect these products, will now be considered.

Products of Weathering—Clay Minerals

Clay minerals which occur in weathering products have three principal sources. They may be (1) inherited from parent material, (2) formed by alteration of parent minerals with slight structural change, for example, muscovite (illite) or biotite (vermiculite), and (3) formed by synthesis from weathering products and structurally different from them. Van Houten (1953) and others have discussed the problem of distinguishing between inherited and authigenic clay minerals, especially in the case of soils derived from sedimentary rocks and unconsolidated sediments.

The specific conditions of clay mineral genesis are poorly known. Laboratory synthesis of clay minerals under conditions comparable to those prevailing in nature has not been accomplished. However, the composition of soil clays, as related to parent material and environmental conditions, permits certain conclusions on the subject of origin. A summary of the environmental conditions leading to the formation of different clay minerals in soil profile development is presented in Table 4-3.

Barshad (in Bear, 1955) argues that the most important chemical and mineralogical changes of parent material during weathering are the result of clay formation from other minerals and the gain or loss of clay minerals by migration in the soil profile.

Cation content and *p*H are the decisive factors in genesis of these clays. Kaolinite forms in laterites under acid conditions, but free percolation is a requisite. Montmorillonite develops in black soils under neutral or slightly alkaline conditions. The difference in the conditions stems from the nature of the removal of the solutes, as will be discussed. With very heavy rainfall, weathering of very porous materials may result in such rapid removal of decomposition products that clay minerals do not form.

As Table 4-3 suggests, regional distinctions may be difficult to make inasmuch as it is the chemical environment, determined by both parent material and hydrology, which ultimately is the controlling factor. Thus, a given rock type may weather to different clay minerals, depending on conditions of drainage. The kind and amount of alkalies and alkaline earths available for weathering reactions also control the type of clay mineral which can form. Thorough leaching of almost any alumino-silicate rock will produce kaolin (or gibbsite, etc.) whereas partial leaching tends to form montmorillonite and illite.

Effects of Removal by Erosion

Ascertaining the genesis of weathered materials observed in the field is complicated by varying rates of removal by erosion. As stated earlier, all minerals have some solubility in water. Minerals continue to dissolve as long as unsaturated solutions move through the weathering zone, and hence there is no "final" or "end product" of weathering. The tendency is to greater stability (lower solubilities) in mineral assemblages resulting from weathering. To illustrate, let the sequence of appearance be

$$Orthoclase \rightarrow illite \rightarrow kaolinite \rightarrow gibbsite$$

in a particular weathering system. If erosion rates exceed the time rate of production of a mineral, the chain may be broken and the "end product" may be any of the minerals in the sequence.

Erosion rates are not well known because erosion is not a uniform removal of a film; rather, it consists of both widespread surface removal and channel erosion. Knowing how much material is removed in a time interval over a known area of the earth is not sufficient for the computation of a realistic rate of removal of the surficial material because the distribution of the material before erosion is not known. It is quite improbable that the distribution was uniform. Variable thicknesses of the weathered zone would be expected simply because erosion is not uniform.

Vegetation is also important in governing the rate of removal of weathering products. Root mats tend to hold soils in place, prolonging the weathering period until the protective plant cover is breached. Failure of the protective root system generally is not uniform under natural conditions, and this leads to uneven erosion (gullying) and consequently an uneven thickness of the weathered material above the bedrock. The stability of vegetation may be more dependent upon extreme conditions or catastrophic events than on long-term mean conditions. A prolonged drought may kill plants. Fire can do the same even during a year of average or greater than average precipitation. A sequence of freeze-thaw events in the Arctic may kill plants in certain years, even though the mean temperature is normal. The brevity of the extreme conditions which may affect the vegetation is in sharp contrast to the long periods presumably represented by accumulation of a thick residue of weathering products. But the extreme conditions may leave no durable record except erosion.

Effects of Weathering Sequence

The order in which minerals are affected by agents of weathering—that is, the weathering sequence—may be specified for fine-grained minerals; it is presented in Table 4-4. This sequence is based on the relative abundance of residual minerals in the clay fractions of soils. Results of other studies show at least partial agreement with this sequence. For example,

Table 4-4.   Weathering sequence of very fine-grained minerals. [According to Jackson et al. (1948).]

Primary minerals	1. Gypsum (halite) 2. Calcite (aragonite, dolomite)	soluble, unweathered or as secondary deposits
	3. Olivine-hornblende (diopside) 4. Biotite (chlorite, glauconite) 5. Albite (microcline, anorthite)	easily and rapidly weathered
Secondary minerals	6. Quartz 7. Illite (muscovite) 8. Hydrous mica intermediates 9. Montmorillonite	slowly weathered
	10. Kaolinite 11. Gibbsite 12. Hematite (goethite, limonite) 13. Anatase (rutile, ilmenite, corundum)	weathered extremely slowly

Ross and Hendricks (1945) state that montmorillonite-type minerals are known to have persisted for long periods but they probably are not the normal end products of weathering. Under acid or neutral leaching and oxidizing environments, kaolin minerals may form from montmorillonite.

Effects of Transport of Materials

The identification of reactants and products should not be the sole result of a weathering study. The disposal of solutes resulting from weathering reactions is of great importance and may affect the course of a reaction, the final products, and the rate at which the reaction proceeds. Two types of movement are possible—continuous and discontinuous transport. During continuous transport the solutes remain in solution throughout the system being studied. Discontinuous transport means that some or all of the solutes are deposited within the system being studied. Discontinuities in transport may result from evaporation or from a change in the composition of the solution due to another reaction which then leads to supersaturation. Continuous and discontinuous transport may be found in all climates, and indeed in portions of the same drainage basin with the same climate and geology. Continuity of transport is controlled by many hydrologic factors of which climate is only one.

Carbonates and sulfates of calcium and magnesium may be concentrated in the soil because of discontinuous transport. Water enters the soil during infrequent rains and reacts with soil particles, especially by solution and base exchange, and carries solutes downward. However, the depth of percolation is limited by the supply of water provided by individual storms, and by losses resulting from evaporation and transpiration. Evaporation may take place within the soil, especially if there are sharp gradients in temperature. Consequently, the dissolved constituents may be precipitated, generally as nodules and irregular aggregates of calcite and gypsum. These precipitates are often called "caliche" in the United States and Spanish-speaking regions. They are known elsewhere by a variety of names.

In calcareous soils, Ca and Mg carbonates commonly constitute 50 to 70% of the total soil mass. Alkali soils often contain sodium salts, which accumulate due to evaporation of soil water. Extensive and long-continued carbonate precipitation may lead to development of thick caliche crusts, such as the "cap rock" of the High Plains in Texas and New Mexico and similar features in other parts of the world.

Effects of Precipitation and Temperature

The local control on the extent and direction of weathering reactions is the total water supply and microhydrology of the weathering zone. Attempts at defining the hydrologic control by studies of precipitation records have not been completely successful because other factors such as relief, permeability, soil temperature profiles, and hydrologic barriers also control the type of transport.

The various constituents of soils are influenced unequally by total rainfall, but no universal generalizations can be drawn from the empirical data. Leaching of certain soluble cations is sensitive to total precipitation. However, variations in quantity of rainfall do not seem to affect importantly the dissolution of organic matter, clay, or iron oxide content. Weathering is impossible without water, and probably, in order to be even moderately effective, requires some as yet undefined threshold amount of precipitation.

It has long been known that the rates of many chemical reactions increase exponentially with rising temperature. Commonly, a 10°C increase in temperature doubles or trebles the reaction rate. Apparently the general relation applies to very slow reactions as well as to fast ones and also to numerous biological phenomena.

The influence of temperature on solubility depends on the heat of solution of the specific salt involved. Most salts absorb heat when dissolved and are, therefore, more soluble at higher temperatures. In the case of NaCl and certain other salts, solubility is almost unaffected by temperature. Still others decrease in solubility as the temperature is raised. The solution of calcite is a special case. Its solubility depends on the solubility of CO_2, and is less at higher temperatures because the solubility of CO_2 decreases.

Jenny (1941) showed that, within belts of uniform moisture conditions and comparable vegetation, the organic matter content of soil decreases exponentially with increasing temperature. The data he presents indicate a doubling of the organic content with each 10°C decrease in temperature, resulting in an accumulation of organic matter in cool climates and accelerated decomposition in warm climates.

Organic Matter

In addition to its effect on pH and ion exchange noted earlier, one of the most important properties of soil organic matter is its power to retain water. It was formerly believed that the organic matter content of tropical latosols is low, owing to high rates of oxidation and bacterial decomposition under continuously high temperatures. However, recent studies in Puerto Rico, Hawaii, Colombia, and the Belgian Congo indicate that organic matter may be present in significant amounts to a depth of 2 to 3 feet, even though surface litter is scant or missing.

Decomposition of organic matter is more rapid in tropical soils, but the total amount present is also greater. Annual production of organic matter in tropical forests amounts to 3,000–11,000 pounds per acre as compared with 800–2,800 pounds per acre in the oak-pine forests of the Sierra Nevada, California (Jenny et al., 1949). In the tropics, practically all the organic matter is within the mineral soil, where decomposition is slow. By contrast, in temperate and cool regions, much of the organic matter rests on top of the mineral soil.

In most high mountain areas and in cold areas of high latitude, including places where the mean annual temperature is below freezing and vegetation is sparse, organic content of the surface soil is commonly 20 to 30%.

In sum, organic matter plays an important part in controlling both the physical and the chemical environment of the weathering zone.

Processes of Physical Weathering

Physical weathering is in-situ disintegration of rocks and comminution of their constituent minerals without major chemical change. Except perhaps in cold regions, these effects are minor by comparison with those resulting from chemical weathering. A significant result of physical weathering wherever it occurs is that it increases the specific surface of the material and thus makes the material more susceptible to chemical attack.

Physical weathering involves stresses generated in the rock by growth of ice crystals, by heating and cooling, wetting and drying, and by organic activity. Because forces from several of these sources may act more or less simultaneously, their individual importance and degree of interaction are not easily evaluated.

The most common example of crystal growth is freezing of water, which produces disruptive effects collectively referred to as frost action. More than half the land area of the world suffers from freezing temperatures, but the intensity of frost action depends on the presence of adequate water as well as on freeze-thaw frequency.

Water freezing at 0°C increases in volume by 9%. If confined, pressures may be generated that exceed the tensile strength of the strongest rock. Although the maximum possible pressures are never developed under natural conditions, the combination of expansion and low compressibility of ice may produce adequate stresses for disintegration of rocks.

The growth of ice crystals, rather than simple freezing, is a major aspect of frost action. Crystals exert pressure in the direction of growth, which is in the direction of heat conduction. As cooling is from the land surface downward, ice crystals grow normal to the surface. Slow cooling allows larger crystals to form. If the supply of water is limited, crystals concentrate near the surface and cause spalling and flaking of rock particles.

Crystal growth other than ice is largely confined to arid and semiarid climates, although it has been suggested that along coasts spray may contribute to salts which influence physical weathering (Kelly and Zumberge, 1961, p. 445). Crystallization of halite, gypsum, calcite, and other salts on or near the surface of rocks produces a minor amount of disruption of mineral grains.

It was formerly believed that natural heating and cooling of rocks would generate stresses sufficient to disintegrate rocks. Such a process was supposed to be most effective in deserts and to cause (1) exfoliation due to expansion, (2) development of cracks perpendicular to rock surfaces due to sudden cooling, and (3) granular disintegration caused by strains arising from differential expansion of constituent materials.

Several experimental studies, made under conditions more severe than those imposed by nature, indicate that stresses generated by natural temperature variations are significantly smaller than the rupture strength of rocks. Although the experimental evidence seems to relegate temperature variations to a minor role in rock disintegration, it is possible that small effects are cumulative over long periods.

Temperature variations are greatest at the surface and decrease rapidly with depth. Diurnal changes are limited to a depth of a few inches, and annual changes to a few feet. For this reason—even if the effects are cumulative—temperature changes cannot form deep residual soils.

The role of organisms in weathering is dominantly chemical rather than physical. However, plant roots are known to widen joints and fractures in rocks. Likewise, burrowing organisms such as ants and many kinds of rodents stir and mix the soil. Termite mounds, for example, are common in the tropics. These activities directly contribute only slightly to rock disintegration, but the increased porosity and permeability greatly facilitate weathering processes.

Products of Physical Weathering

Specific data on the physical properties of debris produced primarily or exclusively by disintegration of rocks are exceedingly scarce, and all examples cited in the literature are deposits formed by frost action.

Size characteristics of some residual frost-riven and mechanically disintegrated rocks are shown graphically in Fig. 4-1. Some of the examples involve only coarse surface material; others represent samples of the entire unconsolidated deposit. In both, the size distributions approximate but deviate somewhat from a log-normal distribution. Apparently frost action usually produces almost no clay sizes and only moderate amounts of silt. That physical disintegration produces few or no particles smaller than about 0.01 mm has been attributed to the depression of the freezing point in very small openings and also to the inherent elasticity of small particles. Krumbein and Tisdel (1940, p. 296) have noted that both mechanically and chemically disintegrated rocks have sorting similar to that of transported sediments but the skewness is reversed. The distributions also follow a law for artificially crushed material which apparently results from random breakage of material.

Fine-grained, moderately permeable rocks such as shales, slates, schists, and some sandstones are especially susceptible to frost action. Highly impermeable and extremely permeable rocks are most resistant. Frost effects may be practically absent on bare cliffs or other sites where there is little water. Vegetation, snow, and ground ice affect water supply, and frost action is most severe where saturation is maintained.

The frequency of fluctuations above and below the freezing point range from more than 200 per year in the high mountains of the middle

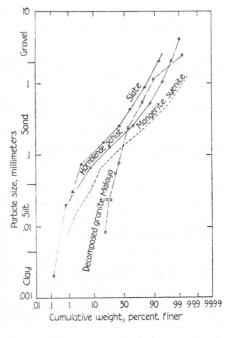

Figure 4-1.

Particle-size distribution of soil material without profile development, caused principally by freeze and thaw, compared with decomposed granite from Malaya. [Schist and syenite sample from Norway (Holtedahl, 1957), shale from Alaska (Taber, 1943).]

latitudes to less than a tenth of this number in the Arctic. The number of freeze-thaw cycles and the temperature in air, however, probably rarely coincide with fluctuations or temperature values in the rock or soil (Kelly and Zumberge, 1961, p. 436). The effects of frost action are greatest at the surface, where freezing and thawing occur most frequently. At depths ranging from a few inches to a few feet, depending on latitude and altitude, freezing and thawing occur only once a year. Deep residual soils extend below the horizon within which mechanical weathering is active.

Growth of ground ice causes heaving and stirring of unconsolidated materials and produces a great variety of microrelief features and mixing of soil materials (see Flint, 1953).

Rates of Weathering

Weathering may be viewed as an interplay between the intensity of the altering agencies and the susceptibility of the materials being weathered. Relatively few generalizations about weathering rates are possible at present because (1) adequate chronologic data generally are lacking, (2) the ranges of variations in conditions during the past are unknown, and (3) chemical and mineralogical changes even under specific environmental conditions are poorly understood. However, it now seems apparent that some of the principal factors which control weathering rates are temperature, quantity of water and its rate of movement, acidity (pH), and properties of minerals and mineral assemblages.

Any attempt to describe rates of weathering processes requires adoption of somewhat arbitrary criteria. Soil scientists refer to the degree and nature of profile development, whereas leaching, oxidation, and changes in mineralogical or chemical composition are considered by geologists to be the most obvious indications of weathering. Whatever the system used, there can be no doubt that determination of rates is vastly more complicated than is suggested by the once-popular slogan, "10,000 years are required to form an inch of soil." Indeed, the tremendous variability is emphasized by Kellogg (1941), who states that building an inch of soil may take anywhere between 10 minutes and 10 million years.

During the last half century, a considerable quantity of field and experimental data bearing on weathering rates has accumulated, much of which was summarized by Jenny (1941). Values vary from several inches in a year to less than 1 inch in 5,000 years. The definition of the degree of weathering is in itself variable, but a review of the data suggests that changes produced by weathering, including the increase in depth of weathered zone, are exponential rather than linear functions of time.

Soils—Introduction

For our purposes a soil may be considered as a unit formed at the earth's surface possessing distinctive zones or horizons characterized by differences in mineralogy, organic content, texture, and structure. Although pedologists and geologists sometimes differ in the definition of a soil, geomorphologists are concerned with the entire weathered zone—from surface down to unweathered bedrock. Characteristic units of the soil

	Horizon	Description
THE SOLUM, the genetic soil developed by the soil processes		Undecomposed organic matter Organic matter, partly decomposed
	A	A dark-colored horizon with high content of organic matter mixed with mineral matter
		A light-colored horizon of maximum removal of materials (eluviation), prominent in Podzolic and faint in Chernozemic soils
	▬▬▬	Transitional zone
	B	Horizons of maximum accumulation of suspended material or of clay; maximum development of blocky or prismatic structure or both
	▬▬▬	Transitional zone
	C	The weathered parent material; includes layers of accumulated calcium carbonate and calcium sulfate found in some soils
	D	Hard rock or any stratum underneath the soil having significance to it

Figure 4-2.   *A hypothetical soil profile that has all of the principal horizons. Not all of these horizons are present in any profile, but every profile has some of them [After Yearbook of Agriculture, 1957.]*

profile are shown in Fig. 4-2, which also gives the general distinguishing features of each of the horizons. Description and classification of soils is in fact based upon profile characteristics.

The soil profile develops as a result of complex chemical and physical processes. Referring to the figure, the B horizon resulting from discontinuous transport is a zone of accumulation of materials leached from the A horizon above. Leaching, primarily of iron, has bleached the A horizon while the accumulation of iron oxides has darkened the B horizon. Similarly, clay has accumulated in the B horizon; the movement and accumulation of clay is known as illuviation. In humid regions hydrous iron and aluminum oxides may accumulate in the B horizon, leaving the A horizon enriched in silica. Such soils are often called "pedalfers." In contrast to the pedalfers are the "pedocal" soils, ones in which calcium carbonate accumulates in the B horizon due to discontinuous transport. Pedocal soils are commonly associated with drier climates.

Morphologically the succession of zones may be from the surface downward, but the sequence of changes may also be manifest from the outside to the inside of individual boulders. These boulders, known as core boulders, are common in deeply weathered zones in the tropics. At greater depths, original rock structure is still preserved and the boulders are less altered.

The great soil groups of pedology are assumed by pedologists to be climatic soil types. According to Thorp and Smith (1949), differences in soils are (a) first order, *zonal* differences, associated with climate and associated biological activity; (b) second order, local or *intrazonal* and *azonal* differences which reflect differences in parent material, relief, biological activity, and age of the soils; and (c) third order, regional and local differences due to activities of man. The major groupings and a few selected examples are given in Table 4-5.

Table 4-5.   Genetic classification of soils with some typical great soil groups.

Major Classification	Type and Location of Environment	Selected Example of a Great Soil Group
Zonal	Cold zone	Tundra
	Light colored soils; arid	Desert
	Dark colored soil; semiarid to subhumid	Chernozem
	Forest-grassland transition	Degraded chernozem
	Light colored podzolized soils of timbered regions	Podzols
	Lateritic soils of forested warm temperate regions	Latosols
Intrazonal		
	Saline and alkali	Saline
	Hydromorphic	Gley
	Calcimorphic	Rendzina
Azonal		
	Little or no soil	Lithosol
	Sands, gravels, debris	Regosols
	Alluvial deposits	Alluvial soils

Table 4-5 indicates that podzols are associated with forested and tundra regions and chernozems with grasslands. In detail, however, as the discussion of weathering demonstrates, the climatic imprint becomes less distinctive as the influence of other variables is magnified in specific localities.

Factors That Affect Soil Formation

Pedologists generally list climate, parent material, biota, and relief as the fundamental factors of soil formation. Because soil properties change, time is also listed. Also implicit in most discussions is the concept that an equilibrium or steady-state condition among these various factors eventually becomes established in a soil. The factors of soil formation are considered individually in succeeding sections and, following Jenny (1941), an attempt is made to segregate the specific effects of each.

Many statements in the literature imply a cause-and-effect relation between the solute content of drainage waters and the climatic environment of weathering. Waters draining silicate rocks in tropical regions, for example, are believed to contain more SiO_2 and less Ca, Na, and K than in temperate and arctic regions. Increased solubility of SiO_2 and rapid removal of the bases are supposed to result from intense tropical leaching. Although sesquioxides of iron and alumina do accumulate in tropical soils, at present it appears that the data are inadequate to justify broad conclusions, mainly because the extent of lithologic control of water composition is unknown. There are sufficient cases of temperate and even subarctic streams with high SiO_2 content—and, likewise, tropical streams with large concentrations of bases—to question the validity of the current concepts relating climate and weathering.

Climate is one control of the hydrothermal conditions of weathering, and thus it has been argued that climate should have a substantial effect on both the nature and intensity of the processes involved. The early Russian school of soil science rated climate as the most important single factor in weathering and soil formation. The most extreme supporters of this opinion maintained that a particular climate would produce a characteristic kind of soil regardless of the parent material involved. This view is gradually giving way to the more realistic belief that under many conditions factors other than climate may exert a dominant influence.

The relations between climate and soil-forming processes are not known because climatological data are not readily applicable to the essential aspects of weathering processes. For example, mean temperatures convey neither information about the actual amount of freezing and thawing, nor of the temperature in the soil. Mean values of precipitation yield limited information on hydrologic properties in the soil. A further difficulty results from the inadequate knowledge of temperature and moisture requirements of the various chemical processes and the ranges of stability of mineral phases formed by weathering.

Mohr and Van Baren (1954) express the opinion that any attempt to group tropical soils on a climatic zonal basis is attended by complications because descriptions of climate do not treat the soil climate. and because of the legacy of past climatic influences different from the present. Influence of climate is only predominant when other factors are most favorable. They cite numerous examples of tropical soils in which factors other than climate, such as parent material or topography, dominate soil development.

A formidable stumbling block to understanding the relation of climate to soil formation is that climate and vegetation have changed considerably, and several times, during the Pleistocene. According to some interpretations, most soils are polygenetic, that is, they bear the imprint of more than one climatic episode. Despite the fact that "great soil groups" seem to occupy fairly specific segments of the "great climatic zones," the widespread belief that soils are adjusted to modern climatic conditions requires careful re-examination.

Paleoclimatic reconstruction is one of the major geological applications of information derived from weathering studies (Bryan and Albritton, 1943; Thorp, Johnson, and Reed, 1951; Simonson, 1954; and many others). In particular, pedocalic and pedalfer-type soils have been used as indicators of arid and humid climates respectively. Broadly speaking, such inferences are probably warranted. Such an interpretation, of course, implies adjustment of soil properties and weathering products to the climatic environment.

Pedalfers and pedocals represent genetic trends that reflect differences due in part to the soil climate. However, many soil climates may occur in the same zonal climate and they may not be causally related. Likewise, very dense, resistant, infertile materials change very slowly in any climate, both because they are only slightly susceptible to weathering and because they cannot support abundant vegetation.

Simonson (1954) suggests that the genetic relationships between factors of soil formation are to some extent interchangeable. Thus, the product of a long interval of time and a low climatic intensity would be comparable to a shorter interval of time and greater intensity of weathering. Red color, which is considered by some workers to be a major indicator of humid-climate weathering, is a property which may follow this rule. Clearly, there is a great need for additional studies of weathering rates, which may lead to distinctions between true equilibria (if any) and cases of apparent equilibrium that result from slow reactions.

Paleoclimatic reconstructions based on weathering phenomena seem to be considerably more complicated than is generally appreciated. The mineralogical and chemical properties of a weathered zone are the result of inheritances of several kinds (parent material, past environments, etc.), its position at or near the surface of the earth, and the environment at that location.

Although there can be no doubt that the role of climate in weathering and soil formation has been overemphasized, it would be a serious error to assume the opposite extreme, that climate is a minor factor in weathering. The current need is for careful field observation and diligent laboratory experimentation directed toward the goal of establishing more precisely the effects of temperature and various other climatic elements on specific weathering processes.

Where other variables, such as parent material and topography, are reasonably constant, a climatic system of soil classification can be used successfully. For example, the climatic conditions of mountain masses change regularly with increasing altitude; that is, precipitation increases and temperature falls and differences in vegetation reflect these differences in climate. Thus a map of soils in the Bighorn Range, Wyoming (Thorp, 1931), shows a gradual transition from desert and pedocalic soils at the foot of the mountains to podzols above 8,000 feet. Similarly, Martin and Fletcher's (1943) investigation of the Mt. Graham area, Arizona, shows well-defined zonation of soils associated with increasing precipitation and decreasing temperature with altitude. Organic content increases and soil *p*H decreases with elevation. Compared with most mountain regions, the parent material is remarkably uniform.

Precipitation

The average amount of precipitation which falls annually on various parts of the earth ranges from essentially none to more than 400 inches. Such figures have little meaning in terms of the moisture available to weathering processes and the rate of removal of solutes. For example, southern Arizona with 3 inches of rainfall is exceedingly arid, whereas northern Alaska with a comparable amount is a land of swamps and muskeg during summer. Similarly, for an area in which a certain quantity of rainfall is sufficient to keep the soil permanently wet, additional amounts probably have little effect on weathering processes. Quite apart from amount is the seasonal distribution of precipitation. Presumably, most weathering reactions are favored by a reasonably equable distribution rather than marked seasonal contrasts, although some physical effects may be enhanced by seasonal variability.

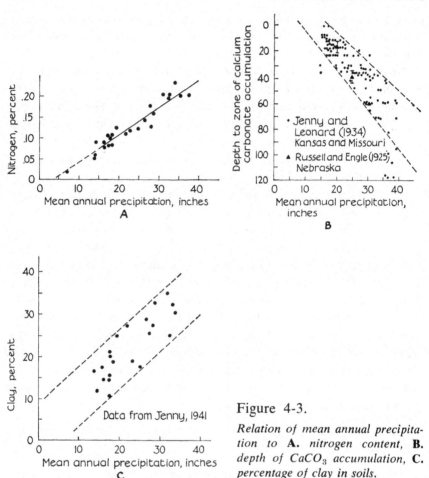

Figure 4-3.

Relation of mean annual precipitation to **A.** *nitrogen content,* **B.** *depth of CaCO₃ accumulation,* **C.** *percentage of clay in soils.*

In many soils the nitrogen content may be taken as an index of the total organic matter. Jenny and Leonard (1934) determined the amount of nitrogen in the surface 10 inches of soils located along the annual isotherm of 52°F (11°C) in the loess belt of the Middle West of the United States. The results, applicable to grassland soils and shown in Fig. 4-3, **A,** indicate a good correlation between soil nitrogen and annual rainfall. In contrast, although there are great variations in depth, a general increase in depth of the carbonate-free zone with increasing rainfall is apparent in the figure (**B**), based on data from loessial soils along a line extending from Colorado to Missouri (Jenny and Leonard, 1934) and on similar information from Nebraska (Russell and Engle, 1925).

Jenny (1941) reviews abundant evidence which shows that soils in humid regions are strongly leached of sodium, potassium, magnesium, and calcium as compared with soils in arid regions.

The sand and silt fractions of soils in humid regions contain on the average $1\frac{1}{2}$ times more quartz than soils of arid regions, indicating apparently a concentration of quartz as a result of weathering of less stable minerals. Jenny and Leonard (1934) showed a positive correlation between clay content and annual rainfall, particularly in the loess belt (Fig. 4-3, **C**). Several writers have shown that the clay content of many tropical soils decreases with increasing rainfall because of the development of concretions.

Jenny and Leonard (1934) also showed that the total exchange capacity of loessial soils in the Middle West increases exponentially with increasing rainfall. At rainfalls greater than 25 inches there is a decrease in exchangeable bases, which is compensated by an increase in exchangeable hydrogen. Certain tropical soils show similar relations for exchangeable hydrogen despite decreasing total exchange capacity with increasing rainfall. Soils of arid regions generally show no exchangeable hydrogen, whereas for humid soils it commonly constitutes more than half of the exchange capacity. The increase of exchangeable hydrogen is reflected in pH values of soils. With increasing rainfall where organic material accumulates, soils become more acid (pH decreases), although in many tropical regions the pH of the soil is 6 or higher.

Studies of Hawaiian soils illustrate the complex interrelationships of rainfall and soil formation. Tanada (1951) and Temura et al. (1953) indicate that the following changes occur as rainfall increases: (1) content of bases and silica decreases, (2) colloid content decreases, (3) percentage of kaolinite decreases, (4) organic matter, gibbsite, and iron oxide content increase; however, rainfall greater than 150 inches produces reducing conditions and decreasing iron oxide content; (5) the contents of illite and montmorillonite-type clays increase with increasing rainfall but reach a maximum at approximately 80 inches.

Parent Material

It might be supposed that if there were a clear correlation between the type of parent material and susceptibility to weathering, there should be a general tendency for the more susceptible rock types to be weathered more deeply than those resistant to weathering. Table 4-6 is a tabulation of available data on depths of weathering of rocks; neither the effects of

Table 4-6.   Depth of weathering.

Location	Rock Type	Depth (feet)	Source[b]
Wakefield, New Hampshire	Granite	10–15	1
Keene, New Hampshire[a]	Granite	10–20	1
Gorham, New Hampshire[a]	Granite	10–15	1
North Conway area, New Hampshire	Granite	up to 30	2
Washington, D. C.	Granite	80+	3
Pikes Peak, Colorado	Granite	20–30	3
Atlanta, Georgia	Granite	95–300	3
Southern California	Granodiorite	up to 200	3
Transvaal, South Africa	Granite	up to 200	3
Hongkong, China	Granite	200+	3
Transvaal-Natal	Granite	65	7
Kwantung, China	Granite	33+	7
Cameroons	Granite-gneiss	65	7
Madagascar	Granite-gneiss-mica-schist	100	7
British Guiana	Granite	16	7
Compos Cerrados, Brazil	Granite	65+	7
Sandwich, New Hampshire[a]	Syenite	2–10	1
Rowan County, North Carolina	Diorite	20	4
Juneau, Alaska	Diorite	20	3
Nicaragua	Dolerite	up to 200	3
British Guiana	Dolerite	33	7
Medford, Massachusetts[a]	Diabase	10+	5
Minnehala County, South Dakota	Diabase	25	3
Rowan County, North Carolina	Meta-gabbro	4	4
Saipan	Andesite	50	6
Cameroons	Andesite	10–65	7
Puerto Rico	Andesite tuff	40+	7
Bintan	Liparite breccia	160	7
Cuba	Serpentine	40–50	7
Saipan	Limestone	6	6
Rio Grande do Sul, Brazil	Shale	390	3

[a] Covered by glacial till.

[b] Sources: 1. Goldthwait and Kruger (1938); 2. Billings (1928); 3. Merrill (1906); 4. Cady (1950); 5. Billings and Roy (1933); 6. McCracken (1957); 7. Mohr and Van Baren (1954).

lithology nor those of climate are especially apparent. If the effects of lithology, climate, and other controlling factors of weathering processes are ever to be segregated and related to observable field conditions, more information on depth of bedrock weathering must be collected and interpreted. A major obstacle to any such effort is the problem of uniformly reliable criteria for distinguishing between weathered and unweathered materials. Among the many examples which illustrate this point is the study by Niggli (1926) of weathering in Switzerland. Certain decomposed gneisses have the appearance of thoroughly weathered rock; yet chemical analysis indicates little change other than hydration and oxidation. In addition, if weathering depth increases with age, a moot point at present, the data will have to be stratified according to age of exposure if valid relations are to be derived.

Other conditions being equal, soils derived from limestone are commonly darker in color and richer in organic matter than noncalcareous soils. But limestone also produces gray, yellow, and red soils, and for reasons which are not yet adequately understood. Red soils on limestones are known in many parts of the world; probably the best described of these is the famous terra rossa of the Mediterranean area. Several red soils are described in the Mediterranean region, however, and it is apparent that past and present climatic conditions may have been quite different in different areas, thus leading to complex and varied relations of soil, lithology, and climate.

On the island of Saipan, where temperatures are 80–85°F all year and the annual precipitation of 80–90 inches is evenly distributed, sites are weathered to a depth of 50 feet or more (McCracken, 1957). The soils are red or yellow, are relatively high in organic content, and commonly contain more than 50% organic matter in the B horizon. Clay in soils derived from volcanic rocks include kaolin, illite, and vermiculite, whereas limestone soils are dominated by montmorillonite, with kaolin next most abundant. There is no significant difference in the free iron oxide content of soils derived from these two types of parent rocks. In contrast to the deep weathering of andesites, the limestones are unweathered at depths greater than 6 feet and the contact between soil and rock is abrupt, though commonly irregular.

At locations separated by a few miles in Rowan County, North Carolina (Cady, 1950), meta-gabbro grades into solid rock at a depth of 4 feet, whereas a diorite is overlain by about 20 feet of saprolite. So far as could be determined, the duration and environmental conditions of weathering are the same in both cases. The weathered products seem to indicate some peculiar relations of the mineral stability series. Feldspar of the meta-gabbro weathers to halloysite before the ferromagnesian minerals are altered appreciably. Furthermore, fresh hornblende and feldspar occur in the sand and silt fractions of all horizons. The weathering products of the diorite in some instances are kaolinite, quartz, and iron oxides, but in others are dominantly gibbsite, chlorite, and allophane.

When the parent material is unconsolidated, the specific effects of rock type are difficult to evaluate, being masked by the effects of permeability and heterogeneity of the material. Jenny (1941) cites several examples of different soil types caused primarily by differences in permeability of the parent materials. Some detailed studies show that remarkable changes in soil properties related to permeability may occur within distances of a few feet.

Composition of parent material affects both the degree of weathering of soil minerals and the development of soil profiles. Krebs and Tedrow (1957), using chemical and X-ray methods to study soils developed on Early Wisconsin tills of different lithologic composition in New Jersey, conclude, in agreement with other workers, that carbonate rock in the parent material retards podzolization. Local differences in weathering environments have had no appreciable effect on the types of clay minerals present in the soil.

Some studies show that illite predominates in soils formed on tills containing abundant granite, quartzite, schist, slate, sandstone, and shale. Soils derived from tills rich in mafic igneous rocks are characterized by predominance of montmorillonite if drainage is poor, or vermiculite if the soil is well drained. Kaolin is less common in soils containing mafic parent material.

In contrast to glacial till, which is heterogeneous in composition and yields a variety of soil types, soils developed on loess and certain wind-deposited sands are exceedingly homogeneous. However, the problems of separating soil properties related to pedogenic influences from those inherited directly from the parent material are as difficult for loess as for other kinds of materials. Thus, it is uncertain whether or not the high clay content of some loesses is the result of in-situ weathering.

Another problem involving unconsolidated materials, and one which has received insufficient attention, is the effect on weathering products and profiles caused by stratification of contrasting materials—for example, till overlain by loess. Distinguishing valid pedogenic horizons and recognizing truncated profiles and buried paleosols are all part of this same problem. More general use of chemical and mineralogical techniques in addition to detailed field examination is clearly indicated.

Topography

The effects of topography on soil formation are associated primarily with surface runoff and subsurface penetration of precipitation. If slopes are very steep the surface erosion may strip away weathered material almost as quickly as it forms. Steep slopes also have more rapid drainage and hence more arid "soil climates" than areas of level topography.

Slope angle and depth of the A horizon for forested loessial soils in Illinois are inversely correlated (Norton and Smith, 1930). On flat areas the surface soil averages 24 inches thick, but decreases to about 9 inches on 14% slopes. Ellis (1938) described similar trends for chernozem soils in Manitoba.

Biologic Factors

Climate and soil properties ultimately determine the effectiveness of the biologic complex in weathering. By contrast with the exothermic oxidizing nature of most inorganic weathering processes, photosynthesis is endothermic and reductive. In addition to carbon, nitrogen, oxygen, and sulfur, most plants and bacteria use and immobilize iron, magnesium, potassium, and certain other elements, thus in part reversing the prevailing inorganic trends.

Natural vegetation of the United States can be classified into three broad types. The fraction of total land area originally occupied by each is given below:

	PERCENTAGE
Desert shrub	15
Grassland	35
Forest	50

Hardwood and coniferous forests are roughly equal in extent.

On the same parent material (loess) and under otherwise comparable environments, organic matter content and base-exchange capacity are much higher under prairie vegetation than under forest. However, exchangeable hydrogen is greater under forest conditions, and the same is true for clay concentration in the B horizon, depth of carbonate leaching, and acidity. Loess soils of Missouri and Iowa have more montmorillonite under grassland than under forest conditions, but the A horizons of the forest soils contain $1\frac{1}{2}$ to 3 times more clay than the prairie soils. On till in the Great Plains of Manitoba, greater decomposition of coarse sediments occurs under forest cover than under grass, and iron in particular is more mobile under forest (Ehrlich and Rice, 1955).

Many pedologists believe that vegetation exerts a major influence on the development of soil characteristics. If soil characteristics reflect conditions over long periods of time, such a view is difficult to reconcile with the fact that vegetation in many parts of the world has changed markedly due to natural causes during the last few millennia, yet the soil does not appear to reflect these changes. As shown by Iversen (1954), Olson (1958), and many others, some changes in vegetation through time may be entirely the result of ecological succession. However, major transitions (such as from conifers to grasses to hardwoods) probably reflect variations in climate and effects of other geographical factors.

Changes of vegetation may influence soil properties rather quickly. Certain properties such as organic content and pH seem to vary in response to changes in vegetation, and a stable assemblage of trees apparently is established within a few hundred years. This appears to be the conclusion of many studies of succession on burned and timbered areas. The time required for the forest floor to attain a constant depth of accumulated organic material is calculated to be less than a decade for tropical forests, 30–60 years for California oak, and 100–200 years for Ponderosa pine (Jenny et al., 1949).

REFERENCES

Bear, F. E. (ed.), 1955, Chemistry of the soil: *Am. Chem. Soc. Mon.* 126, 383 pp.

Billings, M. P., 1928, Petrology of the North Conway quadrangle in the White Mountains of New Hampshire: *Am. Acad. Arts and Sci., Proc.,* v. 63, pp. 67–137.

Billings, M. P., and Roy, C. J., 1933, Weathering of the Medford diabase— pre- or post-glacial? A discussion: *Journ. Geol.,* v. 41, pp. 654–661.

Birch, F., et al., 1942, Handbook of physical constants: *Geol. Soc. Am., Spec. Paper* 36, 325 pp.

Bryan, K., and Albritton, C. C., 1943, Soil phenomena as evidence of climatic change: *Am. Journ. Sci.,* v. 24, pp. 469–490.

Cady, J. G., 1950, Rock weathering and soil formation in the North Carolina Piedmont region: *Soil Sci. Soc. Am. Proc.,* v. 15, pp. 337–342.

Ehrlich, W. A., and Rice, H. M., 1955, Postglacial weathering of Mankato till in Manitoba: *Journ. Geol.,* v. 63, pp. 527–537.

Ellis, J. H., 1938, The soils of Manitoba: Manitoba Economic Survey Board, Winnipeg, Manitoba.

Flint, R. F., 1953, Probable Wisconsin substages and Late-Wisconsin events in northeastern United States and southeastern Canada: *Geol. Soc. Am. Bull.,* v. 64, pp. 897–920.

Goldthwait, J. W., and Kruger, F. C., 1938, Weathered rock in and under the drift in New Hampshire: *Geol. Soc. Am. Bull.,* v. 49, pp. 1183–1198.

Gorham, Eville, 1961, Factors influencing supply of major ions to inland waters, with special reference to the atmosphere: *Geol. Soc. Am. Bull.,* v. 72, pp. 795–840.

Holtedahl, H., 1953, A petrographical and mineralogical study of two high altitude soils from Trollheimen, Norway: *Norsk Geol. Tidsskr,* v. 32, pp. 191–226.

Iversen, J., 1954, The late-glacial flora of Denmark and its relation to climate and soil: *Danmarks Geol. Unders.,* Raekke 2, pp. 87–119.

Jackson, M. L., et al., 1948, Weathering sequence of clay size minerals in soils and sediments: *Journ. Phys. and Coll. Chem.,* v. 52, pp. 1237–1260.

Jenny, H., 1941, Factors of soil formation: McGraw-Hill, New York, 281 pp.

Jenny, H., et al., 1949, Comparative study of decomposition rates of organic matter in temperate and tropical regions: *Soil Sci.,* v. 68, pp. 419–432.

Jenny, H., and Leonard, C. D., 1934, Functional relationships between soil properties and rainfall: *Soil Sci.,* v. 38, pp. 363–381.

Junge, C. E., and Werby, R. T., 1958, The concentration of chloride, sodium, potassium, calcium, and sulfate in rain water over the United States: *Journ. of Meteorology,* v. 15, no. 5, pp. 417–425.

Keeling, C. D., 1960, Measurements of carbon dioxide in the atmosphere: *Am. Geophys. Union Trans.,* v. 41, pp. 512–515.

Keller, W. D., 1962, The principles of chemical weathering: Lucas Bros., Columbia, Mo., 111 pp.

Kelley, W. P., 1948, Cation exchange in soils: Reinhold, New York, 144 pp.

Kellogg, C. E., 1941, The soils that support us: Macmillan, New York, 370 pp.

Kelly, W. C., and Zumberge, J. H., 1961, Weathering of a quartz diorite at Marble Point, McMurdo Sound, Antarctica: *Journ. Geol.,* v. 69, pp. 433–446.

Kessler, D. W., et al., 1940 Physical, mineralogical and durability studies on the building and monumental granites of the United States: *Journ. Research Nat. Bur. Standards, Res. Paper* 1320, pp. 161–206.

Krebs, R. D., and Tedrow, J. C. F., 1957, Genesis of three soils derived from Wisconsin till in New Jersey: *Soil Sci.,* v. 83, pp. 207–218.

Krumbein, W. C., and Tisdel, F. W., 1940, Size distribution of source rocks of sediments: *Am. Journ. Sci.,* v. 238, pp. 296–306.

Martin, W. P., and Fletcher, J. E., 1943, Vertical zonation of great soil groups on Mt. Graham, Arizona: *Ariz. Agr. Exper. Sta. Tech. Bull.* 99, pp. 91–153.

Mason, B., 1958, Principles of geochemistry: Wiley, New York, 310 pp.

McCracken, R. J., 1957, Geology of Saipan Mariane Islands: *U. S. Geol. Survey Prof. Paper* 280-D, pp. 189–206.

Merrill, G. P., 1906, Rocks, rock weathering and soils: Macmillan, New York.

Miller, J. P., 1961, Solutes in small streams draining single rock types, Sangre de Cristo Range, New Mexico: *U. S. Geol. Survey Water Supply Paper* 1535-F, 23 pp.

Mohr, E. C. J., and Van Baren, F. A., 1954, Tropical soils: Interscience Publishers, New York, 498 pp.

Niggli, P., 1926, Die chemische Gesteinsverwitterung in der Schweiz: *Schweiz. Mineralog., Petrog. Mitt.,* v. 5, pp. 322–347.

Norton, E. A., and Smith, R. S., 1930, The influence of topography on soil profile character: *Journ. Am. Soc. Agron.,* v. 22, pp. 251–262.

Olson, J. S., 1958, Rates of succession and soil changes on southern Lake Michigan sand dunes: *Bot. Gaz.,* v. 119, pp. 125–170.

Ross, C. S., and Hendricks, S. B., 1945, Minerals of the montmorillonite group, their origin and relation to soils and clays: *U. S. Geol. Survey Prof. Paper* 205-B, pp. 27–99.

Russell, J. C., and Engle, E. G., 1925, Soil horizons in the central prairies: *Rept. fifth meeting Am. Soil Survey Assoc.,* v. 6, pp. 1–18.

Schattner, Isaac, 1961, Weathering phenomena in the crystalline of the Sinai in the light of current notions: *Bull. Res.* Council of Israel, v. 10G, pp. 247–266.

Simonson, R. W., 1954, Identification and interpretation of buried soils: *Am. Journ. Sci.,* v. 252, pp. 705–732.

Taber, S., 1943, Perennially frozen ground in Alaska; its origin and his-tory: *Geol. Soc. Am. Bull.,* v. 54, pp. 1433–1548.

Tamura, T., et al., 1953, Mineral content of low humic, humic, and hydrol humic latosols of Hawaii: *Soil Sci. Soc. Am. Proc.,* v. 17, pp. 343–346.

Tanada, T., 1951, Certain properties of the inorganic colloidal fraction of Hawaiian soils: *Journ. Soil Sci.,* v. 2, pp. 83–96.

Thorp, J., 1931, The effects of vegetation and climate upon soil profiles in northern and northwestern Wyoming: *Soil Sci.,* v. 32, pp. 283–301.

Thorp, J., and Smith, G. D., 1949, Higher categories of soil classification; order, suborder, and great soil groups: *Soil Sci.,* v. 67, pp. 117–126.

Thorp, J., Johnson, W. M., and Reed, E. C., 1951, Some post-Pliocene buried soils of central United States: *Journ. Soil Sci.,* v. 2, pp. 1–21.

Van Houten, F. B., 1953, Clay minerals in sedimentary rocks and derived soils: *Am. Journ. Sci.,* v. 251, pp. 61–82.

Chapter 5 The Drainage Basin as a Geomorphic Unit

The drainage area may be defined as the area which contributes water to a particular channel or set of channels. It is the "source" area of the precipitation eventually provided to the stream channels by various paths. As such it forms a convenient unit for the consideration of the processes determining the formation of specific landscapes in the various regions of the earth. It provides a limited unit of the earth's surface within which basic climatic quantities can be measured and characteristic landforms described, and a system within which a balance can be struck in terms of inflow and outflow of moisture and energy. The amount of precipitation that falls over a given drainage basin can be measured and, given adequate instrumentation, the quantity of water that flows out of the drainage basin in stream channels, the changes in ground water storage, and the evaporation and transpiration by plants can be estimated. In addition, rates and kinds of denudation may be measured as material transported in solution or as clastic load in stream channels.

The drainage net is the pattern of tributaries and master streams in a drainage basin as delineated on a planimetric map. In theory, the net includes all the minor rills which are definite watercourses, even including all the ephemeral channels in the furthermost headwaters. In practice, the detail of the drainage net is dependent on the scale of the map used to trace the channels.

The network of drainage channels has been variously described as trellis or palmate, and by other terms descriptive of veination of different sorts. The use of such descriptive terms implies an organization of the net, and indeed this organization has been shown to be of unsuspected simplicity. That nature often reveals a simplicity of pattern reflects the

(Nanticoke)

Contour interval 10 feet
Datum is mean sea level

operation of a few dominant physical processes which, when unraveled, can be reduced to known physical or chemical laws.

Quantitative descriptions of various aspects of drainage organization have expanded the number and character of simple relationships. Because the drainage net is intimately associated with the hydraulic geometry of

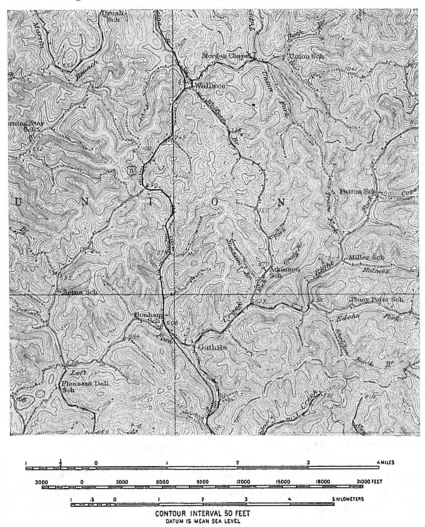

CONTOUR INTERVAL 50 FEET
DATUM IS MEAN SEA LEVEL

Typical examples of drainage density variation. [From United States Geological Survey.] Left: Topographic map of Hurlock quadrangle, showing drainage network on Atlantic Coastal Plain of Maryland. Right: Topographic map of Charleston, West Virginia, quadrangle, showing drainage network on Appalachian Plateau.

stream channels and with the shapes of longitudinal profiles of rivers, few other subjects in the field of fluvial morphology hold more promise for greatly extending basic understanding.

Two topographic maps are shown in Fig. 5-1. Casual inspection reveals obvious differences between them. In the coastal plain of the Atlantic Coast of the United States (Fig. 5-1, **A**) the streams are relatively far apart, there is much area in divide, and the drainageways are relatively sparse—or in reverse terminology, the upland is relatively poorly dissected. On the other hand, the topographic map of the region in the vicinity of Charleston, West Virginia (Fig. 5-1, **B**), shows innumerable stream channels and appears to be considerably more dissected by such channels. The simple statement that basin **A** is poorly dissected and **B** well dissected characterizes one important aspect of these drainage basins.

If the contributing drainage areas of individual stream segments are outlined in each of these topographic maps, further characterization is possible. Such contributing areas are seen to have a variety of shapes. Although most are oval, or perhaps pear-shaped, a number are subrectangular and others almost circular. This adds another measure of description of the drainage characteristics on each of these two topographic maps.

Although such qualitative expressions of dissection and shape tell something about the differences between sample areas, they are of limited use when it comes to investigating the causes of such differences or in distinguishing basins in which differences are less marked. Recent studies have made possible more comprehensive quantitative descriptions that provide useful tools in the study of the origin and processes of land sculpture.

Numbers, Lengths, and Orders of Stream Channels

Among the many ideas contributed by the late Robert E. Horton, an engineer of unique talents, the one which more than any other has caught the imagination of students of fluvial processes is the relationship of stream length and stream number to stream order (Horton, 1945). Defining the last term first, stream order is a measure of the position of a stream in the hierarchy of tributaries. It is best described by a sketch, as in Fig. 5-2, **A.** Given a map of a certain scale, the first-order streams are those which have no tributaries. The second-order streams are those which have as tributaries only first-order channels. However, each second-order stream is considered to extend headward to the tip of the longest tributary it drains. Which tributary to call the headward extension of a given second-order stream, where differences in length are insignificant, is a matter of choice. The one which seems to be the linear extension of

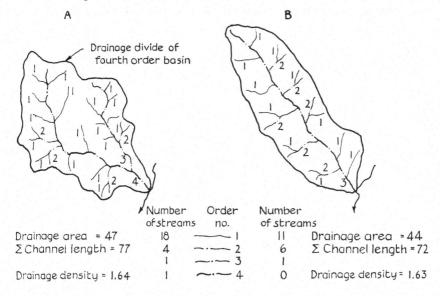

Figure 5-2. *Sketch to show order number of tributaries and other characteristics of the drainage net. Basins **A** and **B** have the same drainage density despite the different shape.*

the second-order stream is usually chosen although, in correlating discharge with drainage basin parameters, the tributary with the largest drainage area is probably most significant.

The third-order stream receives as tributaries only first- and second-order channels, and it also is considered to extend headward to the end of the longest tributary. It can be seen, then, that in practice, after the second-order channels are labeled and third-order channels identified, one of the previously marked second-order streams is renumbered to make it the headward extension of the third-order stream.

Strahler (1957, p. 914) suggested a different way of designating stream order, to overcome the difficulty of renumbering one headwater tributary as necessary in the original Horton scheme. He would restrict the designation of order to stream segments. This simplifies computation but of course shortens the length of the main channel. The laws relating to stream order and number are little affected, however, and the original Horton definition of stream order is retained here.

Where the number of streams of various orders in a drainage basin are counted, their lengths from mouth to drainage divide are measured and averaged. The relation of stream order to number of streams and

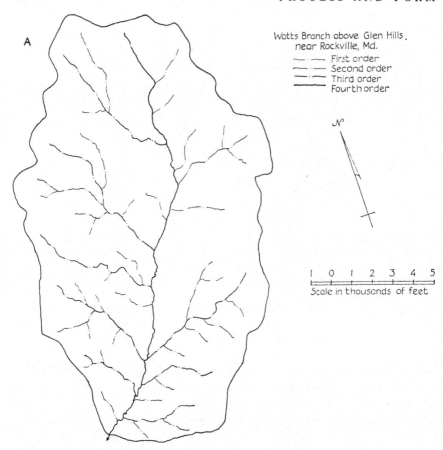

A

Watts Branch above Glen Hills,
near Rockville, Md.
——·—— First order
——·—— Second order
——·— Third order
———— Fourth order

𝒩

| 0 | 2   3   4   5
Scale in thousands of feet

Figure 5-3.   *Watts Branch above Glen Hills, Rockville Quadrangle, Maryland.* **A.** *Planimetric map.*

average length may be plotted. An example is shown in Fig. 5-3 and Table 5-1. The data are for Watts Branch near Rockville, Maryland, a stream where we have done considerable work and therefore one used frequently in this book as a sample small drainage basin. This sample was chosen to provide an idea of the problems encountered in dealing with a small basin having only a few stream orders. Note that the largest stream in the basin is of fourth order and has a drainage area of 6.65 square miles. On the graph of stream length versus order, the alignment of the points is insufficient to provide a good estimate of the relation, though the relation of stream number to stream order is moderately satis-

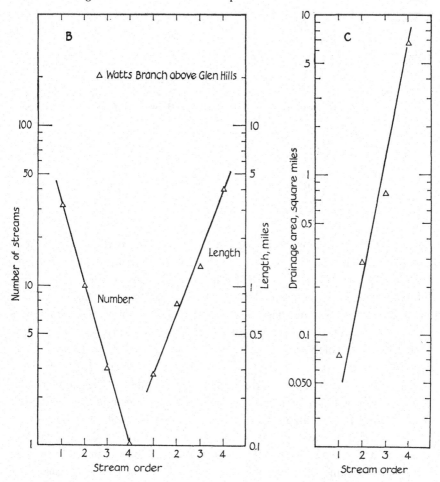

Figure 5-3. *Watts Branch: Horton analysis of drainage networks.* **B.** *Relation of average length and number of streams to stream order.* **C.** *Relation of stream order to drainage area.*

factory. More consistent results are obtainable as the order of the major drainage basin increases.

The three graphs of Fig. 5-3 demonstrate the salient features of what is referred to colloquially as a "Horton analysis." Horton showed that stream order is related to number of streams, channel length, and drainage area by simple geometric relationships; that is, stream order plots against these variables as straight lines on semilogarithmic paper.

Horton introduced the term "bifurcation ratio" to express the ratio of

Table 5-1. Bifurcation characteristics of Watts Branch above Glen Hills, near Rockville, Maryland.

Order	Number of Streams	Bifurcation Ratio	Number of Streams Entering Directly into Next Higher Order	Area of Single Basin (sq mi)	Total Area in Given Order (sq mi)	Percentage of Area of Next Highest Order
1	32		15	.074	2.36	82
2	10	3.2	4	.286	2.86	125
3	3	3.3	3	.765	2.29	34
4	1	3.0		6.65	6.65	
Average		3.2				

the number of streams of any given order to the number in the next lower order. Thus in Fig. 5-3, where there is one stream of fourth order, there are, from the average straight line, 3.2 streams of third order. The bifurcation ratio is, then, the slope of the line relating number of streams to stream order. Among many samples of basins in the United States the bifurcation ratio tends closely to equal 3.5. There are variations, of course. In the examples of basins cited by Horton (1945, p. 290) values of the bifurcation ratio range from 2 to 4. Strahler (1957, p. 915) obtained similar values, and he cites others whose work yielded the same result. Computation of bifurcation ratios for Watts Branch near Rockville, Maryland (annual rainfall about 40 inches), is shown in Table 5-1. The value of the bifurcation ratio is 3.2, whereas that for Arroyo de los Frijoles, near Santa Fe, New Mexico (annual rainfall about 13 inches), is 3.5. Graphs of stream order for the latter basin are presented in Fig. 5-4.

Figure 5-4 demonstrates the effect of map scale on the definition of stream order. On a planimetric map of 2 inches to the mile the smallest tributary shown, and therefore called first order, was of the size of Arroyo Caliente, whose drainage area was 0.027 square miles (17 acres). But a detailed map of this small tributary, presented on Fig. 5-4, **A,** showed a network of ephemeral channels and rills of which the largest was determined to be of fifth order. Thus all stream orders of the network on the 1:24,000 scale map were increased by 4, so that a first-order stream on that map becomes order 5 as demonstrated by the more detailed mapping.

Actual determination from maps of the order of the largest streams in the area studied indicated that two streams were of eleventh order, after

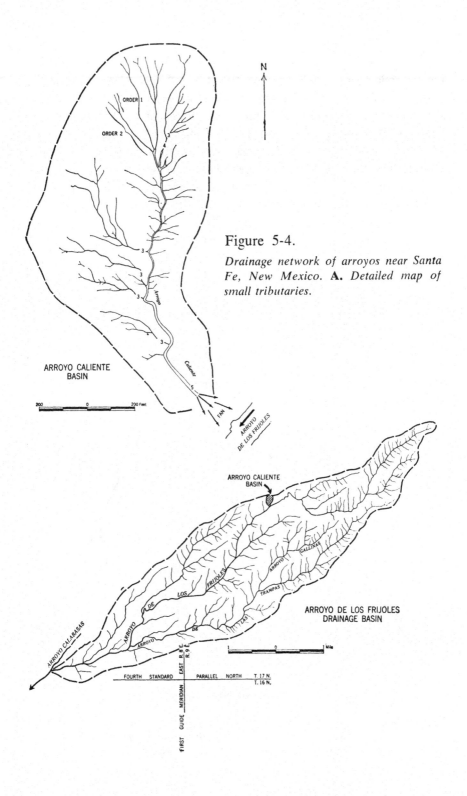

Figure 5-4.

*Drainage network of arroyos near Santa Fe, New Mexico. **A.** Detailed map of small tributaries.*

the correction was made by adding 4 to the smallest streams appearing on maps of scale 1:24,000. It would be laborious to determine on more than a sample area the exact number of streams of the lowest 4 orders. But a good estimate may be made by establishing the slope of the line relating stream order to number of streams, then projecting a line of similar slope through the point representing one stream of order 11. As can be seen in Fig. 5-4, **B,** it is estimated on that basis that the eleventh-order basin contains about 200,000 streams of first order.

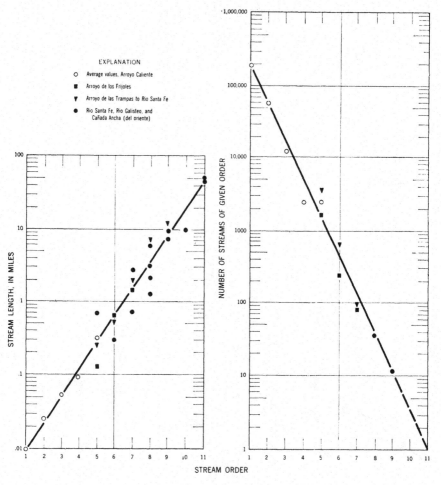

EXPLANATION
○ Average values, Arroyo Caliente
■ Arroyo de los Frijoles
▼ Arroyo de las Trampas to Rio Santa Fe
● Rio Santa Fe, Rio Galisteo, and
    Cañada Ancha (del oriente)

Figure 5-4.   *Drainage network of arroyos near Santa Fe, New Mexico.* **B.** *Relation of stream length to stream order.* **C.** *Relation of number of streams to stream order.*

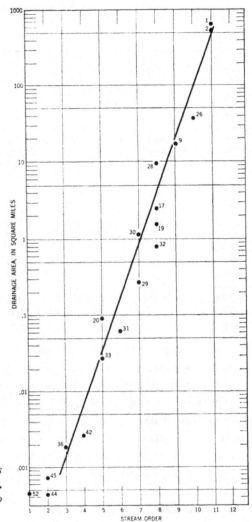

Figure 5-4.

Drainage network of arroyos near Santa Fe, New Mexico. **D.** *Relation of drainage area to stream order.*

Because only for special studies is it useful to include small rills in the drainage net analysis, for most purposes one may restrict consideration only to the net appearing on 1:24,000 scale maps. The choice of map scale, of course, is determined by the objectives of the particular analysis. Neglecting the rills, an extrapolation of the curve relating order to drainage area gives an estimate of the order of the Mississippi River at its mouth as twelfth order. Extrapolating from maps of scale 1:62,500, the Mississippi River is of tenth order, the difference being due to the size of channel designated as order 1.

Table 5-2.   Number and length of river channels of various sizes in the
                  United States (excluding tributaries of smaller order).

Order[a]	Number	Average Length (miles)	Total Length (miles)	Mean Drainage Area. Including Tributaries	River Representative of Each Size
1	1.570,000	1	1.570.000	1	
2	350,000	2.3	810.000	4.7	
3	80,000	5.3	420.000	23	
4	18.000	12	220,000	109	
5	4.200	28	116.000	518	
6	950	64	61.000	2.460	
7	200	147	30.000	11.700	Allegheny
8	41	338	14,000	55.600	Gila
9	8	777	6.200	264,000	Columbia
10	1	1.800	1,800	1.250.000	Mississippi
Total			3.250,000 (approx.)		

[a] The definition is that of Strahler: Order 1 is channel without tributaries; order 2 is channel with only order 1 tributaries, but includes only the length segment between junction upstream of order 1 channels and junction downstream with another order 2 channel.

An estimate of the number of miles of stream channels in a very large area can be made from the same graphs. Reading off the values of stream length and number of streams and multiplying those two parameters, we obtain the results shown in Table 5-2 for the number and total lengths of segments of channel in the United States. Note that order in the table refers to a segment of stream, the definition used by Strahler. This procedure leads to the estimate that there are about 3,200,000 miles of channel in the United States.

Drainage Density and Texture

To define drainage density, let us consider a drainage basin in which there are a number of tributaries and several trunk or major streams. It is a relatively simple matter to measure the cumulative length of all streams shown in this drainage basin. If, then, we also measure the total drainage area, the density of the drainage is given by the quotient of the cumulative length of the stream and the total drainage area:

$$\text{Drainage density} = \frac{\Sigma L}{A_d}$$

Drainage density then is simply a length per unit of area.

A comparison of the maps **A** and **B** of Fig. 5-2 indicates that two areas may well have the same drainage density yet differ in what we would call dissection or, better, texture. It can be seen by the data of Fig. 5-2 that these areas do have about the same length and drainage area, yet one has a greater number of stream channels. Smith (1950) has developed a measure that he calls the texture ratio, which is designed to describe the closeness or proximity of one channel to another. A contour line drawn at about mid-elevation in the drainage basin in Fig. 5-2, **A,** would intercept a number of individual stream channels, whereas one drawn at mid-elevation in Fig. 5-2, **B,** would intercept many fewer drainage channels.

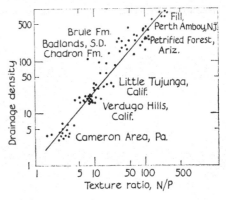

Figure 5-5.

Relation of drainage density to texture ratio, which is obtained by dividing the number of contour crenulations on the most crenulated contour by the perimeter of the basin. [After Smith, 1950.]

To define a texture ratio Smith used the contour with the most crenulations divided by the length of the perimeter of the drainage basin. It is clear that the crenulations of the hypothetical contour or the channel crossings of the contour are a measure of the closeness of the channel spacing. As such, these should be related directly to the drainage density. This relation is shown in Fig. 5-5. Descriptive textural terms were applied to the examples provided in Fig. 5-5. These can be made more precise by using quantitative measures. The descriptive classifications are simply generalizations of the quantitative measurements, and it is to be expected that as data are amassed with which to complete the graph in Fig. 5-5 the discontinuities that now exist will presumably disappear. It should be noted that drainage density ranges from one to 1000 in nature.

Stream length for given orders varies directly with order. In addition, it has been found that the *total* stream length of a given order (as well as the average length) is inversely proportional to the stream order. This relation, however, is a power function, for it is a straight-line plot of the logarithm of the total length of each order and the logarithm of the order (Strahler, 1957, p. 915).

Let us make the transformation from a purely geometric description of a

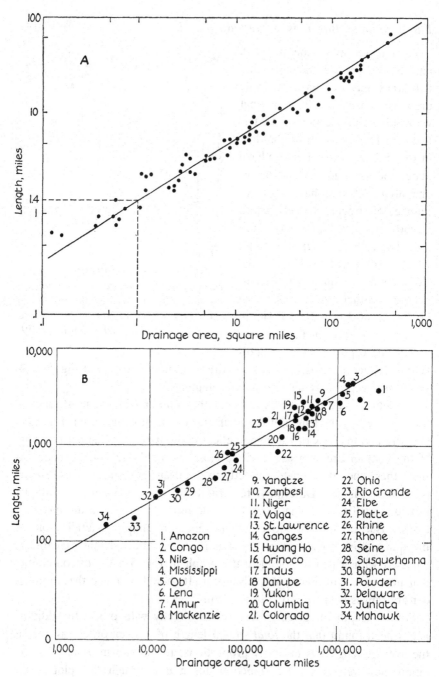

Figure 5-6.   *Relation of channel length in miles to drainage area, square miles. A. Shenandoah Valley, [After Hack, 1957.] B. Various basins including some large rivers of the world.*

drainage basin, such as we have considered so far, to the possible relations between the geometry and the erosional processes which may be the cause of the observed geometry. Two relationships involving channel length and drainage area appear to be of special interest.

If we measure successive distances from the headwaters along a watercourse (or their horizontal projection on a topographic map) and the total contributing drainage area at each of these points, we can draw a graph to show the relation between progressive increase in drainage area with increasing channel length. Such a graph is shown in Fig. 5-6, **A**, for streams described by Hack (1957) in the Shenandoah Valley and adjacent mountains in Virginia, and for various large rivers of the world in Fig. 5-6, **B**. There is relatively little scatter of points, and they define a curve whose equation is

$$L = 1.4A_d^{0.6},$$

where L is the channel length in miles and A_d the drainage area in square miles. In the northeastern United States the coefficient averages about 1.4 and ranges between 1 and 2.5. On the average, a drainage basin of 1 square mile will contain a channel 1.4 miles long. The exponent varies somewhat from one geographic region to another, but values generally lie between 0.6 and 0.7 for several regions in the United States.

Because large increments of drainage area are added at each point where a tributary joins the principal watercourse, the curve defined in Fig. 5-6, **A**, if it were drawn for a given drainage basin, would actually be composed of successive uniform segments broken by discontinuous upward steps. Important increments of drainage area contribute directly to the main channel. But the largest increments of drainage area are those which are added at tributary junctions.

If, as basins of larger size are considered, average basin width and length increased in the same proportion, the relation of length of the longest stream would increase as the square root of basin area. That the exponent is not 0.5 but 0.6 to 0.7 indicates that with increasing size, drainage basins elongate; length increases faster than width. This is true through a wide spectrum of basin size.

The position of the intercept defined by the length of channel at a drainage area of 1 square mile is less complex and of major interest when considering the process of channel formation and maintenance. The existence of a particular length of channel in a unit area implies several things.

1. The distance between two adjacent channels is by definition the reciprocal of the drainage density or A_d/L. One-half of this distance represents the length of overland flow from a divide to a channel. The length of overland flow at a drainage area of 1 square mile is then given by the intercept of the length-area relation.

2. The length of overland flow is an important parameter affecting the quantity of water required to exceed a threshold of erosion. Thus, on the average, 1 square mile of drainage area is sufficient to produce and maintain 1.4 miles of channel, as shown in Fig. 5-6, **A.**

Such a concept of channel maintenance is perhaps more easily visualized where the coordinates are reversed. In Fig. 5-7, cumulative drainage area is related to cumulative channel length. From the intercept of this graph, Schumm (1956) has defined a "constant of channel maintenance." When, as in the example, the cumulative stream length on the filled land at Perth Amboy, New Jersey, is 1 foot, the basin area is 8.7 square feet. Similarly, when the stream length is 1 foot in Chileno Canyon in the San Gabriel Mountains of California, the drainage basin area is 316 square feet. Assuming that these truly represent the smallest channels existent in these two locations, we may say that the surface area required to maintain a foot of channel—that is, the constant of channel maintenance—is 316 square feet at Chileno Canyon and 8.7 square feet at Perth Amboy.

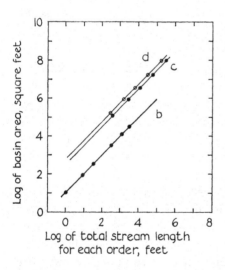

Figure 5-7.

Relation of drainage basin area to cumulative channel length of each order. [After Schumm, 1956.] b. Perth Amboy, New Jersey. c. Chileno Canyon, California. d. Hughesville, Maryland.

Thus far this discussion has concerned itself only with demonstrations of orderliness in the seemingly random distribution of a drainage network. In Chapter 10 the possible mechanism leading to this design will be discussed.

It was noted earlier that drainage area increases abruptly at tributary junctions, while length, of course, increases continuously. It is significant

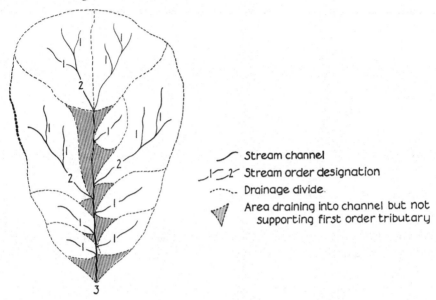

Figure 5-8.   *Hypothetical basin showing drainage contributing directly to main stem of master stream.*

however, that about half the streams of a given order enter directly into channels of two or more orders higher. This generalization holds for a number of samples inspected for this characteristic, but needs further checking to explore under what circumstances it is not true. For the most part these lower-order streams draining directly into streams of several orders higher will enter them in their lower rather than in their headwater reaches.

In the hypothetical basin sketched in Fig. 5-8, note the small wedge-shaped areas that drain directly into the master stream of third order, but are individually of such small size that they do not maintain even a first-order channel. The percentage of total area which so drains directly into the main channel is surprisingly high. It comprises about 20% of the total drainage area in basins of third and fourth orders. What percentage it represents in basis of still higher order has not been computed.

There is another closely related aspect of drainage basin configuration that is important for flood-control engineering. From an inspection of the numbers and drainage areas of various orders of streams, it will be seen that the totality of basins of a given order constitute less of the total area of a basin of larger order than might be expected. As an example, if all tributaries of about 10 square miles area within a basin of several hundreds

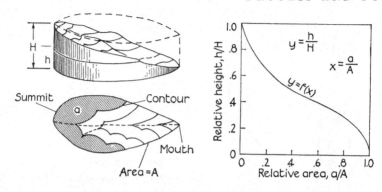

Figure 5-9.   *Nondimensional hypsometric curve. [After Strahler, 1957.]*

of square miles in total area were to be dammed, the area upstream from the dams would total only about 50% of the larger basin within which they are located.

To give another example, consider the basin of eleventh order in New Mexico described by the graphs of Fig. 5-4. The eleventh-order basin has a drainage area of 500 square miles. Within it there are about 10 basins of ninth order, each of which has a drainage area of 20 square miles. Thus the total area within the 10 ninth-order basins is 200 square miles, only 40% of the area of the eleventh-order basin in which they occur.

In the same eleventh-order basin there are about 150 sub-basins of seventh order, each having an area of about 1 square mile. Their sum is only 150 square miles, or 30% of the total drainage area of the eleventh-order basin within which they occur.

This is one of the reasons that adequate flood protection for a valley downstream cannot be achieved by the construction even of a very large number of dams on small upstream tributaries. Too large an area would remain uncontrolled by the many small dams (Leopold and Maddock, 1954, p. 37).

To summarize, drainage nets are so organized that there is a characteristic length, drainage area, and number of channels of each order; as size of basin increases, length of basin increases faster than width; the small-order streams draining into large master streams, particularly in downstream portions of the basin, account for a relatively large part of the drainage area of any master stream; areas too small to maintain even a first-order stream also drain directly into master streams, and these areas also constitute an appreciable part of the drainage area; drainage density ranges widely but bifurcation ratios have a limited range, 3.5 being a common value.

Description of the Drainage Basin in Cross Section

Geologists and engineers are familiar with the two-dimensional graph relating elevation and distance, the longitudinal profile drawn along a river channel. The relief is defined as the difference in elevation between valleys and the tops of mountains. In a description of erosion by running water, the relief ratio—the quotient of relief and channel length (the average slope of the longitudinal profile measured from headwater to the point of interest along the channel)—can in some instances be related to the quantity of clastic material that is eroded from a drainage basin. Neither relief nor relief ratio, however, deals with the three-dimensional picture of a region as does the hypsometric curve (after Strahler, 1957), illustrated in Fig. 5-9. Elevation is measured relative to the maximum height **H,** and area above a given contour is measured relative to the total area obtained by passing a horizontal plane through the mouth of the drainage basin. The shape of the curve shows the proportion of the total area which exists at various

Table 5-3.   Descriptive parameters of drainage basins.

Name	Symbol Used in This Text	Dimensions	Definition or Derivation
Stream order	R	None	An integer designation of a segment of a channel according to the number and order of tributaries
Stream number	N	None	Number of streams of a given length
Stream length	L	Length	The distance along a stream channel
Average length of streams of a given order	L_R	Length	Where R is the given order
Bifurcation ratio	r_b	None	Average ratio of number of streams of a given order to number in next higher order, N_1/N_2
Length ratio	r_e	None	Average ratio of average length of streams of a given order to average length of streams of next lower order, L_2/L_1
Drainage area	A_d	Length^2	Basin area contributing precipitation to a given channel segment
Texture ratio	T	$\dfrac{1}{\text{Length}}$	Ratio of maximum number of channels crossed by contour to basin perimeter

elevations. Inasmuch as landform development under exogenous processes may be manifested in changes in the form of hypsometric curves, this index is useful in describing the possible evolution of landforms.

There are of course any number of additional parameters which we might devise to describe the various characteristics of drainage basins. If our concern is with slope or the slope of a particular landform, we could describe it in terms of cumulative curves, of changes of slope, of tangents expressing the slope at a given point, or in other graphic terms. Strahler has developed a large number of measures of potential use in describing drainage basins. Because they cover a wide range of possible problems, many of which are not discussed in this book, some of them are given in Table 5-3.

REFERENCES

Hack, J. T., 1957, Studies of longitudinal stream profiles in Virginia and Maryland: U. S. Geol. Survey Prof. Paper 294-B, pp. 45–97.

Horton, R. E., 1945, Erosional developments of streams and their drainage basins; hydrophysical approach to quantitative morphology: Geol. Soc. Am. Bull., v. 56, pp. 275–370.

Leopold, L. B., and Maddock, T., 1954, The flood control controversy: Ronald Press, New York, 278 pp.

Leopold, L. B., and Miller, J. P., 1956, Ephemeral streams—hydraulic factors and their relation to the drainage net: U. S. Geol. Survey Prof. Paper 282-A, pp. 1–37.

Schumm, S. A., 1956, Evolution of drainage systems and slopes in badlands at Perth Amboy, New Jersey: Geol. Soc. Am. Bull., v. 67, pp. 597–646.

Smith, K. G., 1950, Standards for grading texture of erosional topography: Am. Journ. Sci., v. 248, pp. 655–668.

Strahler, A. N., 1957, Quantitative analysis of watershed geomorphology: Am. Geophys. Union Trans., v. 38, pp. 913–920.

Wolman, M. G., and Leopold, L. B., 1957, River flood plains: some observations on their formation: U. S. Geol. Survey Prof. Paper 282-C, pp. 87–107.

Chapter 6 Water and Sediment in Channels

.... Niagara has power and it has form and it is beautiful for thirty seconds, but the water at the bottom that has been Niagara is no better and no different from the water at the top that will be Niagara. Something wonderful and terrible has happened to it, but it is the same water and nothing at all would have happened if it had not been for an aberration in one of nature's forms. The river is the water's true form and it is a very satisfactory form for the water and Niagara is altogether wrong.

GERTRUDE STEIN

QUOTED BY JOHN HYDE PRESTON IN

"A CONVERSATION WITH GERTRUDE STEIN"
ATLANTIC, *August 1935*, p. 192.

Introduction

Rivers drain water from the continents to the ocean and are the principal routes for transporting the products of weathering. Gravity provides the force by which both excess water and movable debris are brought from higher to lower elevation.

In accomplishing this transfer the water which flows off the land toward the ocean forms and maintains a highly organized system of physical and hydraulic features. So complex are the details of interrelations in this organized system that to describe adequately any single portion tends to make one lose sight of other equally important features. As in any part of the natural environment, the interrelations in the system make it difficult to visualize all of it simultaneously. Yet it is these interrelationships which constitute the most distinctive and pervasive characteristic of rivers.

In this discussion we attempt to keep this feature before the reader, even though it seems to complicate the subject at times. The complications arise in large part because of gaps in our present understanding and not as a result of any mysterious properties of the system. As one would expect, the fundamental aspects of rivers and their hydraulic processes turn upon basic principles of physics or chemistry. Infinite variations seem possible because of different local conditions of lithology, topography, climate, or vegetation. But it is the systemization, the interrelationships, and the basic mechanical principles which seem to us most important for an understanding of rivers.

Forces Acting in Channels

A few fundamentals of hydraulics are necessary to introduce the subject of the geometric and hydraulic properties of stream channels. These will perhaps be common knowledge among many readers, but may be helpful to some.

Water flowing in any open channel is subject to two principal external forces. Gravity exerts the force which propels the water downslope; friction between the water and the channel boundaries tends to resist the downslope movement.

The force of gravity is directed vertically. With respect to any plane inclined to the horizontal, there is a tangential component of the gravitational force. Let F be the gravitational force exerted on a body situated on an inclined plane, the force being equal to the weight of the body. In a direction along an inclined plane, the component of the gravitational force is $F \sin \beta$, where β is the angle of inclination of the plane.

If the body is at rest it is obvious that a frictional resistance exerted by the surface of the inclined plane against the object is larger than the downslope component of the body's weight. If there were no friction, then the body would immediately begin to move downslope under the influence of the tangential component of its weight, and would accelerate at the gravitational rate.

If the body is moving down the inclined plane at a uniform speed—that is, without acceleration—then a frictional force equal and opposite to the tangential downslope component of the body weight must be acting on it.

This same principle applies to water flowing in a channel. The slope of the water surface is comparable to the inclined plane. The force exerted by gravity, tending to move the water downslope, is the tangential component of the weight of the water.

The resisting force exerted on the water by the bed and banks of the channel is a shearing stress. An analogous case for laminar flow would involve two thin flat plates separated by a thin layer of liquid, the two plates being drawn past one another. The force required to draw one plate past the other is a function of the surface area in contact, the viscosity of the liquid, and the square of the relative velocities of the plates. Similarly, the resisting stress per unit area of streambed is proportional to the square of the velocity of the water moving over the bed.

If there is no acceleration, the downchannel or tangential stress which water exerts on the channel boundary is equal to and opposite to the parallel resisting stress exerted by the bed on the moving fluid. The shearing stress is transmitted from one layer to another through the medium of viscous or turbulent exchange of momentum as a result of a gradient of velocity.

The ability of flowing water to carve a channel, transport debris, and thus ultimately to degrade the landscape, depends on these forces—the gravitational impelling force, and the resistances offered to it. The effects of lithology and topography on the ability of flowing water to carve and transport are exerted principally through their relation to the resisting forces. For this reason a knowledge of some of these hydraulic relations is becoming increasingly important to the geomorphologist.

Velocity and Its Distribution

Shear stress, τ, is expressed as

$$\tau = K\frac{dv}{dy},$$

where v is velocity, y is depth, and K is a coefficient. In viscous flow K is the molecular viscosity while in turbulent flow K becomes the coefficient of eddy viscosity. The gradient of velocity or change of velocity in a certain distance is a shear because it expresses the displacement of one part of a body relative to another.

At the bed of a channel velocity will be zero. Above the bed the velocity is greater, usually increasing with distance above the bed. The rate of increase of velocity, dv/dy, is governed by the way in which mixing takes place between slower-moving elements of the flow nearer the bottom and faster-moving elements above.

In nonturbulent or laminar flow, mixing is molecular and by viscous forces. Velocity varies linearly with depth and shear stress is proportional to velocity. In turbulent flow, mixing is done by turbulent eddies and shear stress is proportional to the square of the velocity. Experimental studies have shown that the laminar or turbulent character of the flow can be expressed in terms of a dimensionless number, the Reynolds number

$$\frac{vd}{\mu/\rho},$$

where v is velocity, d is depth, and μ/ρ is the kinematic viscosity (viscosity divided by the density).

It has been shown from considerations of momentum transfer that the velocity in an open channel should decrease toward the bed in a logarithmic curve. A relation in which velocity is proportional to the logarithm of depth does describe the change of velocity in the vertical quite well in most river channels. Two sets of velocity measurements by current meter are shown in Figs. 6-1 and 6-2, in which the same data are plotted on Cartesian and on semilogarithmic paper.

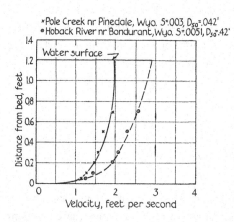

Figure 6-1.

Current-meter measurements of velocity in rivers, plotted as functions of depth; the examples chosen are identical in depth, but differ in mean size of bed material, D_{50}, and in slope. Data plotted on Cartesian coordinates.

Owing to the logarithmic change of velocity with distance from the bed, there is a depth at which the local velocity equals the mean velocity of the curve as a whole. This position averages about 0.6 of the distance from the surface to the bed. It is known empirically that the mean velocity can also be closely approximated by the average of the velocity occurring at a distance from the surface 0.2 of the total depth and that occurring at 0.8 of the total depth (see Fig. 6-3). These relations are used in stream-gaging practice. If the current meter is placed at a depth 0.6 of the distance from surface to bed, the

velocity measured will be approximately the mean for the vertical
section. If measurements are
taken at 0.2 and 0.8 depths, the
average is a close approximation
of the mean value.

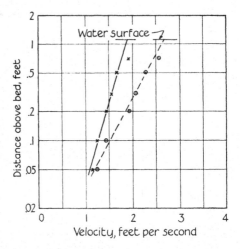

The shape of the velocity distribution curve depends on the
roughness of the channel bed. For
a given depth of flow, the larger
the roughness elements or projections from the bed, the greater the
loss of turbulent energy at the
bed, and this results in a steeper
gradient of velocity toward the
bed. The actual velocity profiles
of two channels having equal
depth but different bed roughness
are shown in Figs. 6-1 and 6-2.

Figure 6-2.

*Current-meter measurements of velocity
in rivers, plotted as functions of depth;
the examples chosen are identical in
depth, but differ in mean size of bed material, D_{50}, and in slope. The same data
as in Fig. 6-1, plotted on semilogarithmic
paper.*

The usual reference to velocity
in natural streams is not to a
velocity at a point but rather to a
mean velocity for the channel as a whole. The rate of flow in terms of volume of fluid passing a given cross section per unit of time is called discharge. Thus discharge is the product of the cross-sectional area of the
flowing fluid times a mean velocity. In this context of open channels of
natural streams, velocity is equivalent
to speed. Velocity is usually defined as
a vector having both direction and magnitude, but as used here it implies
speed in a direction perpendicular to
the cross section.

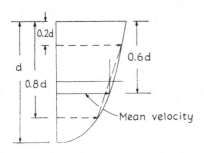

Velocity of water flowing in an open
channel is dependent on several factors,
among which are energy gradient (usually approximated by water-surface
slope), depth, and roughness. Though
in engineering practice velocity is computed by one of several semiempirical

Figure 6-3.

*Relation of position of mean ve-
locity to the distribution of velocity
with depth in a river.*

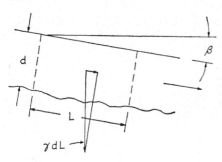

Figure 6-4.

*Definition sketch: flow in channel seg-
ment of unit width.*

formulas which contain those variables, it should be recognized that velocity varies from one part of a given cross section to another and is the integrated result of complex interaction of water moving at different speeds in different parts of the channel.

Factors Controlling Flow Velocity

The manner in which the boundary shear exerts its influence on velocity is best seen by considering the most commonly used equations applicable to open-channel hydraulics, the equations of Chezy and Manning.

Consider a channel segment of unit width w, length L, and depth d, as in Fig. 6-4. The force acting on the unit of water in the direction of flow is the downslope component of its weight, or

$$F = \rho g d L w \sin \beta,$$

where ρ is the mass density. For small angles,

$$\sin \beta \doteq \tan \beta = \text{slope } s.$$

The resisting stress is the stress per unit area, τ, times the boundary area over which the stress is applied:

$$F = \tau(2d + w)L.$$

Because there is no acceleration of the water, the impelling force equals the resisting force, or

$$\rho g d L w s = \tau(2d + w)L.$$

Since $dw = $ area of cross section, A, therefore

$$\rho g A s = \tau(2d + w),$$

or

$$\tau = \rho g s \frac{A}{2d + w}.$$

The ratio $A/(2d + w)$ is defined as hydraulic radius R. Note that $2d + w$ is simply the wetted perimeter; in a very wide channel, where width is very much greater than depth, the hydraulic radius is equivalent to the mean depth A/w. We may now compute the shear stress as:

$$\tau = \rho g R s = \gamma R s,$$

where γ = specific weight in lbs./cu. ft. Its dimensions are

$$\frac{MLT^{-2}}{L^3} = ML^{-2}T^{-2}.$$

Thus, the resisting force per unit area of channel boundary is proportional to the product of hydraulic mean depth (hydraulic radius) and slope. Slope in this context expresses the downstream rate of loss of potential energy through friction and is the slope of this energy line, known as the energy gradeline. In many hydraulic computations this slope is approximated by the slope of the water surface, as the two are parallel where the flow is uniform.

In hydraulics there is a resistance to flow which is comparable to the resistance of a wire carrying electric current. This flow resistance may be defined as

$$\text{Resistance} = \frac{\tau}{v^2},$$

in which, as before, τ is the shear stress per unit area of boundary and v is the mean velocity of flow in the conduit. It has been shown experimentally that in turbulent flow the resistance is proportional to the square of the mean flow velocity, provided the boundary does not change as velocity is varied. Therefore

$$\tau = k_1 v^2,$$

where k_1 is the flow resistance. Substituting in the previous formula for τ,

$$\tau = k_1 v^2 = \gamma R s$$

or

$$v^2 = k_2 \gamma R s.$$

Let $\sqrt{k_2 \gamma}$ be called C, the Chezy coefficient; then

$$v = C\sqrt{Rs},$$

or mean velocity is proportional to the square root of the product of hydraulic radius and slope. In wide shallow channels where R is almost equivalent to mean depth, d, then mean velocity is proportional to the square root of the depth-slope product.

A variation of this is the most widely used formula for mean flow velocity in American engineering practice. The formula, based on field and experimental determination of the values of the resistance coefficient is

$$v = 1.49 \frac{R^{2/3} s^{1/2}}{n},$$

and is known as the Manning equation. Experimental values of n, Manning's n, vary from about 0.01 for smooth metal surfaces to 0.06 for natural channels with many rocks, protuberances, or irregular alignment. As can be seen from these values, n is a conservative quantity. However, n is not independent of R.

It is truly surprising that engineering practice has depended to such an extent on a formula as empirical as this one, derived nearly a century ago. Many engineers have become very proficient at estimating the value of n to apply to a given channel. The U. S. Geological Survey has developed a series of stereophotographs of different channels for which the value of n has been computed from measurements of velocity, depth, and slope. This file of comparative pictures is a useful tool. Because n must be estimated, its determination is a matter of judgment and not an independent measure of specified physical attributes of the channel.

Recall that Reynolds number, vd/v, is a measure of flow turbulence, in which v is velocity, d is depth, and v is kinematic viscosity. If the channel boundary is characterized solely in terms of simple roughness elements uniformly distributed, the resistance to flow provided by the boundaries under different flow characteristics can be expressed as a function of Reynolds number, flow depth, and height of the roughness elements. A well-known diagram (for example, see Rouse, 1950, Fig. 79) expresses this function of resistance in dimensionless terms. High values of Reynolds number indicate flow in the turbulent range, and low values indicate flow in the laminar range. In laminar flow the resistance decreases as Reynolds number increases. Beyond a transition zone in the region of fully developed turbulent flow, resistance is no longer related to the Reynolds number, but is instead a function of the ratio of the depth of flow, d, to the height of roughness elements, k; the greater d/k, the less the resistance.

Because flow in natural rivers is—nearly without exception—fully turbulent, the skin resistance, or resistance produced by the boundary surface, is not a function of Reynolds number but depends only on the ratio of depth to size of the roughness elements. This ratio is called the relative roughness. There is in addition resistance offered by a variety of bed and bank features such as bars or riffles, undulations of the channel bank, and bends in the channel. Flow resistance is computed in hydraulic data from observed relations among velocity, depth, and slope, rather than being judged independently by measurement of the actual roughness elements. The computed value of flow resistance, then, includes the effects of all types of roughness.

It is of particular interest, then, insofar as data from natural rivers are concerned, to determine whether the total resistance computed from natural river data bears the same relation to relative roughness as occurs in pipes or experimental channels. Further, what is the relation of actual bed particle size to the size of uniformly distributed roughness elements which would offer the same resistance?

To answer these questions requires a set of river data more complete than is usual at any ordinary measurement location, certainly a variety of measurements not available at the usual stream-gaging station. For flows of different magnitude it is necessary to have observations of slope, depth, and velocity, and for the same reach an objective measurement of bed material size.

Such data are available for five locations on Brandywine Creek, Pennsylvania (Wolman, 1955, pp. 17 and 51–53). The measure of bed particle size used for the present computations is D_{84}, the diameter which is equal to or larger than 84% of the bed particles. The choice of the 84% value is arbitrary; it is two standard deviations larger than the mean size, assuming a normal distribution. Experience of many investigators has shown that particles larger than the median size play an important role in flow resistance, and therefore a single parameter to describe bed particle size should be some size larger than the median.

The Brandywine Creek data plotted in the form usually used for presentation of analogous data for pipes are presented in Fig. 6-5. Theory and experiments yield a relationship of the form

$$\frac{1}{\sqrt{f}} = 2 \log \frac{d}{D} + \text{constant.}$$

The nondimensional factor, **f**, known as the Darcy-Weisbach resistance coefficient is proportional to gds/v^2. Note that

$$\frac{C}{\sqrt{g}} = \frac{v}{\sqrt{gds}},$$

where C is the Chezy coefficient. The quantity $\sqrt{gds}$ has the dimensions of velocity and is known as the shear velocity, v_*.

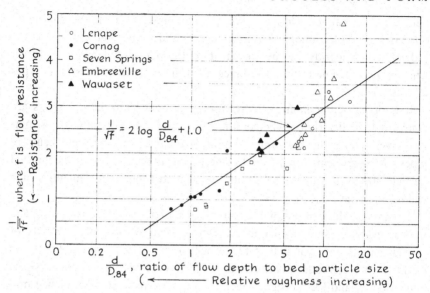

Figure 6-5.   *Resistance to flow as a function of relative roughness, the ratio of flow depth, d, to the height of roughness elements, D_{84}, Brandywine Creek, Pennsylvania. [Data from Wolman, 1955.]*

The relation defined by the river data plotted in Fig. 6-5 is

$$\frac{1}{\sqrt{f}} = 2 \log \frac{d}{D_{84}} + 1.0.$$

This indicates that, at least in the Brandywine data, the bed particles exert an effect on flow similar to uniformly distributed skin resistance, in that total flow resistance is a function of relative roughness.

Energy Losses in Streamflow

Thus far, a simple characteristic of the boundary surface has been taken as the key element in the resistance to flow. Indeed, in natural streams a principal effect of the geology of an area on the rivers which drain it lies in the influence of lithology on the soils and on the debris of the hillslope. This is reflected in the size, character, and quantity of debris contributed to the river. The debris forms not only the burden of the river but determines the character of the channel boundaries. The effect of the debris and in turn the river's modification of the debris are hydraulic phenomena reflected in the resistance to flow. In natural channels the total resistance expressed by a parameter such as the Manning n includes not only the effect of particle size per se but such factors as vegetation, curvature, obstructions, and the effect of changing forms on the bed, generated by movement of sand and silt in waves or other configurations. Effect of bed form on resistance is discussed elsewhere in this chapter. It can be seen here as elsewhere in geomorphology that the interaction of the surface and a geomorphic agent is registered or measured in terms of resistance. For this reason, resistance or energy loss is of particular concern in geomorphology.

Energy is a quantity which, irrespective of the form it may assume, can be measured as the product of a force (mass times acceleration) times the distance through which the force acts. The energy of a discrete body therefore has the dimension ML^2T^{-2}. The energy per unit volume has the dimension $ML^{-1}T^{-2}$, which is also the dimension of a stress—force per unit area. When energy is considered in terms of elevation, the force assumed is that of gravity, Mg acting on a body, or ρg acting on unit volume. The decrease in energy along the direction of flow is the energy converted from kinetic or potential form into thermal form. The energy available is the sum of potential energy (head or elevation), and kinetic energy ($v^2/2g$). In steady uniform flow, where the kinetic energy is the same at various points along the channel—that is, where there is no change in velocity with time at a point (steady flow) nor a change in velocity with distance along the channel (uniform flow)—the slope of the water surface is parallel to the slope of the energy grade line and is a direct measure of the energy loss.

The decrease in total energy along the direction of flow is an indication of the fact that energy is converted from potential or kinetic form into thermal form. The energy loss is caused by boundary friction and ultimately is converted into heat.

There is an upward flux of turbulent energy away from the boundary and it is in the body of the flow that most of the final degeneration into heat takes place. It is the migration of turbulent energy away from the boundary that distinguishes shear turbulence from simple homogeneous turbulence. This migration of turbulent eddies maintains the upward force necessary to support a load of suspended sediment, as will be explained.

The fact that flow energy is converted into heat is apparent from the fact that a meltwater stream on the surface of a glacier, even though at a constant temperature of 32°F, can progressively melt its way downward through the ice by the heat generated by the fluid friction of the flow.

The elements of resistance causing the energy loss in turbulence include the roughness of the boundary surface and discrete areas of energy dissipation in bends, junctions, or protuberances. Each causes distortion of the flow, for it deflects the flow or part of it from its former direction. Such deflection creates energy dissipation by eddying, secondary circulation, and increased shear rate.

The energy loss, or head loss, of each of these resistances is, in closed pipes, dependent on the square of the flow velocity, because within a pipe there is no change in the flow boundary as velocity varies. But in open channels deformation of the boundary is possible, owing to changes that can occur in the free water surface. Thus, not all types of resistance in open channels will cause energy dissipation strictly in proportion to the square of the flow velocity.

With this in mind, resistance in open channels may be classified into three types. *Skin resistance,* as we have seen, depends, for any given shape and size of cross section, upon the square of the flow velocity and the roughness of the boundary surface. In a straight and uniform pipe the whole of this resistance may be assumed to be uniformly distributed along the flow and is expressed as a force per unit boundary area. In river channels this is the resistance most affected by the size and character of material in bed and banks.

A second type is *internal distortion resistance,* caused by discrete boundary features that set up eddies and secondary circulations. In pipes this is exemplified by bends, fixtures, or junctures. In river channels it is represented by bars, bends, individual boulders, undulations of the bed, and by bank protuberances. This type may be even greater than skin resistance in open channels, particularly where channel bends are involved. Internal distortion resistance is also proportional to the square of the flow velocity.

A third type is *spill resistance,* which occurs locally at particular places in open channels under some conditions. The energy is dissipated by local waves and turbulence when a sudden reduction of velocity is forcibly imposed on the flow. Spill resistance is associated with local high velocities, as when water backs up behind an obstruction and spills into lower velocity flow. Blocks of bank material slumped into a channel cause such spills, as do some channel bends of tight curvature. Spill resistance departs from the velocity-squared relation and varies with velocity to higher powers. Such resistances cause local foci of intense energy dissipation.

Little is known about the localization and amounts of these different types of energy dissipation, despite the fact that longitudinal profiles of rivers, profiles of the water surface during floods, and many other significant features of rivers are highly dependent on rates of energy dissipation. There has been little investigation of the relative effects of various bed materials, or random bank projections, of pool and riffle sequences, and of river bends on energy losses.

In natural rivers carrying sediment, resistance due to bars, ripples, and dunes is a function of sediment transport rate (Einstein and Barbarossa, 1952). The resistance caused by channel alignment and curvature usually cannot be separated from the sum of all resistances computed from data on natural rivers. However, laboratory experiments with sinuous and straight channels have shown that channel curvature alone can cause energy loss of the same order of magnitude as that due to skin friction. Where curves are tight (small radius of curvature relative to channel width), the resistance due to curvature may be double that due to skin resistance.

Some of the characteristics and causes of energy loss in a natural channel are shown diagrammatically in Fig. 6-6, which includes (above) a hypothetical plan view of a river channel, and (below) a graph showing the longitudinal profile of the energy grade line, the water surface, and a straight line representing the average energy slope in the same reach. The diagram has been constructed to show the following phenomena. At the distance of

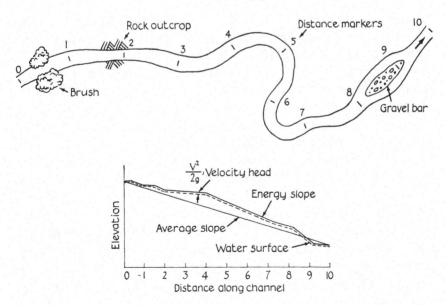

Figure 6-6. *Relation of channel features to profile of water surface and energy profile; note effects of obstructions, curves, and central island.*

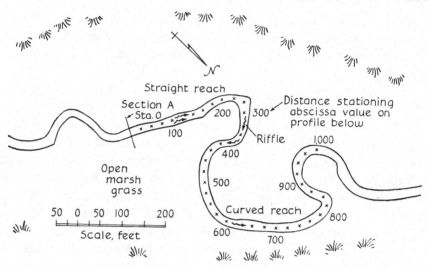

Figure 6-7.   *Planimetric sketch map of a reach of Rio Grande del Ran-*
chos near Taos, New Mexico. In Fig. 6-8 are shown profiles
of the bed and of the water surface and energy gradeline for
two discharges. The higher represents near-bankfull flow
and the lower about half bankfull.

0.5 units on the distance scale, brush is indicated encroaching on the channel, and at this point there are a steepening of the water surface and a
slight increase in the velocity head—that is, the kinetic energy of flow. The
latter is indicated by the vertical spread of the lines representing the water
surface and the energy grade line and is equal to the quantity $v^2/2g$.

The energy grade line steepens again where the channel is narrowed by
a rock outcrop. Note, in each of these places of local steepening, that velocity increases above its average value. The smallest slope occurs in the
nearly straight reach between stations 2 and 4. The scale of the sketch is
too small to show the usual alternating steep zone over a riffle (submerged
gravel bar) and pool or deep in the nearly straight reach.

Then the slope steepens somewhat as the channel describes the two
curves between stations 4 and 7 because, when discharge and bed material
remain the same, more energy is used up in a curved channel than in an
otherwise comparable straight one, owing to the energy loss due to the
curvature. The steepest place is where the channel divides around an island, stations 8 to 9.

Some idea of the magnitude of the kinetic energy head in relation to the
channel slope may be obtained by considering the conservatism of velocity

downstream in relation to increasing width and decreasing slope downstream. At high stage, velocities in most streams average between 3 and 6 feet per second, and so the respective velocity heads are 0.14 foot and 0.56 foot. A creek with a gravel bed and a width of 20 feet may have an average slope of 0.007 or .14 foot in a distance equal to one channel width. If its mean velocity is 3 feet per second, the downstream slope provides a fall equal to the velocity head in a distance about equal to the width. But a great river a mile wide with a velocity of 6 feet per second may have a slope of .00002. Its fall will equal one velocity head in a distance of 5.6 widths.

A field example is given in Figs. 6-7 and 6-8. In addition to the planimetric map (Fig. 6-7), Fig. 6-8 shows the thalweg profile, water surface, and energy grade line for a medium stage and a bankfull stage on the Rio Grande del Ranchos near Taos, New Mexico. The water-surface data are the mean of measurements on the two sides of the channel. Energy head is computed from current-meter measurements of velocity at each individual point plotted on the graph.

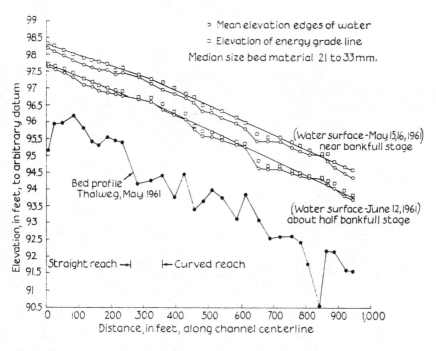

Figure 6-8.   *Rio Grande del Ranchos near Talpa, New Mexico, showing pool-riffle sequence, water-surface profiles, and thalweg.*

Measurement of Velocity and Discharge

The preceding discussion of velocity distribution dealt essentially with the variation of velocity in a two-dimensional case above a given boundary. For the river cross section as a whole, a full three-dimensional velocity distribution is needed. Figure 6-9 shows the velocity distribution in a natural river channel in a straight reach. The velocity decreases toward the bed and walls, and theoretically the maximum occurs at the surface in the center-line of the channel. In a very wide channel the velocity distribution throughout the middle two-thirds, or perhaps even nine-tenths, of the section may be nearly uniform. The depression of the maximum velocity below the surface is often observed in small rivers and in river curves.

Because discharge is the product of velocity and cross-sectional area, measurement of discharge requires measurement of velocity and area. Be-

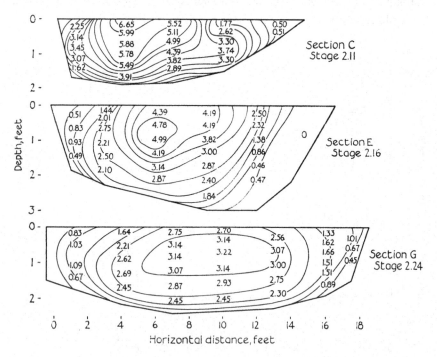

Figure 6-9. *Velocity distribution in a natural river channel in a straight reach; three examples for the same stream, Baldwin Creek near Lander, Wyoming, at different sections within a single reach of a few hundred feet; flow is near bankfull.*

cause both velocity and depth vary throughout a section, in standard engineering practice velocity is usually measured at perhaps 30 positions across the flowing stream. In the United States observations are made at either one or two depths at each position, in accord with the generalized logarithmic velocity curve described earlier. In many countries in Europe more points are measured in each vertical. At each lateral position the depth of water is recorded and the cross-sectional area so determined is multiplied by the mean velocity in the vertical to obtain a discharge for each subarea. These discharges are then summed to give the total for the cross section. In English units discharge is expressed as cubic feet per second.

In United States practice a cup-type current meter is used to measure velocity. When a current meter is not available or cannot be used, floats of various kinds may be timed over measured distances, using the average of the rate measured by two or three floats at various positions across the channel. Such measurements are crude but they are far better than guesses and are particularly valuable during floods when equipment is either unavailable or unusuable. A stick weighted by a rock or a spike at one end so that it floats upright is easy to make in the field and provides a serviceable float. In the nonexistent "average" river an estimate of mean velocity in a given vertical position is obtainable by timing the rate of travel of an upright float and multiplying this rate by about 0.80 (Matthes, 1956).

Mean velocity of rivers in flood varies from about 6 to 10 feet per second. The mean velocity attained in large rivers is generally slightly higher than that in small ones. There are, of course, many local situations where, owing to constrictions or rapids, velocity attains greater values. The figures cited above would include a large majority of river channels in reaches that have no unusual features.

A tabulation was made by the U. S. Geological Survey of the largest measured values of velocity at a single point (not the average for the whole river cross section). The maximum point velocity usually is on the order of 25–50% greater than the average velocity for the cross section. Out of 2,950 measurements included in the sample, the median value was 4.11 feet per second, the mean 4.84 feet per second, and less than 1% of the total exceeded 13 feet per second. One of the highest velocities ever measured by current meter by the U. S. Geological Survey was 22 feet per second in a rock gorge of the Potomac River at Chain Bridge near Washington, D. C., during the flood of March 1936. Velocities of 30 feet per second (20 miles per hour) are known, but none greater than that value.

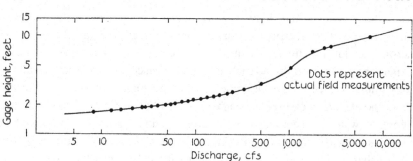

Figure 6-10.   *Rating curve; plot of gage height versus discharge, Seneca
Creek at Dawsonville, Maryland, drainage area 100 sq. mi.
The mean annual discharge at this location is about 100 cfs.*

A gaging station involves two particular kinds of observations: (a) the
maintenance of a graphic or digital continuous record of water stage (water
level in the river), and (b) velocity measurements from which the stage
record can be converted into discharge. To utilize a series of individual
velocity measurements made at various times, a rating curve is established
for each station, relating discharge to stage (Fig. 6-10). The stability of
this relationship is primarily a function of the stability of those physical
conditions which control the height of the water at a given discharge.
Changing bed elevation, variation in vegetation, and channel pattern all
may affect the control. For this reason, in rivers of moderate size a concrete
weir (called a control) is often placed across the channel to provide a stable
control free from the vicissitudes of scour and fill and changes in vegetation and debris.

At any known river stage, then, the corresponding discharge may be determined by reading off the rating curve. The total quantity of water flowing
past a gaging station in a given period of time is determined by summing the
products of discharge and time for appropriate intervals. Total water quantity in a month or year is usually expressed in acre-feet; 1 cubic foot per
second for 1 day is approximately 2 acre-feet (other conversions are shown
in Appendix A, p. 505).

Because many research problems in river morphology require such measurement data, it is well to point out that there are available thousands of
direct observations of velocity, width, depth, discharge, and stage at a variety
of river sections. There are more than 7,300 gaging stations in the United
States and its territories and many in Europe but many fewer in the less
developed areas of the world. It should be noted, however, that other fac-

tors needed for morphologic studies are not required in discharge measurement and therefore are in short supply. This applies particularly to measurements of water-surface slope, size and character of materials comprising river bed and bank, the gage height of the bankfull stage, and other physiographic features. There are very few detailed observations of the distribution of turbulence, maps of flow trajectories or streamlines, variations of water-surface profile with stage, or profiles across the channel and along it. Also lacking are detailed measurements of physical and hydraulic parameters in certain kinds of channels, particularly in ephemeral rills, in tidal estuaries without inflow from the land, and in various channel patterns.

The Debris Load of Rivers: Introduction

Although chemical and hydraulic problems can be dealt with in terms of clear water and fixed boundaries, the fundamental aspect of river mechanics is the interrelationships between the flow of water, the movement of sediment, and the mobile boundaries. These relationships are complex in detail and the mechanical principles governing their behavior are not yet adequately explained. In addition, there is a chicken-and-egg uncertainty about the interrelationships which makes simple statements of cause and effect uncertain and misleading. This discussion first considers principles of movement of sediment particles and the form and composition of the beds of alluvial channels.

A particle on the surface of the bed of a channel exerts on the particles on which it rests a vertical force equal to its immersed weight. A natural channel has some slope and therefore the immersed weight can be resolved into a component normal to the bed surface and a component tangential to the bed and directed downslope.

To move a given grain, a force must be exerted on it, the normal upward component of which is equal to the downward normal component of the immersed weight. The force exerted can be visualized as a torque or couple exerted by the flowing water dragging over the exposed top of the grain, or as a direct force of the water impinging on the area exposed to the flow. In either event, the force exerted is usually thought of as a drag stress proportional to the exposed area of the grain. In natural streams, which are usually fully turbulent, the minute turbulent eddies near the bed cause a fluctuation of the local flow velocity at any one point. This gives some statistical or random chance that a given grain will move rather than neighboring grains.

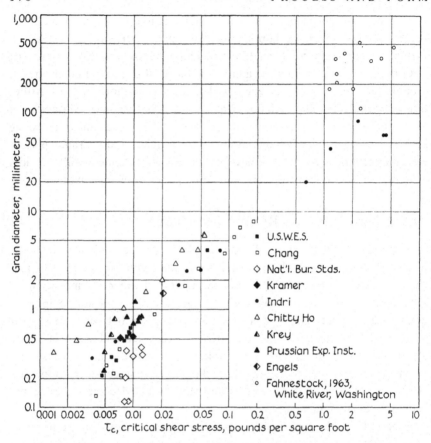

Figure 6-11.   **A.** *Laboratory and field data on critical shear stress required to initiate movement of particles.*

As demonstrated in the derivation of the Chezy formula (p. 157), the fluid stress acting on the bed may be expressed as a function of the product of water depth and water-surface slope. Mean flow velocity is not a parameter directly influencing applied stress, but because velocity is partly dependent on depth and slope, fluid stress on the bed is correlated with velocity.

Similarly, discharge is proportional to velocity and depth, and therefore has some correlation with fluid stress on the bed.

When the applied stress is low enough, no grains move. If it is increased, a condition is reached where a few grains begin to move. This condition corresponds to the critical stress or critical tractive force commonly referred to in literature on grain transport.

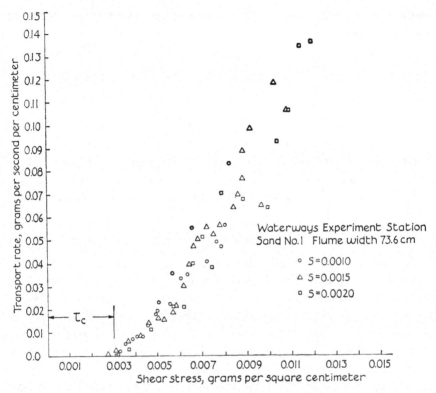

Figure 6-11. **B.** *A sample to show how a value of critical shear stress τ_c, can be derived from data on relation of transport rate to mean bed shear; data from Waterways Experiment Station. [Johnson, 1943.]*

If τ is the mean stress per unit area of the exposed grain, then the total force on a grain is the product of mean stress times the area per grain. The number of grains in a unit area of bed is

$$\text{Number of grains} = \eta \, \frac{\text{area}}{D^2},$$

where D is grain diameter and η is a packing coefficient. Thus the exposed area per grain is D^2/η and the fluid force on a grain is $\tau(D^2/\eta)$.

The gravitational weight of a grain is its volume times its immersed weight per unit volume, or

$$\frac{\pi}{6} \, D^3(\sigma - \rho)g,$$

where σ is mass density of grain and ρ is mass density of fluid. When a particle is about to move, the force tending to move it is just balanced by the immersed weight:

$$\frac{\pi}{6}(\sigma - \rho)g D^3 \sin \beta = \tau \frac{D^2}{\eta}.$$

Then

$$\tau = \eta \frac{\pi}{6}(\sigma - \rho)g D \sin \beta,$$

where β is the angle of inclination of the bed relative to the horizontal. When β is small, $\sin \beta$ equals $\tan \beta$. Let $\eta(\pi/6)$ be a constant K and $\tan \beta \doteq$ slope s. Then the bed stress at the time of incipient motion of the grains is

$$\tau_c = K(\sigma - \rho)g D s.$$

Because τ_c is the critical stress necessary for grain movement—*competence* in the terminology of geomorphology—it may approximate values of τ_c, experimentally determined in flumes. However, as White (1940) emphasizes, this experimental value is subject to a wide variation from the value derived from the simplified assumptions above. Figure 6-11, **A,** presents a sample of laboratory and field data, showing relation of grain diameter to critical shear stress or tractive force.

Among other factors it is important to note that the statistical variation in velocity in turbulent flow encompasses fluctuations at least 2 times the mean. Since shear stress is proportional to the velocity squared, this amounts to a four-fold increase in shear stress. From a practical standpoint the most satisfactory estimate of τ_c would be one based on plotting measured rate of transport against mean bed shear for a variety of conditions and extrapolating the empirical relation to the intercept where transport rate equals zero (Fig. 6-11, **B**).

Because of the importance of sediment transport in geomorphic problems, the discussion here is somewhat extended and includes an initial consideration of some fundamental principles governing transport. This discussion, based primarily upon the work of R. A. Bagnold, is useful in demonstrating the nature of the forces involved in transporting sediment but does not purport to be a derivation of a practical formula for computing rate of sediment movement. In a later section reference is made to several equations frequently used to estimate the rate of transport of sediment in rivers.

The Nature of Fluid Force and Its Relation to Debris Movement*

In any mechanical system having no mean acceleration there must be an equilibrium of applied forces. This is true for balls bouncing on the ground, airplanes flying through the air, or solids immersed in a fluid. In the last case the excess weight of the solids must be transmitted to the ground by (a) direct solid contact, continuous or intermittent, (b) flux of fluid momentum, or (c) by both agencies.

A group of bouncing balls exerts the same weight on the ground as if the balls were in stationary contact. An aircraft is supported by the upward dynamic reaction of an equal downward flux of momentum diffused in a cone to a wide area on the ground.

In the case of solids being transported by a fluid the total mass M of sediment has an excess weight (weight less buoyant force) of

$$\frac{\sigma - \rho}{\sigma} Mg,$$

where σ is the mass density of the grain and ρ that of the fluid and g is the acceleration of gravity. The total mass M is made up of the bed-load mass, M_b, and suspended load, M_s. In respect to the bed-load mass M_b, its excess weight, $(\sigma - \rho/\sigma)M_bg$, is supported by direct, though intermittent, contact with the bed. The part of the load which is suspended, having a mass M_s and an excess $(\sigma - \rho/\sigma)M_sg$, is supported by a flux of fluid momentum. Whereas an aircraft creates its own downward flux, the suspended load experiences an upward flux of eddy momentum created by the passage of the fluid past the channel boundary.

In ordinary solid friction, the frictional stress opposing motion, T, is proportional to the perpendicular force $P = Mg$, equal to the weight of the object (Fig. 6-12, **A**).

The coefficient of friction is defined as the ratio of these forces, T/P, at incipient motion, and this ratio is also equal to the tangent of the angle of repose or angle of slip, α:

$$\frac{T}{P} = \tan \alpha,$$

---

* This explanation is drawn from an unpublished manuscript by R. A. Bagnold, 1960, "The physics underlying sediment transport." The explanation has been abbreviated and paraphrased; the authors are responsible for this present form.

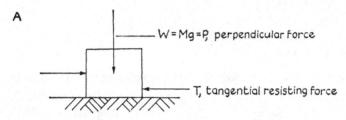

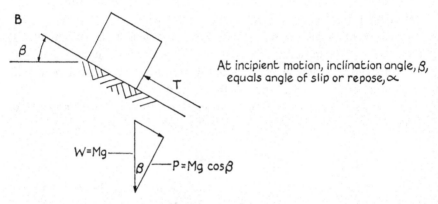

Figure 6-12.   *Relations of applied and resisting forces on an object resting on a supporting surface. A. Subject to an outside tangential force. B. At incipient motion on an inclined plane, inclination angle, β, equals angle of slip or repose, α.*

as shown by Fig. 6-12, **B,** or

$$T = P \tan \alpha.$$

When the object rests on an inclined rather than on a horizontal surface, the force resisting motion in the tangent direction, T, is proportional to the normal component of the body weight:

$$P = Mg \cos \beta,$$

where β is the angle of inclination. Hence,

$$T = Mg \cos \beta \tan \alpha.$$

The force needed to initiate motion is T minus the downhill component of the body's weight, or

$$Mg \cos \beta \tan \alpha - Mg \sin \beta.$$

It can be seen that the force needed to initiate motion becomes zero when the angle of inclination, β, is made equal to the angle of friction, α.

The relations discussed above are quite general. They apply equally well to a solid grain immersed in a fluid, in which case the appropriate weight of the object of mass M is its immersed weight,

$$\frac{\sigma - \rho}{\sigma} Mg.$$

Thus the force necessary to initiate motion of an immersed object is

$$\frac{\sigma - \rho}{\sigma} Mg(\cos \beta \tan \alpha - \sin \beta).$$

To maintain motion at a velocity V it is necessary to exert a force constantly in order to provide the energy that is degraded into heat during the motion. The rate at which the work is performed, or energy supplied, is the product of the force exerted and the velocity maintained, inasmuch as force × velocity $= \dfrac{\text{force} \times \text{distance}}{\text{time}} = $ power. Thus, if a bed particle travels with relative velocity, V_b, the power required is the product of the force exerted times this velocity, or

$$\text{power} = \frac{\sigma - \rho}{\sigma} M_b g V_b (\cos \beta \tan \alpha - \sin \beta).$$

The coefficient of friction, $T/P = \tan \alpha$, refers to solid contact transmitted from solid to solid. In fluids, a comparable fluid or flow coefficient (discussed on page 157) relates a tangential fluid stress, τ, to a fluid velocity v in the same direction. The perpendicular stress, P, has no counterpart in the fluid case.

The counterpart of the coefficient of friction is, for the fluid case, the flow coefficient, v/v_*, which (as shown on page 159) is equal to a function of the Darcy-Weisbach resistance factor $\mathbf{f}$, in the form

$$\frac{1}{\sqrt{\mathbf{f}}} = \frac{v}{v_*}.$$

The resisting stress exerted by a channel bed to a fluid with bed load is transmitted partly by solid stress, T, and partly by fluid stress, τ. In equilibrium the resisting tangential stress, $T + \tau$, must equal the total applied stress, $\Im$, or

$$\Im = T + \tau,$$

where

$$T = \frac{\sigma - \rho}{\sigma} M_b g \tan \alpha.$$

The total applied stress ჳ, composed of the solid stress and fluid stress, is the total measurable shear stress. Where neither bed nor suspended load is in motion, the total applied stress consists only of fluid stress uninfluenced by grains. An increase in bed load increases the solid stress component, T. An increase in suspended load decreases the fluid stress component, τ, by damping the turbulence. As grain concentration increases, the solid stress T not only comprises an increasingly large part of the total, but the total stress, ჳ, becomes very much greater than the fluid stress in the absence of grains (Bagnold, 1956, pp. 240–242).

Discharge of water is expressed as the movement of a volume of water a unit distance downstream in a unit of time. This represents a rate of flow. Similarly, a transport rate of sediment is the movement of a mass of sediment a unit distance in a unit of time, with dimensions MLT^{-1}. This rate can be expressed as the product of a mass times a velocity.

For a mass immersed in a fluid, a dynamic transport rate, I, may be defined in terms of the immersed weight, $(\sigma - \rho/\sigma)Mg$, times the velocity of the grains:

$$I = \frac{\sigma - \rho}{\sigma} MgV.$$

This transport rate has the dimensions ML^2T^{-3}, which are the dimensions of rate of doing work or power.

For the movement of a bed-load mass, M_b, for which the corresponding dynamic transport rate is I_b, the rate of work done, or power, is

$$\text{power} = I_b(\cos \beta \tan \alpha \pm \sin \beta),$$

where

$$I_b = \frac{\sigma - \rho}{\sigma} M_b g V_b.$$

The sign of sin β depends on whether the bed slopes downward or upward in the flow direction. As Bagnold points out, this type of expression is general and would apply to transport by road, rail, dogsled, or bed sediment.

The previous derivation applies to grains transported as bed load in which the weight of the grains is supported by the grain bed. In contrast, the weight of the suspended grains is carried by the flow and ultimately by the fluid in the interstitial spaces of the bed.

Where the transport is by fluid alone, as in the case of suspended sediment load or a fleet of aircraft, power is necessary to maintain the supporting flux of fluid momentum. This again is a product of the appropriate mass and a velocity. The velocity here, however, is the vertical component of the fluid velocity relative to the load. In the case of suspended debris, it is approximated by settling velocity, v_s. Thus,

$$\text{power} = \frac{\sigma - \rho}{\sigma} M_s g v_s.$$

When the channel is inclined to the horizontal at an angle, β, the required vertical component of velocity, v_s, is in part provided for or augmented by the vertical component of the downslope velocity of the particle V_s:

$$\text{power} = \frac{\sigma - \rho}{\sigma} M_s g (v_s \pm V_s \sin \beta).$$

As previously, defining a dynamic transport rate of suspended load as

$$I_s = \frac{\sigma - \rho}{\sigma} M_s g V_s,$$

then

$$\text{power} = I_s \left(\frac{v_s}{V_s} \pm \sin \beta \right)$$

When the ratio v_s/V_s equals $\sin \beta$, the necessary power is zero. In suspended load under these conditions, the falling sediment provides its own power or would suspend itself.

The power discussed above is that necessary to transport the solid objects by the fluid, either with solid contact at the boundary or by fluid alone. In mechanical systems the energy used is less than the total available; that is, there is always some loss. The part used is the product of an efficiency factor, e, and the total available power, Ω. For solid support,

$$I_b(\cos \beta \tan \alpha \pm \sin \beta) = e_b \Omega_b,$$

and for fluid support,

$$I_s \left(\frac{v_s}{V_s} \pm \sin \beta \right) = e_s \Omega_s,$$

where subscripts b and s refer respectively to conditions of bed-load and suspended transport, and $\overline{V}_s$ is the average downslope velocity of suspended grains.

Power is the rate of doing work and is equivalent to a force times velocity. In a stream of flowing water the appropriate force is the gravitational tractive force, $\rho g A s$, where A is the cross-sectional area of flowing water and s is the slope of the energy grade line. The velocity appropriate is the mean flow velocity of the fluid v. Because Av equals the discharge, Q, then

$$\text{available power} = \rho g Q s.$$

This is the rate of doing work or the rate of degradation of potential energy into heat by the flowing water. Only part of this energy is available at the bed to move the bed load along.

The rate of doing work, or power, of a stream per unit of width—that is, per unit of bed area—is

$$\frac{\Omega}{w} = \omega = \frac{\rho g Q s}{w},$$

where w is channel width and ω is power per unit width. Because the total applied stress per unit area is $\mathfrak{I}$, the power per unit area is also

$$\omega = \mathfrak{I} v.$$

The mean flow velocity, v, can also be expressed in terms of unit stress through the flow coefficient or resistance factor:

$$\frac{v}{v_*} = \frac{v}{\sqrt{\mathfrak{I}/\rho}}.$$

If c_b is the flow coefficient appropriate to the effective mean height above the bed where the fluid is pushing the grains at their average velocity, let

$$c_b = \frac{v}{\sqrt{\mathfrak{I}/\rho}}$$

or

$$v = c_b\sqrt{\mathfrak{I}/\rho}.$$

Now substituting this expression for v, the power per unit area available for bed load is

$$\omega_b = \mathfrak{I} c_b\sqrt{\mathfrak{I}/\rho}.$$

The relation of transport rate to available power may be written for bed load:

$$I_b(\cos \beta \tan \alpha \pm \sin \beta) = e_b \Omega_b$$

Per unit of width, let

$$\frac{I_b}{w} = i_b$$

and

$$\frac{\Omega}{w} = \omega_b.$$

Then a transport equation for bed load takes the form

$$i_b(\cos \beta \tan \alpha \pm \sin \beta) = e_b \omega_b$$
$$= e_b c_b \mho \sqrt{\mho/\rho}.$$

A comparable expression can be derived for the suspended sediment. The sum of the bed load and suspended load equals the total load. The coefficients e_b and c_b can be evaluated from experiment by measuring the other parameters in the equation. Thus far this has not been done for most field and laboratory data.

Some important aspects of the derivation can be readily summarized. Transport rate of debris, whether as bed load or suspended load, can be expressed in terms of the force required to maintain movement at a given velocity. The product of force and velocity has the dimensions of power. The particles being submerged, their appropriate weight is the submerged weight. In the case of bed load, the power required is affected by the coefficient of friction and the angle of the bed to the horizontal, because the necessary applied stress is provided in part by the downslope component of the immersed weight of the particles. The appropriate velocity is the downstream velocity of the particles relative to the bed.

Bagnold (1954) points out that "the available power on the right hand side of the bedload equation is the bed stress times the fluid velocity at the effective distance from the bed at which the fluid pushes the bed material along. Since this height is a few sediment diameters only, the relevant fluid velocity is smaller than the mean flow velocity, and considerably smaller in the case of a big river.

"In the same units the available power to transport suspended load is the bed stress times the mean flow velocity. Hence, all else being the same, the ratio

available bedload power
——————————————————
available suspended-load power

becomes smaller the bigger the river, and may be misleadingly larger in the case of small flumes."

The power necessary to carry suspended load is also a product of a force or weight times velocity, but the velocity applicable here is the settling velocity of the particle. The required vertical velocity component may be partly provided, or it may be increased by the vertical component of the downslope velocity of the particle.

The derivation brings out the difference between solid friction exerted from solid to solid and tangential fluid stress exerted between fluid and solid. The resistance exerted by channel bed to a fluid with bed load is partly solid friction and partly fluid stress.

Grains in motion near the bed exert a force that tends to maintain grains on the surface in position against the stress provided by the flowing fluid-grain mixture. A grain swept off the bed by an initial impulse tends to be pulled back to the bed by gravity. When sufficient grains are in motion, the probability that this accelerated grain will collide with another is very great, and the faster moving of the two transmits to the slower an impulse tending to knock the latter up, down, or aside. When enough of these collisions are occurring there will be some grains knocked upward and some downward. Some of these downward blows are exerted against the bed. The net effect of these blows delivered to the bed can be considered a force exerted by the moving grains against the bed, and this force may be resolved into a normal (perpendicular) component and a tangential component. The ratio of these components is dependent on the mean conditions of encounter. The tangential component transmits a shear stress from fluid to bed and vice versa, which is additive to, and may greatly exceed, the shear stress transmitted by the sheared fluid surrounding the bed-load particles.

The normal component, under steady conditions, amounts to the immersed weight of the moving bed-load particles.

A static grain bed cannot be sheared without some degree of dilation or dispersion. This dispersion must be upward, against the downward or normal component of the body force pulling the grains toward the bed. Thus all grain flow requires the exertion of some kind of dispersive stress between the sheared grain layers and the bed. Following Bagnold (1954), who enunciated the concept, this stress is called *dispersive grain stress*.

Bed Load and Suspended Load

The separation of bed and suspended load has often been quite arbitrary. Workers in the field are agreed that suspended load in rivers is that portion of the load borne by the upward momentum or flux of momentum in turbulent eddies in the flow. Bed load, although variously used in engineering practice, applies to the sediment that moves by sliding, rolling, or saltating on or very near the bed. Bed material includes the sizes of material found in the bed. These may be transported anywhere within the flow (Colby, 1961, p. D-2). As conditions of flow change, particles making up the bed load at one instant may become suspended at the next and vice versa.

Some workers make still another distinction: that load which because of its fine size has such a small settling velocity that it would be held in suspension as colloidal particles, they call "wash load." Thus wash load is comprised of exceedingly fine particles having vanishingly low rates of settling; hence these particles pass through the river system relatively unrelated to the hydraulic conditions in a given reach. Clay minerals and sizes in appreciable concentrations, however, may affect the configuration of the bed and through it the overall rate of sediment movement.

We believe that the Bagnold definition states better than any other the difference between bed load and suspended load. Though in engineering practice it does not give a practical basis for differentiating quantitatively the sediment in the two modes of transport, it does provide a rational and physically sound basis for understanding the difference. Bagnold defines "bed load" as that portion of the total load whose immersed weight is carried by the solid bed—that is, borne by the bed grains. "Suspended load" is that part whose immersed weight is carried by the fluid and thus finally by the interstitial fluid between the bed grains.

Computation of Sediment Load

The computation of suspended load is usually based upon a sediment diffusion equation which for two-dimensional, steady, uniform flow is given as

$$\bar{c}v_s = -K_s \frac{d\bar{c}}{dy},$$

where $\bar{c}$ is the mean concentration at a distance y above the bed, v_s is the appropriate fall velocity of the sediment, and K_s is the sediment diffusion coefficient (for example, Rouse, 1950, p. 799). The equation is similar to the diffusion equation for turbulent flow,

$$\tau = K \frac{dv}{dy},$$

in which K is the eddy viscosity. The equation states that the upward moving eddies supply a flux of momentum which supports the sedimentary particles tending to fall under the influence of gravity. Because concentration decreases with distance from the bed, there is a net upward transfer of sediment from layers of higher to layers of lower concentration, although there is no net flow of water.

Integration gives an equation of the form

$$\frac{\bar{c}}{\bar{c}_a} = f\left(\frac{y}{a}\right) \frac{v_s}{K\sqrt{\tau/\rho}},$$

which relates the concentration $\bar{c}$, at any height y, to the concentration $\bar{c}_a$, at a reference elevation a. Note that here again the distribution of sediment is related to the shear stress, τ, and to the particle size expressed in terms of the settling velocity, v_s. The equation shows that the relative concentration of particles of a particular fall velocity will increase as the shear stress increases. Similarly, for a given shear stress, the smaller the particle (in general, the lower its fall velocity), the more uniform will its distribution be in the vertical section. Large particles will be found at higher concentrations near the bed. Empirical data from field and laboratory agree well with the theoretical formulation.

Equations for the computation of bed load have, to considerable extent, been of the type first suggested by Du Boys, in which transport rate depends on a coefficient, the bed shear ($\tau = \gamma RS$), and a critical shear derived as shown in Fig. 6-11, **B.** The coefficients in these equations are derived primarily from laboratory studies, and the application of the equations to field studies poses many yet unsolved problems. A summary of a number of such equations is given in Rouse (1950, p. 194).

Many equations have been developed to estimate total sediment discharge. Based in part on theoretical analyses and in part on laboratory experiments in flumes, in final form all of them include some wholly empirical relations. The best known is the Einstein bed-load function (Einstein, 1950). This complex equation shows the dependence of sediment transport rate on shear stress and particle size. Several engineering equations use a mean particle diameter, but in the Einstein method a rate of transport is computed for each size and summed to give a total. Based on the excellent and original work of Einstein, an adaptation of this general method was developed by Colby and Hembree (1955) which they entitled the "Modified Einstein computation of sediment discharge." The modified method is simpler in computation and uses parameters more readily available from actual stream measurements. This modified Einstein method is presently the one most used in engineering work in the United States.

Like many important variables in stream morphology, mean velocity is intercorrelated with other dynamically significant variables. Figure 6-13 presents a typical relationship between the measured mean velocity in the Rio Grande and the sediment transport, computed by the modified Einstein technique. The graph includes several points in which there was a high concentration of fine sediment, and these points appear to fall above the general curve. The empirical relations of transport rate and mean velocity confirm Gilbert's (1914) early experimental findings that velocity is highly correlated with sediment discharge.

As discussed previously, transport rate is related to several intercorrelated parameters, some of which are independent and some dependent. The relation between transport rate of sands, mean velocity, and depth (Fig. 6-14) represents an empirically useful relationship, but one in which the basic mechanics are complex and poorly understood. Here, for low values of velocity, increasing depth is associated with decreasing rate of transport of sand. At velocities of about 2.5 feet per second the transport rate of sands is independent of channel depth in channels having the same mean velocity. Channels having a higher value of mean velocity have higher rates of transport of sand with increases in water depth.

Colby (1961, p. D-8) attributes these apparently anomalous effects of depth to the fact that at low velocities most of the sediment is being moved near the bed. An increase in depth at constant mean velocity reduces the velocity near the bed, and because of the low velocity

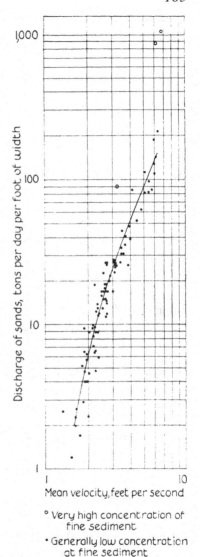

° Very high concentration of fine sediment

• Generally low concentration of fine sediment

Figure 6-13.

Discharge of sands computed by the modified Einstein technique, plotted against mean velocity for the Rio Grande in New Mexico and for some diversions from it; data are unadjusted for water temperature. [After Colby, 1961.]

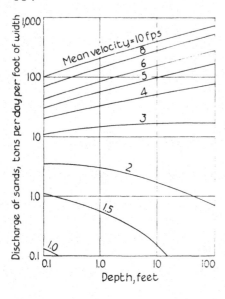

Figure 6-14.

Empirical relation between transport rate of sands and mean velocity as a function of water depth; water temperature 60° F; median diameter of bed sands 0.30 mm. [After Colby, 1961.]

throughout the vertical the reduced concentration near the bed is not compensated for by higher concentrations at greater depths. In contrast, at high velocities sediment is distributed more generally throughout the vertical. An increase in depth may lower concentrations near the bed but high velocities will continue to prevail throughout the vertical, resulting in greater total transport in the section.

The high degree of correlation between transport rate and mean velocity can be useful, particularly in dealing with field data, where the need is for a rough and workable way of estimating total transport rate. For such purposes a few rates of sediment transport can be computed by the modified Einstein method and these transport rates plotted against the correlative mean velocities, establishing an empirical relation from which further estimates of transport rate can be read directly.

Estimates of the rate of sediment transport in natural channels based on existing equations, however, may be as much as 100% in error. In practice, several are often used as checks. Each has merits or demerits for use in different kinds of streams, but detailed comparisons are beyond the scope of this book. In addition to those already mentioned more detailed references can be found in Meyer-Peter and Muller (1948), and Laursen (1958). The number of empirical equations itself testifies to the continuing need of a more fundamental knowledge of the basic processes governing sediment movement. Because of the variables involved it appears likely that major advances will be made primarily through advances in theory and critical experiments rather than by amassing volumes of additional data.

Measurement of Sediment Load

The standard method of measuring suspended load in the United States employs a sampling "fish" of brass or aluminum, containing an ordinary pint milk bottle. The nozzle protruding forward from the nose has an opening of about 0.3 inch in diameter and is connected to the bottle and to an external air vent so designed that the entrance velocity to the nozzle is always the same as the ambient velocity of the stream (Nelson and Benedict, 1951, p. 897). Neither the sediment moving nearest the bed nor the suspended material larger than the diameter of the nozzle can be sampled. It is unusable for measuring debris larger than coarse sand. Thus the network of sediment-measuring stations in the United States, 184 river locations in 1961, does not include any of the mountain streams or headwater channels where gravel is the dominant size of material on the bed. Sediment load in cobble- and boulder-covered channels is unmeasured.

In streams carrying fine gravel to sand, bed-load traps developed in Europe have given useful results but of unknown validity. A recent review of methods of bed-load measurement, however, was not encouraging (Hubbell, 1963). Although troughs placed across and beneath the beds of some sand-bed streams (Einstein, 1944) have proved workable, this method is limited to favorable locations on small streams and is too elaborate for general use. There is thus a need for new techniques and instruments for measuring debris loads of rivers under natural conditions. So important is this to geomorphology, it is a major challenge to men entering the field.

Where material carried as bed load is thrown into suspension, as at narrows or falls in natural streams, the total load can sometimes be measured with a suspended-load sampler. The required conditions have also been created artificially by constructing a grid of small baffles on a concrete apron on the bed of a short section of the channel (Benedict, Albertson, and Matejka, 1955).

In practice, the suspended-load sampler is used to obtain depth-integrated samples at several places across the stream cross section. By depth-integrated is meant that the sampling fish is lowered into the water with a uniform motion until it touches the bed, then raised again to the surface at the same speed. While submerged, the water-sediment mixture enters the nozzle at a rate representative of the local velocity, and thus through time the sampler integrates the sample of load at various distances from below the surface in proportion to the local velocity.

Samplers of similar design are used to measure sediment concentration at any given depth by controlling opening and closing of the nozzle.

Physically, the concentration of sediment in a vertical column of water-sediment mixture, or in a unit height of the column, is the ratio of sediment to fluid or to the mixture of fluid and sediment in the volume. If one could instantly enclose the volume of interest, the concentration in the enclosed volume would be a physicist's concept of true concentration. This might be called a "static" concentration. The static concentration determines: (1) the weight of sediment supported on the stream bed; (2) the density of the fluid mixture in the stream; and in part (3) the effect of concentration on the fall velocity of particles of sediment in the fluid.

The standard stream-sampling procedures just described do not, however, provide a "static" concentration. Byrnon Colby suggests the following illustration. Consider an 8-foot length of a conveyor system of four parallel belts 1 foot wide. These belts are carrying flat plates 1 foot square. The first and second belts have one plate every 8 feet of belt; the third has one every 4 feet, and the fourth has one every 2 feet. The concentration or ratio of area occupied by the plates is 12.5, 12.5, 25, and 50% for belts 1 to 4 respectively. The average concentration for all 4 belts is 25%. This is a static concentration similar to the concentration in a sample of fluid-sediment mixture trapped instantaneously. The average concentration and the load on the conveyor system are the same, regardless of how the 8 plates are distributed on the belts.

If we assign respective velocities of 4, 3, 2, and 1 feet per second to the belts, the static concentration does not change. However, in 8 seconds the first belt moves 32 feet and 4 plates past any given reference point; the second belt moves 24 feet and 3 plates; the third belt moves 16 feet and 4 plates; and the fourth belt moves 8 feet and 4 plates. The total is 80 feet of belt and 15 plates, giving an observed concentration of 18.8%. This is a discharge-weighted concentration, one in which, so to speak, high velocity may compensate for low concentration, and vice versa.

The concentration of suspended sediment in a stream may be visualized as the concentration in the volume of flow enclosed between two cross sections 1 foot apart. This "stream" concentration is the concentration that determines the density of the water-sediment mixture and determines the effect of sediment concentration on fluid drag and particle fall velocity. The stream concentration is not determined directly in routine sedimentation measurements. Usually, for fine sediments (smaller than 62 μ) the concentration of suspended sediment in the stream is approximately equal to the concentration of suspended sediment in the stream discharge. For coarser sediments the suspended-sediment concentration normally increases, but the flow velocity decreases from surface to bottom of the stream. Lower than average sediment concentrations are transported in higher than average stream velocities. Therefore, the suspended-sediment concentration in the stream discharge (the concentration that would be collected in a reservoir) is smaller than the concentration in a cross section of the stream.

A static concentration by volume is needed for analysis of the mechanics of sediment transport, but only a discharge-weighted concentration in parts per million by weight is usually published. The problem remains of converting a discharge-weighted concentration into a static concentration. This can be done only if one of the following applies: (1) the concentration is uniform throughout the stream depth, (2) the velocity is uniform throughout the depth, or (3) the distribution of both velocity and concentration is known throughout the depth. The moving belts show these limitations. If all belts move at the same speed, the static and discharge-weighted concentrations are equal; if all belts have the same concentration, the static and discharge-weighted concentrations are equal; otherwise, to determine the relation of static and discharge-weighted concentration one must know the speed of each belt and the spacing of plates on each belt.

The U. S. Geological Survey publishes the concentration of suspended sediment in parts per million (ppm) by weight. The concentration is computed as one million times the ratio of the weight of sediment to the weight of water-sediment mixture. In a complete sediment measurement the concentration is obtained from analyses of samples that give the ratio of sediment discharge to water discharge: (1) at a point in a stream cross section, (2) as an average in a stream vertical, or (3) as an integrated sample representative of the entire stream cross section. Samples at a point are weighted with stream discharge to obtain a discharge-weighted concentration for the stream vertical, and concentrations in a vertical are weighted with discharge to obtain a discharge-weighted concentration for the entire stream. Usually, complete sediment measurements are taken only at intervals of several days, but they form the basis for records of sediment concentration. Therefore, published concentrations are discharge-weighted concentrations and they are the concentrations that, divided by one million and multiplied by stream discharge by weight, give suspended-sediment discharge by weight.

Laboratory operations on the samples collected consist generally of analysis of sediment concentration and for some, but by no means all, of the samples, an analysis of the particle-size distribution. In the United States sediment-concentration data and size analyses made for the sites at which sediment is measured are published in Water-Supply Papers of the U. S. Geological Survey. A procedure widely used, but the implications of which are by no means understood, is the dispersion of the clay fraction of the sample before the size analysis is made. The effective size of clay material, so far as the transportation by river processes is concerned, probably is the size in flocculated condition. By dispersion of these flocs the true size is obtained. Although the dispersed size may be descriptive, it may not be most meaningful in terms of transport and deposition in nature. More must be known about the action and theory of suspension before it is known whether the present practice is the best. For some studies, size analysis is made in native water without dispersion.

The Stream Bed and Its Description

The characteristics of a stream bed are of interest to the geomorphologist on several counts. First, as a part of the channel they are a feature of the landscape which enters into the description of the landforms of an area. Second, materials making up the bed of a river may form part of the geologic or stratigraphic record. Their recognition and interpretation can be used in unraveling the geologic history and in reconstructing the paleogeography. Third, the materials on the bed of the river and their behavior are an integral part of the dynamics of river mechanics. An understanding of this dynamic interaction is in fact the essential ingredient in interpreting the geologic evidence provided by the deposits. For many aspects of rivers and river mechanics no discussion extant in the literature is better than the book by Leliavsky (1955); this is particularly true with regard to his discussion of the stream bed. His emphasis is solely on river mechanics, whereas in the present volume we are attempting to relate river processes to geomorphic problems. For a discussion of the stream bed, a student seeking other details should consult Leliavsky's work.

Omitting vegetation, the channel bed consists essentially of two different elements: discrete particles whose effect on the flow is primarily a function of the sizes and shapes of the grains or particles themselves, and aggregates of particles which form definite structures on the channel bed. Dynamically these aggregations create a form drag on the flow which is in addition to the effect of the grains themselves (Einstein and Barbarossa, 1952). Consider first the discrete particles on the river bed rather than the differentiated forms which they may comprise.

The customary mode of sampling and describing aggregates of discrete particles is by size analysis based upon the percentage by weight of the different size components. Ideally a sample of a stream bed would cover a representative surface area of the bed perhaps one grain diameter thick. It is usual to combine or to composite surface samples taken at a number of different points. Unless otherwise specified, published descriptions of the mean diameter or other statistical measures of particles on the bed of a channel refer to such a composite surface sample. The thickness or depth of the sample, however, is more likely to be 6 inches to a foot, depending upon the character of the bed. Bulk samples of sand are sieved in the laboratory, while silt and clay fractions may be separated by the visual accumulation tube or pipette methods. A typical example showing the size distribution of river-bed sediments is given in Fig. 6-15. Some summary data applicable to the samples presented in this figure are shown in Table 6-1.

Figure 6-15 and the data of Table 6-1 indicate that grain size tends to decrease in size relatively rapidly in the first few miles and much more

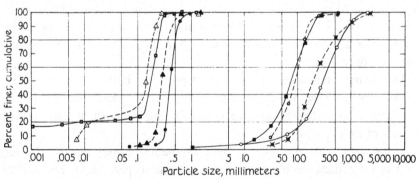

□ Mississippi R. at Head of Passes, La.
● Mississippi R. at Mayersville, Miss.
△▲ Missouri R. at Omaha, Neb.
○ Rock Creek at Roberts, Mont.
○ W. Fork Rock Creek nr. Red Lodge, Mont.
✴ Rock Creek nr. Red Lodge, Mont.
■ Yellowstone R. at Billings, Mont.

Figure 6-15. *Samples of size distribution of bed material, Yellowstone-Missouri-Mississippi River System.*

Table 6-1.   Samples of size distribution of bed material, Yellowstone-Missouri-Mississippi River System.

River and Location	Bankfull Q^1 (cfs)	Slope	Bed Material D_{50} (mm)	Distance from Headwater (miles)
West Fork Rock Creek near Red Lodge, Montana	260	0.035	270	15
Rock Creek near Red Lodge, Montana	600	.021	210	23
Rock Creek at Roberts, Montana	—	.0086	90	36
Yellowstone River at Billings, Montana	26,000	.0015	80	75
Missouri River at Omaha, Nebraska	110,000	.000012	.29–.15	1,300
Mississippi River at St. Louis, Missouri	436,000	.000010		2,400
Mississippi River at Mayersville, Mississippi	2,070,000^2	.000076	.40	3,070
Mississippi River at Head of Passes, Louisiana			.16	3,500

[1] 1.5-year recurrence interval.

[2] Mean depth 43 feet, stage 120.3 at Mayersville, and near bankfull.

slowly thereafter. This difference is accentuated by differences in sampling technique, as the very coarse materials upstream were described by counting. On the Missouri and below, sieve samples were used. (These techniques are discussed below.) Also, the materials become moderately well sorted after a relatively short distance of transport, and thereafter sorting increases only slowly.

The changes in grain size in the Mississippi River in the 1,091 miles from Cairo, Illinois, to the mouth, have probably been as well documented as in any large river of the world. The graphs of Fig. 6-16 show some salient aspects of these changes.

In the distance of more than 1000 miles the mean grain size remains for most of the distance in the range of coarse sand to pea gravel. The muddy Mississippi is generally pictured in terms of the appearance of the water, which is influenced by the suspended fraction of the load, and it is perhaps not generally realized that the bed material is usually sand rather than silt or clay. This is clearly shown both in graphs *A* and *C*.

Considering the sand fraction only (graph *B*), the decrease in size downstream appears to be nearly linear, although it should be kept in mind that the sand size embraces a larger and larger percentage of the bed material as the mouth is approached. Even near New Orleans (graph *C*) more than 70% of the bed material is of sand size.

Inspection of these examples of downstream change of bed-material size in rivers illustrates certain impressions which we have gained from experience with such data. Good data on the downstream change in grain size are meager. Usually they consist of samples taken by a variety of techniques and are therefore lacking in consistency (as is Fig. 6-15). Few if any set of data extant are as consistent and voluminous as those from which Fig. 6-16 is derived; more such examples are needed, including extensions into headwater areas of larger grain size.

The exponential decrease of particle size downstream—shown by Sternberg (Hack, 1957) and widely quoted—may well represent the ideal case, where there are no or few tributaries introducing sediment of different history and origin into the master stream. But in many river systems—if the reach of river studied is long enough to show the effects of abrasion—tributary entrance usually complicates the picture by introducing a variety of new materials. More field studies are needed on processes of abrasion, in which observations are made not only on particle size but also grain lithology. Hack's (1957) work is almost unique in this respect and at present is the most instructive in the literature.

An exponential decrease downstream of particle size implies a rapid decrease in size in the most headwater reaches, the rate of decrease diminishing downstream. This seems to be a general characteristic. When the predominant grain size is sand, the size-frequency distribution tends to be log-normal; that is, the logarithm of particle size plots as a straight line on probability paper (Blench, 1952). Geologists have recognized that river sediments are not well sorted, at least as compared with sediments deposited by some other geomorphic agents. Size distribution graphs in Fig. 6-15 are generally quite typical.

Because much remains to be learned about abrasion and sorting of river debris, methods of measurement deserve considerable attention because measurement itself presently impedes progress in knowledge.

There are two important difficulties connected with large bulk samples. First, the spatial distribution of the surface materials rather than the sizes alone may be most important, and compositing may mask such differences. Second, if the particles on the bed are very large they are difficult to remove. As Wentworth (1931) has shown, if the size distribution of the al-

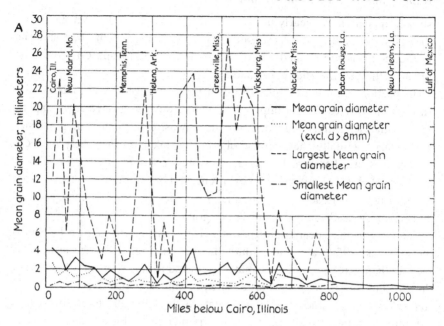

Figure 6-16.   *Changes in bed-material size along the main stem of the Mississippi River from Cairo, Illinois to the mouth, based on 615 bed samples. [After Mississippi River Commission, 1935.]* **A.** *Downstream change in size of certain classes of grains. The variation in size of maximum grain is large; there is no significant downstream decrease in maximum size through 800 miles, in part presumably owing to tributary entrance.* **B.** *Downstream change in mean particle size. The trend toward finer size is distinct but the quantitative change in size is small.* **C.** *Downstream changes in size distribution showing increase in percentage of sand.*

luvial materials approximates a log-normal distribution, samples of very large volume are required to provide a truly representative description of sediments containing large particles. Most studies of coarse alluvial materials probably have not satisfied this requirement. Plumley's (1948) study of terrace gravels, however, provides an excellent illustration of field methods which can be used in overcoming some sampling problems.

To circumvent some of the problems posed by sampling coarse gravels in stream beds (particles greater than 4 mm), we use a method of sampling based on measurements of the diameter of individual particles on a stream bed. In streams flowing on gravel, roughly 100 pebbles are measured. A

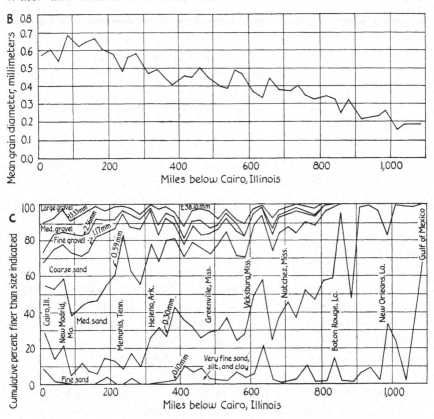

grid pattern locating the sampling points can be paced, outlined by surveys, or designated by floating bobbins. As a rule the sorting in the pebble-count samples is likely to be somewhat better (that is, more uniform) than in the grab samples. Brush (1961) found that statistically there was no significant difference between results obtained by different operators making pebble counts, nor within the same traverse, but differences between traverses were significant at the 1% level. On the basis of these tests he used 60 pebbles as the smallest number necessary to give reproducible results.

The method works satisfactorily for determining size distributions—even up to boulder size—but is has the disadvantages of not being applicable for sizes smaller than pea-gravel and cannot be used unless the stream can be waded. If the stream is not wadable, an estimate of size may be made by sampling along the stream margins or on exposed bars.

Coarse particles in the gravel-size range may develop bed forms similar to those associated with finer sands and silts. Thiel (1932, p. 455) described a series of gravel dunes formed as water burst from a lake and flowed initially at depths of perhaps 40 feet and at estimated velocities of more than 10 feet per second. The gravel in the dunes was several inches in diameter, and the dunes had wavelengths of about 40 feet and amplitudes of 8 feet. Blench (1961, p. 223) observed dunes in fine gravel (1 mm) at high discharges on the bed of the channel in an experimental flume several feet wide.

By far the most common form in gravel, of course, is the gravel bar and the associated pool and riffle sequence seen in many river channels. The geometric scale, however, including the spacing of the pool and riffle sequence, appears to differ from that of the dunes. The spacing of pools and riffles, discussed in connection with river patterns, is on the order of 5 to 7 widths, while bars and dunes are much smaller, 2 to 100 occurring in a reach equal to one stream width.

Bed forms or features in alluvium composed of sand and silt-size sediments are far more important both in distribution and in dynamic significance than are those which occur in coarser gravels. These forms in the finer sediments range from small features, such as ripples on a sand bed, with amplitudes measured in parts of an inch and wavelengths or spacings of a foot or less, to larger features, including dunes with dimensions of tens of feet. Carey and Keller (1957) describe sand waves in the Mississippi River as much as 34 feet high with wavelengths up to 2 miles. Smaller sand waves or dunes ride on the backs of these larger ones. Sundborg (1956, pp. 208, 271) describes transverse bars spaced roughly 0.05 widths apart, about 30 feet, which he views as different in kind and origin from ripples, but which are comparable to the dunes described by many investigators. The size of ripples and dunes is determined both by flow conditions (larger amplitudes with greater depths) and by size of the sand particles. There are theoretical reasons why dunes should be absent or at least relatively uncommon in sediments coarser than 0.6 mm (Bagnold, 1956, p. 261).

In laboratory flumes the amplitude of ripples and dunes appears to be related to flow depth and velocity. The scale relations under the full range of conditions found in rivers, however, is not yet known. River-bed forms change with changes in hydraulic factors and with the character of the sediment load. These changes are described in more detail in the discussion of channel changes at a given cross section associated with changing discharge (pp. 215–222).

The problem of sampling sand beds poses the same questions of variability in space and depth met in sampling coarser sediments. The cliche of the statistical and sampling literature—that the nature of the question being asked must determine the sampling procedures—again applies. If the problem is one of estimating rates of sediment movement in sand-bed streams, weighed samples are clearly preferable inasmuch as they give a measure of volume and weight appropriate to the units required in transportation. Even here, however, although the customary descriptive statistical measure is usually given as the mean, geometric mean, or median diameter, the significance of sorting coefficient or standard deviation in sediment transport is not yet clear. Einstein (1950) believes that the D_{65} (the diameter exceeded by 35% of the grains) is the important parameter in considerations of bed-load transport—an empirical choice, because it gave the best results in his equations.

In coarse material, where surface roughness of large particles is important, measurement of individual pebbles may be the only practicable method of showing both spatial and size variation. Our limited data indicate that the larger size fractions—for example, the diameter at which 84% of the distribution is finer—may correlate best with a measure of the hydraulic resistance (see Fig. 6-5). In many geomorphic studies the largest sizes alone may be of interest, or sizes beyond a specific threshold may be of particular engineering significance.

Obviously, many of these descriptive measures will not be correlative with one another. Certain classes of problems, however, will prove amenable to specific methods of description. As data on alluvial channels accumulate, they can be appropriately classified. There are at present relatively little comprehensive data available on size distributions and the related hydraulic environments in alluvial channels.

REFERENCES

Bagnold, R. A., 1954, Some flume experiments on large grains but little denser than the transporting fluid, and their implications: *Inst. Civil Engrs. Proc.*, Paper No. 6041, pp. 174–205.

Bagnold, R. A., 1956, The flow of cohesionless grains in fluids: *Roy. Soc. London Phil. Trans.*, ser. A, no. 964, v. 249, pp. 235–297.

Benedict, P. C., Albertson, M. L., and Matejka, D. Q., 1955, Total sediment load measured in turbulent flume: *Am. Soc. Civil Engrs. Trans.*, v. 120, pp. 457–489.

Blench, T., 1952, "Normal" size distribution found in samples of river bed sand: *Civ. Engr.*, v. 22, p. 147.

Blench, T., 1961, Discussion of resistance to flow in alluvial channels: *Am. Soc. Civil Engrs. Proc., Hydr. Journ.*, pp. 221–228.

Brush, L. M., Jr., 1961, Drainage basins, channels and flow characteristics of selected streams in Central Pennsylvania: *U. S. Geol. Survey Prof. Paper* 282-F, pp. 145–181.

Carey, W. C., and Keller, M. D., 1957, Systematic changes in the beds of alluvial rivers: *Am. Soc. Civil Engrs. Proc.*, v. 83, Paper 1331, *Journ. Hylr. Div.*, 24 pp.

Colby, B. R., and Hembree, C. H., 1955, Computations of total sediment discharge, Niobrara River near Cody, Nebraska: *U. S. Geol. Survey Water-Supply Paper* 1357, 187 pp.

Colby, B. R., 1961, Effect of depth of flow on discharge of bed material: *U. S. Geol. Survey Water-Supply Paper* 1498-D, pp. 1–10.

Einstein, H. A., 1944, Bed-load transportation in Mountain Creek: *U. S. Dept. Agric. Soil Cons. Serv. Tech. Publ. 55*, 47 pp.

Einstein, H. A., 1950, The bedload function for sediment transportation in open channel flows: *U. S. Dept. Agric. Tech. Bull.* 1026, 70 pp.

Einstein, H. A., and Barbarossa, N. L., 1952, River channel roughness: *Am. Soc. Civil Engrs. Trans.*, v. 117, pp. 1121–1146.

Gilbert, G. K., 1914, The transportation of debris by running water: *U. S. Geol. Survey Prof. Paper* 86, 363 pp.

Hack, J. T., 1957, Studies of longitudinal stream profiles in Virginia and Maryland: *U. S. Geol. Survey Prof. Paper* 294-B, 97 pp.

Hubbell, D. W. 1963, Apparatus and techniques for measuring bed load: *U. S. Geol. Survey Water-Supply Paper* 1748 (in preparation).

Johnson, J. W., 1943, Laboratory investigations of bedload transportation and bed roughness: *U. S. Dept. Agric., Soil Cons. Serv. Paper 50*, processed.

Laursen, E. M., 1958, Sediment-transport mechanics in stable channel design: *Am. Soc. Civil Engrs. Trans.*, v. 123, pp. 195–206.

Leliavsky, S., 1955, An introduction to fluvial hydraulics: Constable, London, 257 pp.

Leopold, L. B., and Maddock, T., Jr., 1953, The hydraulic geometry of stream channels and some physiographic implications: *U. S. Geol. Survey Prof. Paper* 252, 57 pp.

Matthes, Gerard, 1956, River surveys in unmapped territory: *Am. Soc. Civil Engr. Trans.*, v. 121, pp. 739–758.

Meyer-Peter, E., and Muller, R., 1948, Formula for bedload transportation: *Internatl. Asso. for Hydr. Structures Res. Proc.*, 2nd Meeting, Stockholm, pp. 39–65.

Mississippi River Commission, 1935, Studies of river bed materials and their movement with special reference to the lower Mississippi River: *U. S. Waterways Exper. Sta.*, Paper 17, Vicksburg, Miss., 161 pp.

Nelson, M. E., and Benedict, P. C., 1951, Measurement and analysis of suspended load in streams: *Am. Soc. Civil Engrs. Trans.*, v. 116, pp. 891–918.

Plumley, W. J., 1948, Black Hills terrace gravels; a study in sediment transport: *Journ. Geol.*, v. 58, pp. 554–558.

Rouse, H., ed., 1950, Engineering hydraulics: Wiley, New York, 1033 pp.

Sundborg, A., 1956, The river Klarälven, a study of fluvial processes: *Geografiska Annaler*, v. 38, pp. 127–316.

Thiel, G. A., 1932, Giant current ripples in coarse fluvial gravel, *Journ. Geol.*, v. 40, pp. 452–458.

Wentworth, C. K., 1931, The mechanical composition of sediments in graphic form: *Univ. Iowa Studies in Nat. Hist.,* v. 14, no. 3, 127 pp.

White, C. M., 1940, Equilibrium of grains on bed of stream: *Roy. Soc.* *London Proc.,* Ser. A, v. 174, pp. 322–334.

Wolman, M. G., 1955, The natural channel of Brandywine Creek, Pennsylvania: *U. S. Geol. Survey Prof. Paper* 271, 56 pp.

Chapter 7

Channel Form and Process

Rain added to a river that is rank
Perforce will force it overflow its bank.

SHAKESPEARE

Venus and Adonis

Shape of the Channel

The shape of the cross section of a river channel at any location is a function of the flow, the quantity and character of the sediment in movement through the section, and the character or composition of the materials making up the bed and banks of the channel. In nature the last will usually include vegetation. Because the flow exerts a shear stress upon the bed and banks, one "adjusted" or stable form which the channel can assume is one in which the shear stress at every point on the perimeter of the channel is just balanced by the resisting stress of the bed or bank at each point. Consider, for example, a channel in uniform noncohesive coarse sand at a constant discharge.

Particles of sand on the bed will be at an incipient stage of motion when the shear stress τ is equal to the immersed weight of the grain. The critical stress is

$$\tau = \frac{\sigma - \rho}{\sigma} \, Mg(\cos \beta \tan \alpha - \sin \beta),$$

where the angle of inclination of the channel is β, the angle of friction is α, and $(\sigma - \rho/\sigma)Mg$ is the immersed weight.

In addition to the shear stress on the sloping side of the sand channel,

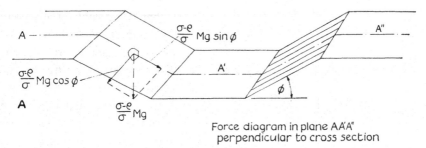

Force diagram in plane AA'A"
perpendicular to cross section

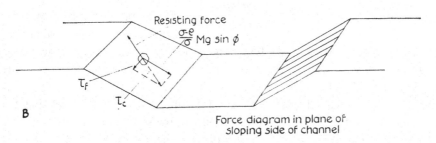

Force diagram in plane of
sloping side of channel

Figure 7-1.　*Resolution of forces on submerged particle on sloping channel bank.* **A.** *No flow.* **B.** *Flow.*

there is a component of the gravitational force directed parallel to the side slope, which also contributes to the instability of the particle. This force, acting on the particle on the sloping bank, is shown in Fig. 7-1, **B**. In still water (Fig. 7-1, **A**), equilibrium of the particle is maintained when the component of the immersed weight is balanced by the frictional force:

$$\frac{\sigma - \rho}{\sigma} Mg \sin \phi = \frac{\sigma - \rho}{\sigma} Mg \cos \phi \tan \alpha,$$

where $\tan \alpha$ is the coefficient of friction equal to the tangent of the angle of repose and ϕ is the angle of the side slope.

When water is flowing along the side slope, then the component of the immersed weight of the particle must at equilibrium equal τ_c, where τ_c is the resultant of the fluid shear stress τ_f and the component tending to roll the particle down the side slope (Fig. 7-1, **B**). When the side slope is zero, then the shear stress required for movement equals that on the bed. The steeper the angle of the side slope, the less is the fluid shear stress, τ_f, required to move a given size particle from the bank.

Let us consider the profile of a cross section. The angle ϕ which the

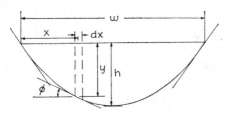

Figure 7-2.

Sinusoidal shape of channel in noncohesive sand. Perimeter at all points at "threshold" of erosion. [Modified after Koechlin, 1924.]

side slope will assume at any point is a function of the angle of repose of the particles and the ratio of the tractive force or shear stress of the flow and the critical tractive force. In this ideal sand channel, at each point on the perimeter the resisting force due to the weight of the particle is just balanced by the applied stress. The maximum angle of the bank will approach the angle of repose, α, near the surface, where the shear stress provided by the flow approaches zero. As shear stress increases toward the center or deeper part of the channel, the inclination of the side slope must decline.

It will be recalled from the analysis of fluid flow in a channel that the shear stress of the fluid, τ, is proportional to the square of the flow velocity, and the velocity in turn is a function of the roughness of the channel boundary, the slope, and the depth of flow, as in the Chezy equation. In the stable or equilibrium sand channel, then, the channel must be able to transmit the flow, but the shear stress associated with the flow must equal that required for stability of the bed and banks. By equating expressions for these two stresses and expressing the tangent of the side slope as a differential or change of height with change in lateral distance, an expression for the channel shape can be derived. The resultant cross section (Fig. 7-2) has a sinusoidal shape close to that of a parabola. Such a derivation is given in Koechlin (1924, pp. 94–102). Similar ones have been made more recently by Lane and others (see Lane, 1955; Scheidegger, 1961, p. 161).

This analysis brings out an essential point. In the simplest stable natural channel with movable bed and banks, two conditions must be satisfied simultaneously—the transmission of the flow and the stability of the banks. Such a channel has been called "threshold" (Henderson, 1961, p. 112), describing the fact that each point on the perimeter is at the threshold of movement. In this hypothetical condition a channel could not transport sediment because the required increase in stress would cause erosion of the banks. In actuality a natural channel not only carries sediment but migrates laterally by erosion of one bank, maintaining on the average a constant channel cross section by deposition at the opposite bank. In this case the condition of no bank erosion is replaced by an equilibrium between erosion and deposition. The form of the cross section is "stable," meaning constant, but position of the channel is not.

There are, of course, relatively few channels in nature which are composed of uniform noncohesive sands and flowing at constant discharge. Small channels in the laboratory as well as individual distributary channels of braided rivers often are in nearly noncohesive materials, and these do have sinusoidal or parabolic shapes very similar to those derived in the analytical example. Figure 7-3, for example, shows some cross sections of laboratory and distributary channels which conform closely to the idealized case. For a given roughness and slope, the size of the channel is a function of the flow. Increasing the discharge alters the overall size, but the shape of the channel margins tends to remain contant. Because the banks in the noncohesive sand are unable to withstand an increase in shear stress associated with an increase in depth, as discharge is increased the banks give way and the central portion of the channel, *ABCD* in Fig. 7-3, simply enlarges.

From the ideal case, the effect that changes in bank material will have on channel form can be postulated. This in essence is what geologists have long

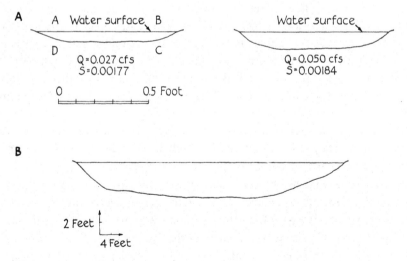

Figure 7-3.   **A.** *Cross sections of laboratory channels developed in noncohesive sand; mean diameter 0.67 mm. Sections show similarity in shape of channels at different discharges.* **B.** *Cross section of anabranch of braided channel, showing similarity of channels in laboratory and field in noncohesive material.*

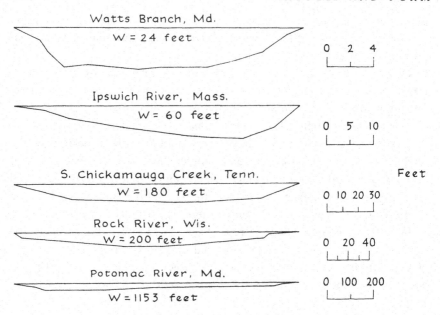

Figure 7-4. *Cross sections of some natural rivers scaled so that the width is the same.*

referred to as the effect of relative resistance of bed and bank material. As the threshold of erosion of the bank material increases, whether by addition of coarse or cohesive sediments or by the presence of vegetation or bedrock, with no change in bed material or discharge, the channel will be narrower. Thus channels with cohesive silty banks and sand beds will be narrower than comparable ones with sand banks and sand beds (Schumm, 1960).

Most rivers in cross section are not parabolic and they are certainly not semicircular. They tend more to be generally trapezoidal in straight reaches but are asymmetric at curves or bends. The appearance of rectangularity increases somewhat as the river gets larger downstream, since width increases downstream faster than does depth. Some typical cross sections of rivers are shown in Fig. 7-4, where they have been drawn at different scales so that the width is the same for each on the printed page. When cross sections are drawn without vertical exaggeration, the shapes tend to bear a resemblance to those of channels in cohesionless materials, as shown in Fig. 7-3. The relatively large width-to-depth ratio for the biggest river is apparent. The asymmetry of the cross section in bends is described in a later section when meanders are discussed.

Riffles and Bars

A straight or nonmeandering channel characteristically has an undulating bed and alternates along its length between deeps and shallows, spaced more or less regularly at a repeating distance equal to 5 to 7 widths. The same can be said about meandering channels, but this seems more to be expected because the pool or deep is associated with the bend, where there exists an obvious tendency to erode the concave bank. The similarity in spacing of the riffles in both straight and meandering channels suggests that the mechanism which creates the tendency for meandering is present even in the straight channel and that this mechanism is associated with some form of wave phenomenon.

The alternating pool and riffle is present in practically all perennial channels in which the bed material is larger than coarse sand, but it appears to be most characteristic of gravel-bed streams—whether the gravel is pea-size or the size of a man's head. There appears to be a latent tendency for the development of pools and riffles even in boulder-bed channels.

Measurement of lengths of individual pools and riffles is partly a matter of judgment. With this qualification, we may say that we found in Seneca Creek that the average length of one repeating distance is 324 feet, which is 5.1 times the mean width of the bankfull channel. The ratio 5.1 in Seneca Creek compares favorably with the corresponding ratio, 5, for meanders in streams of comparable size. In Seneca Creek the average length of pool is 1.6 times the length of riffle. No corresponding figure is available for meanders.

Gravel bars that form the riffles in rivers generally are lobate in shape and slope alternately first toward one bank and then toward the other. The low-water channel then bends around the low point or nose of each and thus tends to have a sinuous course even within the banks in a reach which is generally straight. This tendency for alternating pitch of gravel bars bears a close analogy to the "skew shoals" of Quraishy (1944) and to the alternating dunes in the flume studies of Wolman and Brush (1961). Its significance is not yet known.

The material in the bed tends to be somewhat larger on the riffles or gravel bars than in the pools. This is often obvious in streams having boulders in the gravel bars, for the adjoining pools may be quite free of any big rocks. The variation of grain size is more subtle in channels carrying cobbles or fine gravel.

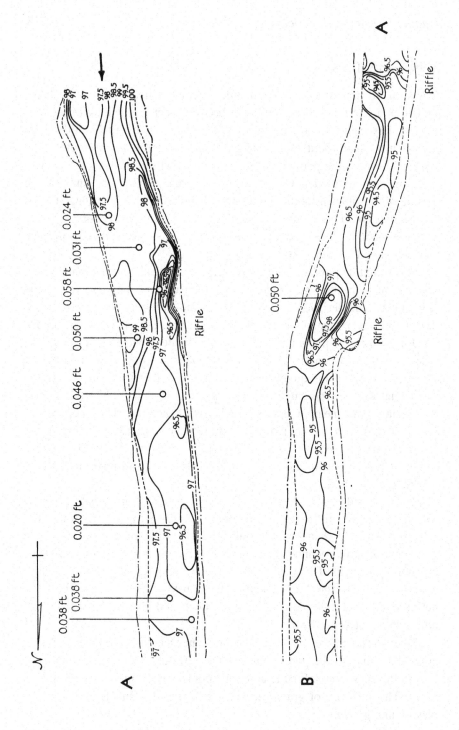

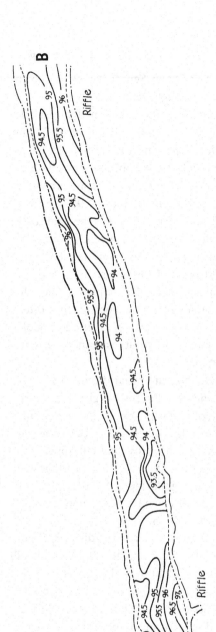

B

Contour interval 0.5 foot; datum arbitrary

Scale, feet

50 25 0 25 50 100 150

Figure 7-5.   A detailed topographic map of the bed of the channel of Seneca Creek near Dawsonville, Maryland. The river is shown also in the photograph of Fig. 3-24. The alternation of pools and riffles can be seen, as well as the tendency for the thalweg or deepest section to wander from one side of the stream to the other. Circles show location of measurements of bed-material size; figures keyed to circles show median size in feet.

Local variation of median size of bed gravel in Seneca Creek is shown by the data plotted in circles in Fig. 7-5. The slight tendency for larger size on the riffles than in the pools can be seen in the data. The median particle size on the riffles varied from 0.058 to 0.046 feet, 90% being smaller than 0.17 feet. In the pools the median varies from sand to 0.038 feet. There is a tendency for grain size to grade gradually from pool toward riffle bar.

In streams even in the same locality, where bedrock outcrops are more common than in the reach of Seneca Creek studied, Hack (1957, p. 55) found that the material composing the riffle bars may not be markedly coarser than that of the pools. Where outcrops provide a ready source of coarse material in or near a reach there is apparently some tendency for coarse fragments to collect in the deepest parts of the pools.

At low flow the water surface over a pool and riffle sequence tends to consist of alternating flat reaches of low gradient and steeper reaches often involving white water. This appearance of smooth water over the "pool" and "riffles" over the bar—terms well known to trout fishermen—led us to use these terms in describing the feature.

As the water rises during flood, the difference in appearance of the water over pool and riffle tends to disappear, and at sufficiently high flow, which we believe to be about bankfull, the longitudinal profile of the water surface tends to become nearly straight. The riffle is then said to be "drowned out," no longer influencing the water surface. This drowning out appears to occur at a smaller discharge in a meandering river than in an otherwise comparable straight reach of channel. The significance of the stepped or nonlinear profile at bankfull stage will be discussed when comparison is made with the water-surface profile of a meandering reach.

Over a pool, water-surface slope increases with increasing discharge, but over a riffle it decreases. Depth increases with discharge over both pool and riffle. The depth-slope product, which in a uniform channel is proportional to bed shear, varies with discharge, depending upon the relative rate of change of slope and depth. Observations over the riffle are inconclusive, but it is clear that shear increases with discharge in the pool; this increase in shear is more rapid over the pool than over the riffle, at least for stages below bankfull.

The periodic spacing of pool and riffle was noted by Stuart; in his work on the ecology of salmon and trout, he had found that water flowing through the gravel of a riffle provides aeration essential to the incubation of fish ova (Stuart, 1953, p. 408). Being concerned with the effect of diversion and realignment of certain gravel streams in Scotland on their ability to maintain trout, Stuart noted that new stream beds dredged by a dragline were, when just constructed, of uniform depth and without pools and riffles. With the aim of producing the usual pool and riffle sequence, he directed the operator of the dragline to leave piles of gravel on the stream bed at intervals appropriate to riffles—that is, 5 to 7 widths apart. After a few flood seasons these piles had been smoothed out and presented to the eye a picture that in all respects appeared natural for a pool and riffle sequence. Moreover, the riffles so formed have been stable over a number of years of subsequent observation.

Miller (1958, p. 46) showed that high mountain streams in the Sangre de Cristo range in New Mexico do not exhibit alternating pools and riffles. Although there are occasional deep reaches, the alternation of deeps and shallows with periodic spacing related to channel width is absent. He argued that the coarse bed material derived from steep cliffs, Pleistocene glaciation, and frost action was too large to be moved by even the high flows of the channel under the present climatic regimen.

Pole Creek, a mountain stream that has incised itself into a moraine of Wisconsin age near Pinedale, Wyoming, has a coarse gravel bed derived from the moraine. Through this reach the stream averages 80 feet in width and 3 to 4 feet in depth at bankfull stage. It exhibits alternating deeps and shallows, which in form are typical pools and riffles, but their spacing is variable and not clearly related to any function of width.

In many of the pools boulders were conspicuously absent, and the bed material was fine enough to be counted by picking up and measuring individual pebbles. Median grain diameter in pools varied between 0.04 and 0.4 foot, and local channel gradient from 0.002 to 0.013. The rapids, on the other hand, were composed principally of boulders, which were measured individually in place on a sampling grid. A comparative median diameter was 1 to 2 feet and average slope through the rapids was 0.02.

To obtain a quantitative measurement of how boulders tend to be concentrated in the rapids or riffles, the number of boulders equal to or greater than 3 feet in diameter were counted in sample reaches of pool and riffle. The average number per 100 square feet of stream was zero in the pools sampled. In the nearby riffles the average numbers were 0.18, 0.27, 0.38, 0.65.

This same sampling method could be applied to boulders seen on the surface of the moraine into which the stream was incised. The average number of boulders per 100 square feet on the moraine was 0.24.

Although these measurements are crude, they tend to support the conclusion that in Pole Creek the pools have a relative dearth of large boulders compared with the source material and that boulders have been concentrated in the riffles by stream action. Thus, boulders must have been swept out of incipient pools and collected in incipient riffles.

Downstream from the nose of the moraine we found no large boulders in the channel. In fact, the change in character of the bed material from coarse to fine was striking. No rocks larger than 1 foot in diameter were found downstream from a point 150 feet below the nose of the moraine. The absence of large rocks in the channel was accompanied by a decrease of average channel slope to about 0.004.

The absence of large rocks as soon as the stream left their moraine implies that disintegration of boulders in the riffles must proceed hand-in-hand with downcutting. This confirms the observation of Hack (1957, p. 83) that hard rocks of large size disintegrate in the channel very near their original source and, once broken down to requisite size for transport, become part of the bed material through a long reach downstream.

The bed material in riffles is of somewhat larger size than that of pools. At some stages of flow, material must be evacuated from pools but cannot all be transported over the riffle; thus it tends to accumulate there.

On Seneca Creek, Maryland, we have painted all the individual pieces of gravel (one-quarter to 6 inches in diameter) lying at the surface of a gravel bar during low flow when the bar is exposed. During a subsequent high flow, all the painted particles moved but the bar itself was subsequently the same height and topography as before. Some of the painted pebbles were found on the next riffle downstream. In our studies at Seneca Creek, the movement of gravel of the median size on the riffle requires a discharge that fills the channel about 0.75 full (depth equal to 0.75 bankfull depth), which has a recurrence interval of about one year.

Stuart (1953, p. 21), in his studies of trout spawning in small streams, noted that high floods seemed to disturb the gravel less than smaller sharp rises or initial and falling stages of higher flows. He suggested that the gravel movement was in the nature of creep rather than occurring by sudden shifts. Major floods can, of course, rework and modify the channel bed. Some details of the effect of a great flood on sand and gravel streams have been described by Wolman and Eiler (1958) in the Connecticut River basin.

In gravel-bed channels during periods of observations extending up to 7 years, we found no indication that the bars comprising riffles move downstream with time. Movement of gravel bars or riffles appears to be relatively slow.

One of the requirements for the existence of pools and riffles in non-meandering streams appears to be some degree of heterogeneity of bed material size. Channels that carry uniform sand or uniform silt appear to have little tendency to form pools and riffles. In part this might be attributed in sand channels to the concomitant high width-to-depth ratio associated with relatively noncohesive bank materials, and this ratio in turn tends to promote a braided pattern. Pools and riffles are less well developed in braided than in the straight but nonbraided channels.

In sandy ephemeral channels viewed in the usual dry state, pools and riffles are generally absent, though careful observation or detailed mapping discloses an analogous feature—thin surface accumulations of coarse material in the form of gravel bars. Their distribution, shown by the planimetric map of a reach of Arroyo de los Frijoles near Santa Fe, New Mexico (Fig. 7-6), is strikingly reminiscent of the occurrence of pools and riffles in gravelly perennial streams, for they tend to be spaced at 5 to 7 widths along the channel length and remain with minor change from year to year. In these bars by far the majority of the cobbles are at or very near the surface, and the sand below is quite free of rocks and cobbles. Such gravel

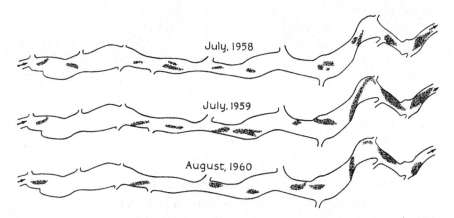

July, 1958

July, 1959

August, 1960

Figure 7-6. *Surface accumulations of gravel in the channel of an ephemeral wash resemble riffles or bars in a perennial stream. Map shows gravel accumulations in a reach of Arroyo de los Frijoles near Santa Fe, New Mexico, observed in successive years. The gravel moves during individual flows but the zones of accumulation change little.*

Figure 7-7.

Large rocks at surface in the sandy channel of an ephemeral wash. Arroyo de los Frijoles near Santa Fe, New Mexico. Section was exposed when a minor channel shift caused local cutting into a "gravel bar."

bars, then, are mere surface features, though we presume they are caused by the same general process which accounts for riffles and bars in perennial rivers.

That the large rocks accumulate at the surface of sandy ephemeral washes is particularly surprising in view of the fact that the channel bed scours at high flow and fills again to approximately the same level when the flow ceases. The photograph of Fig. 7-7 shows the surface and a stratigraphic section of the bed of Arroyo de los Frijoles, exposed when a minor channel shift caused the water to cut into a gravel bar.

Further investigations by the authors indicate that the accumulation of coarse particles on the bed also occurs in gravelly stream beds of the sub-humid areas. In Watts Branch near Rockville, Maryland, for example, the largest size class (3- to 4-inch diameter) invariably is found at the surface of the bed and not below. In the deposits laid down on point bars during the observation period, very few cobbles larger than 2 inches were found and none in the largest size class.

Table 7-1. Number of pebbles of different sizes at various depths, Seneca Creek near Dawsonville, Maryland.

Depth Level Below Stream-bed Surface (inches)	Number of Pebbles in Each Size Class			
	Size Class in Inches			
	2.0	1–2	$\frac{1}{2}$–1	Total
0–2	19	154	625	798
2–4	4	115	481	600
4–6	4	73	334	411
6–8	2	72	316	390
8–10	1	66	422	489

As another example, Table 7-1 presents a count of number of particles of various sizes at different depths for a gravel bar comprising a riffle in a pool and riffle sequence of a straight reach. It will be noted that in the top 2 inches, 2.4% of the pebbles were greater than 2 inches in diameter. At no deeper level do pebbles of this size exceed 1% of the sample.

In rotating-drum experiments conducted by Schoklitsch to test the rate of abrasion of particles, he found that the large particles were generally on top of the mixture (Schoklitsch, 1933, p. 4).

In the case of the natural stream bed it might be argued that the immediate surface layer of cobbles on a stream bed contains a dearth of sand and fine gravel because the finer grains are winnowed away due to differential transportation. This is certainly true in some instances, but in others field observations indicate that near-surface layers contain practically all the largest particles, a difference between surface and underlying layers that cannot be explained by winnowing.

The explanation of the phenomenon appears to be in the effect of the intergranular dispersive stress. The dispersive stress increases as the square of the particle diameter (Bagnold, 1954, p. 62), and hence differential stress on the larger particles may be enough to force them to the surface.

where the dispersive stress is zero. The same phenomenon occurs in dry granular material that flows under gravity. When a truck dumps dry gravel in a pile, the largest particles come to the surface, roll down the face of the conical pile, and thus tend to segregate themselves at the base of the pile.

The full significance of this phenomenon is not known. The concentration of the largest movable particles near the surface of the stream bed seems to occur in a variety of channels in quasi-equilibrium. The phenomenon can probably have but little significance in a stream that is aggrading except that it may tend to cause a random alternation of large and small particles—in other words, lenses or layers of different size material. In any event, interpretation of alternating beds of sand and gravel in sedimentary deposits should be made keeping in mind this tendency for reversal of bedding.

Gravel bars or riffles also consist of a concentration of units—in this instance individual cobbles or pebbles—which are replaced during periods of high flow. Thus the gravel bar is made up of particles that lodge on the bar only temporarily; yet the bar is an entity that may change shape or position. The gravel bar, then, may be regarded as a kind of kinematic wave in the traffic of clastic debris.

A kinematic wave is a wave consisting of a concentration of units, particles, or individuals, but through which these units may move. A concentration of cars on a highway is an example. A concentration of cars behind a traffic signal is a wave through which individual cars move. The wave, or zone of traffic concentration, may move in the same or opposite direction as the traffic, or may remain fixed in space. A flood wave moving down a river is a kinematic wave—in this case, a wave that travels downstream faster than the water particles themselves.

The analysis of kinematic waves (in Lighthill and Whitham's theory of traffic flow) has been shown to depend on two principal factors: flux or transport rate and linear concentration (Lighthill and Whitham, p. 319). Flux is measured by the number of units (automobiles) passing a point in a given time. Linear concentration at any place is measured by the mean number of automobiles per unit of distance. Flux is directly analogous to transport rate as defined in studies of river sediment. Linear concentration, as applied to gravel movement, would be the mean grain density. It has been used in sediment studies only by Bagnold (1956, p. 241).

The relation of transport rate to linear concentration, illustrated in Fig. 7-8, provides important information about the action of kinematic waves.

When the linear concentration approaches zero, the distance between traffic units approaches infinity, and it can be seen, then, that transport rate becomes zero. Thus the plot goes through the origin. Experience also indicates that when linear concentration becomes very large, traffic units are very close together and finally jam. In bumper-to-bumper traffic the transport rate approaches zero. Thus the plot goes through some value of $y = 0$, and $x > 0$. The slope of the tangent to the curve, line AB in Fig. 7-8, is the velocity of the ki-

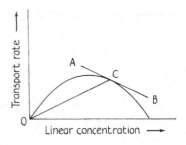

Figure 7-8.

Diagrammatic relation of transport rate to linear concentration in a kinematic wave.

nematic wave itself, negative slopes indicating that the wave moves upstream or in a direction opposite to that of the traffic. At the maximum ordinate value of the curve the wave is stationary.

The slope of the chord OC joining the origin to any point on the curve indicates the velocity of the individual traffic unit, the speed of the automobile at that point in the flow described by the coordinate values at Point C. An individual traffic unit in passing through a kinematic wave experiences different values of speed as linear concentration changes, because the speed of any traffic unit depends in part on the interaction between units. It is common experience that drivers of automobiles traveling at high speed stay farther apart (smaller linear concentration) than when traveling at slower speed. In road traffic the interaction of vehicle speed and linear concentration is caused by driver reaction in consideration of braking distance. In the transport of debris by flowing fluid, interaction among individual units must take place. Close juxtaposition of particles decreases the effectiveness of fluid propulsion of the particles. Debris being moved as bed load cannot move as fast as the ambient fluid. The speed of particles relative to fluid decreases as linear concentration increases.

A "traffic jam" occurs in particle motion in a fluid when the linear concentration reaches some high value. As observed by Bagnold (1954, p. 203), this condition in a flume may result in the particles jamming up so that they slide along the bed of the flume as a solid block.

It is visualized, then, that when debris particles are propelled along by the fluid (but with a slippage between fluid and particle), a tendency for increased linear concentration will cause a slowing up of particle speed relative to the water. In a small flume in which $\frac{1}{8}$-inch glass beads were transported by water in a single "lane of traffic"—that is, in a flow where particles are not passing one another—Langbein* demonstrated the decrease of particle speed with decreased distance between particles. If no other factors were changing, this would lead to a condition of still further concentration of particles, for the effect of the initial disturbance is to increase the magnitude of the disturbance. Such a condition of disequilibrium is analyzed in the theory of queues. The steady-state condition is achieved when the effect of decreased particle speed due to increased linear concentration is balanced by increased fluid force on the particle. This might be provided by increased velocity gradient associated with a decreased depth of water.

Our data obtained during 5 years of observation of painted rocks in Arroyo de los Frijoles, New Mexico, involved about 14,000 individual rock movements during storm runoff. The results agree with those of the bead movements in Langbein's miniature flume. We found that, for a given rock size, increasingly close spacing of individual rocks requires a geometrically increasing discharge to begin motion. Both in the miniature flume and in the ephemeral stream, particles spaced at distances greater than 8 diameters do not interact—that is, do not affect each other in regard to initial movement. Thus any initial grouping of rocks, as in a gravel bar, tends to preserve the concentration of rocks and thus promotes the formation of a gravel bar.

When the bed debris is heterogeneous in size, the possibility of interaction between grains is apparently enhanced. A local perturbation which would tend to increase temporarily the linear concentration would also make the particles there slow up, and this leads to the cessation of movement of grains, building up a mound of grains on the bed. But the growth of the bar shallows the water over it, probably resulting in an increased local slope of the water surface. These changes probably increase the gradient of velocity near the bed. The net result, apparently, is to increase locally the transporting power, and there is established an equilibrium between factors enhancing transport and the tendency for local deposition.

The result is a mound of gravel and sand, whose surface particles during the duration of high flow are constantly being traded for new particles arriving from upstream. The bar is then a queue in which some particles are at rest, sooner or later to be plucked off the bed and moved downstream, only to be replaced by others. The movement of an individual grain therefore must be jerky, perhaps moving more continuously through the

* Personal communication.

pool and waiting or temporarily stored in the gravel bar. An average speed, taking into account the stops and interruptions in movement, must be greater in the traverse of the pool than over the bar or riffle.

The hypothesis sketched above is believed to be correct in outline, but many important questions remain unanswered. The more or less regular spacing of bars, averaging 5 to 7 channel widths apart, is not explained by the hypothesis. Also, the position of most gravel bars remains nearly fixed in space, though sand bars and dunes tend to move downstream. Presumably the erosion of the upstream slope of the sand dune and deposition on the steep downstream face has no counterpart in the gravel bars of pool and riffle sequence. In the Elbe River, sand bars spaced on alternate sides of the channel have been observed to move downstream at a rate of roughly 300 meters per year (Gierloff-Emden, 1953, pp. 298–306). The theory of kinematic waves indicates that the wave which does not move upstream or downstream is associated with maximum transport rate; where the tangent to the curve in Fig. 7-8 is horizontal, the ordinate value (transport rate) is maximum. Probably this is significant in the mechanism, though at the present time one can only guess that there is some unknown relation to rates of energy expenditure.

Variation of Hydraulic Characteristics at a Given Cross Section

Most rivers experience a wide range of flows. Because a reach of channel or a given cross section must transmit varying amounts of water passed into the section from the channel upstream, at different discharges the observed mean velocity, mean depth, and width of flowing water reflect the hydraulic characteristics of the channel cross section. Graphs of these three parameters as functions of discharge at the cross section constitute a part of what Leopold and Maddock (1953) called the hydraulic geometry of stream channels.

The graphs plotted in Fig. 7-9 demonstrate some characteristics which describe many natural river cross sections. In many sections the graphs tend to be straight lines on logarithmic paper. With increasing discharge at a given cross section, the width, mean depth, and mean velocity each increase as power functions:

$$w = aQ^b, \qquad d = cQ^f, \qquad v = kQ^m,$$

where w is width, d is mean depth, v is mean velocity, Q is discharge,

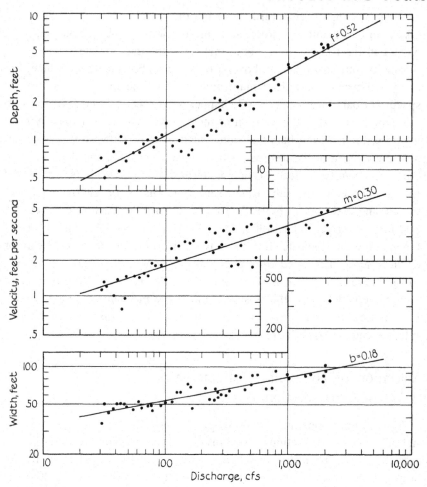

Figure 7-9. *Changes of width, mean depth, and mean velocity with discharge at a river cross section, Seneca Creek at Dawsonville, Maryland; drainage area 100 sq. mi.*

and a, c, k, b, f, and m are numerical coefficients. Graphs in which values of various channel characteristics are plotted against discharge at a given river cross section are termed *at-a-station curves,* to differentiate them from similar graphs describing how the parameters change downstream as discharge increases by successive contributions of tributaries.

The product of width and mean depth is the cross-sectional area of flowing water. Discharge is the product of mean velocity and cross-sectional area of flow. Thus $w \times d = A$, and $w \times d \times v = Q$. It follows that $aQ^b \times cQ^f \times kQ^m = Q$, or $b + f + m = 1$, and $a \times c \times k = 1.0$.

These relations show that when two channels have the same rate of increase of width with discharge (same value of exponent b), if one has a larger rate of increase of depth (larger value of f), it must then have a smaller rate of increase of velocity with discharge.

Many analyses of river cross sections have been made since the original suggestion was published, and they have provided a clearer indication of the range of values of the exponents in these equations. A summary of average values of the exponents for different groups of data are shown in Table 7-5 (p. 244).

The data necessary for graphs of the type shown in Fig. 7-9 are sometimes available from current-meter measurements at river sections. Measurements made for the determination of discharge near a gaging station are not always taken at the same cross section. This variation in the sites of measurements leads to considerable scatter of points on the graph, but when observations are all taken at the same cross section the scatter is reduced.

The exponents b, f, and m essentially describe both the geometry of the channel and the resistance to erosion associated with the character of bed and banks. For example, (1) a wide dish-shaped channel would have a rapid rate of increase in width with increasing discharge; (2) a boxlike channel with straight steep sides—such as one might expect to find in cohesive materials—would have a low value for b and a high value for f. The laboratory channels in cohesionless sand typify channel (1), and the cross sections of Brandywine Creek, Pennsylvania, are representative of channel (2).

The values of exponents in relation to the resistance of bank and bed to erosion are classified, with examples, in Table 7-2. When the materials comprising the channel are moved at all stages or discharges, as in the laboratory channels, the rate of change of width is large, the rate of change of depth approaches zero, and velocity increases very slightly. If some flows are below the threshold of erosion, the channel form is set by the higher discharges, which are capable of eroding bed and banks. Thus on Brandywine Creek, with cohesive bank material, width changes very little with discharge and adjustments occur in velocity and depth. The relation of these changes at a station to those changes which take place in the downstream direction are discussed later in this chapter.

Table 7-2.   Relations of discharge to width, depth, and area of cross section in channels having cohesive and noncohesive bank materials. [From Wolman and Brush (1961), p. 206.]

| | NONCOHESIVE BANK MATERIAL | | COHESIVE BANK MATERIAL | |
	At a station	Downstream Direction	At a station	Downstream Direction
At or above critical point of incipient motion	$A \propto Q^{0.9}$ $w \propto Q^{0.9}$ (Flume in 2.0 mm sand)		$w \propto Q^{0+}$ $d \propto Q^f$ $v \propto Q^m$ $f = m$	$A \propto Q^{0.9}$ $w \propto Q^{0.75}$ or $w \propto Q^{0.5}$ (Many rivers; Leopold and Maddock, 1953)
Below critical point of incipient motion	$A \propto Q^{0.9}$ $w \propto Q^{0.75}$ (Platte River at Grand Island, Nebraska)	Same as for cohesive material in downstream direction	(Brandywine Creek, Pennsylvania)	Determined by above; that is, at or above point of incipient motion

We have studied the geographic distribution of values of the exponents b, f, and m over the United States. Clear-cut patterns do not exist, but it is evident, among the channels studied, that those in the humid eastern United States and those in the wet mountain area of the central and north Rocky Mountains tend to have lower values for the exponent b than channels in semiarid Southwest or parts of the High Plains. In other words, the ephemeral streams in semiarid areas tend to increase more rapidly in width as discharge increases than streams in more humid areas, as surmised above from consideration of bank materials. There appears to be some tendency for values of m, the rate of change of velocity, to be low throughout the Great Plains area as compared with both eastern streams and mountain western ones. High sediment loads might account for this, but the data are not sufficiently complete to make any generalization.

Exploring further the relations among the exponents in the equations, we may use the fact that mean depth increases approximately in direct proportion to water-surface elevation or stage, so long as the flow is within banks—that is, at discharges less than bankfull. Thus, we can say that the equation $d = cQ^f$ is an approximation of a discharge rating curve, a plot of water stage and discharge. A value of about 0.4 for f indicates in mathematical terms what was discussed previously—that discharge increases much faster than depth of water in a channel. When the mean depth doubles, the discharge increases in proportion to $2^{2.5}$, or is increased nearly 6 times.

In addition, the equation $d = cQ^f$ indicates that a rating curve should plot approximately as a power function (as a straight line on log paper), and indeed such plots are widely used in stream-gaging practice. Cross-sectional area, $A = wd$, increases as the function $A \propto Q^{b+f}$. Using the average value for 158 river sections, $A \propto Q^{.57}$. Thus, doubling the cross-sectional area of flowing water means an increase of discharge of $2^{1.8}$ or 3.5 times.

Study of river data suggested that rating curves of various stations are so similar in form that they could be generalized into a nondimensional form. A group of 13 gaging stations distributed over the eastern half of

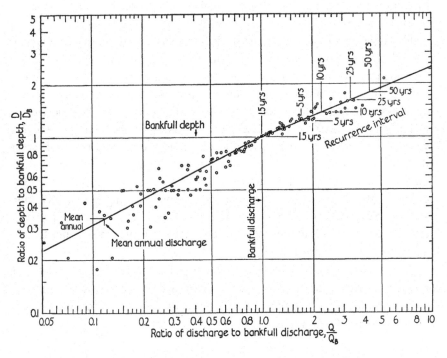

Figure 7-10.   *A nondimensional rating curve for 13 gaging stations in the eastern half of the United States. Depth is expressed as ratio to mean height of streambanks, and discharge as ratio to bankfull discharge.*

the United States was chosen for special field surveys in order to obtain uniform estimates of the stage at which bankfull flow occurred. The data from these stations are plotted in Fig. 7-10. The figure shows that bankfull stage has a recurrence interval averaging 1.5 years, a conclusion in agreement with other data (see p. 320).

Analysis of the data indicates, as a broad general average, that a flood plain will be inundated to a depth equal to $\frac{8}{10}$ the mean height of the banks by a flood recurring once in 50 years. It is also interesting to note that a flow equal to average annual discharge fills the channel to about $\frac{1}{3}$ the bankfull depth, or about $\frac{1}{4}$ full.

The ratio m/f can be related to the transportation of load through the interdependence of the various channel factors. The ratio may be defined as

$$\frac{m}{f} = \frac{\text{rate of increase of velocity with discharge}}{\text{rate of increase of depth with discharge}}.$$

Measurement data indicate that the higher this ratio, the more rapid the increase of measured sediment load with increase of discharge.

The relation of measured debris load to measured discharge may be plotted directly from observational data. However, all correlations of measured load of natural streams with other factors is influenced by the limitations of these measurements. As discussed earlier, measurements of sediment load are available in quantity only on that portion of the load caught in the samplers now in use. These do not measure the sediment moving very close to the bed, nor as a rule do they measure moving material larger than coarse sand. Where total load and the load caught in a modern sampler have been compared, it has been shown that the device catches on the order of two-thirds of the total load in streams carrying principally sand and silt. Nevertheless, studies of channels indicate that even this percentage of the load does provide an index to the manner in which channel factors and sediment load are related.

A plot of suspended load as a function of discharge gives a rough correlation, as shown in Fig. 7-11. This relation may be approximated through at least a major part of the range of discharge by a straight line on logarithmic paper. Such a line is defined by the expression

$$G = pQ^j,$$

in which G is sediment load in tons per day and p and j are numerical constants. Values of j typically lie in the range 2.0 to 3.0. This relation then becomes part of the hydraulic geometry.

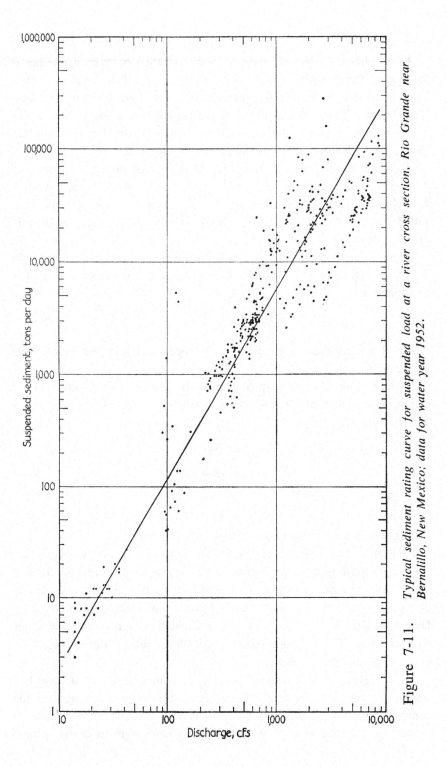

Figure 7-11.   *Typical sediment rating curve for suspended load at a river cross section, Rio Grande near Bernalillo, New Mexico; data for water year 1952.*

As previously discussed, there is an interrelation between a number of factors in the hydraulics of flow in an open channel—velocity, depth, slope, and roughness, which appear in the Chezy or Manning equation. Velocity, depth, width, and discharge are interconnected by definition, and debris load is related to the same factors. Thus, to complete the hydraulic geometry, both slope and roughness must be included.

As discharge increases at a station, water-surface slope generally tends to remain nearly constant. From the direct observations of the concomitant changes in the other variables, the change in resistance may be computed, and it is found that the total boundary resistance decreases with increasing discharge.

Consider an open channel in which the roughness on bed and on banks is similar. The Darcy-Weisbach nondimensional expression for total flow resistance is

$$\mathbf{f} \propto g\, \frac{Rs}{v^2},$$

where R is hydraulic mean depth, s is slope, and g is gravitational acceleration.

Using $R = d$, which is approximately true for wide channels, and assuming that water-surface slope is approximately constant for discharges less than bankfull, then

$$\mathbf{f} \propto \frac{Q^f}{Q^{2m}}$$
$$\propto Q^{f-2m}.$$

if $\mathbf{f} = .45$ and $m = .43$, then

$$\mathbf{f} \propto Q^{-.41}$$

or channel resistance decreases with increasing discharge.

Resistance or roughness in a hydraulic sense cannot be measured directly but must be computed from observations of velocity, depth, and slope. Direct measurements of all of these factors at a specific cross section at various discharges are available in relatively few places on natural rivers. Data from Brandywine Creek, Pennsylvania, show that resistance decreases with increasing discharge at a station and with the same order of magnitude of the exponent shown above.

The behavior of the hydraulic parameters is more complex in sand-bed streams, where changes in discharge are associated with changes in the configuration of the bed. Few if any sand beds in their natural state are smooth. If they are smooth initially, sand grains begin to move as flow moves over them, and ripples and then dunes are formed on the bed, which scour on the upstream gently sloping face and deposit on the steeper downstream face, and thus move downstream. With increasing velocity the dunes enlarge to the point where the transport rate at the crest is such that deposition and hence growth of dune height is no longer possible. Smaller dunes, or ripples, may be superimposed on the dunes. At higher velocities the dunes may be erased over much of the bed and the bed will become sensibly flat or plane. With further increases in flow, waves form on the surface of the stream when the water velocity exceeds the celerity of a simple harmonic wave of the same wavelength as the bed form. In this condition antidunes are formed on the bed in phase with the surface waves. The antidune is a bed wave which accompanies surface waves. The latter move upstream for short distances in trains, then collapse, reform, and migrate upstream, only to break up again. Antidunes also move progressively upstream and are usually transitory, continually forming, dissipating, and reforming. Although antidune formation often occurs when the Froude number of the flow has a value at or near unity (Langbein, 1942), their formation and growth are also related to the rate of sediment movement. No comprehensive criteria are yet available to specify all conditions for their formation. Because cross-bedding is produced by dune movement, if antidune cross-bredding is preserved in stratigraphic sections it will "face" upstream.

The subject of bed configuration is one on which a large literature exists, particularly reporting hydraulic experiments in flumes. Our aim here is merely to outline the main considerations and to suggest in what ways bed configuration is important in river morphology and in channel equilibrium.

First, to give a picture of the hydraulic effects, Fig. 7-12 presents a set of data relating to a particular sand size and includes two or three observations related to each of the principal bed forms. From the sketches at the bottom of the figure it will be noted that the sequence of bed forms—rippled or flat, dunes, plane, and antidunes—is correlated with increasing velocity. In nature the initial flat bed is rarely if ever observed.

Change from flat bed to ripples or dunes is accompanied by an increase in velocity despite an increase in flow resistance. The increased rugosity (physical roughness) of the bed as dunes form would logically be expected to result in an increase of resistance.

At a higher velocity the dunes are washed out and give way to a plane bed or one on which a few low dunes exist on a generally flat bed. This plane bed takes on the appearance of a distant dust storm viewed from

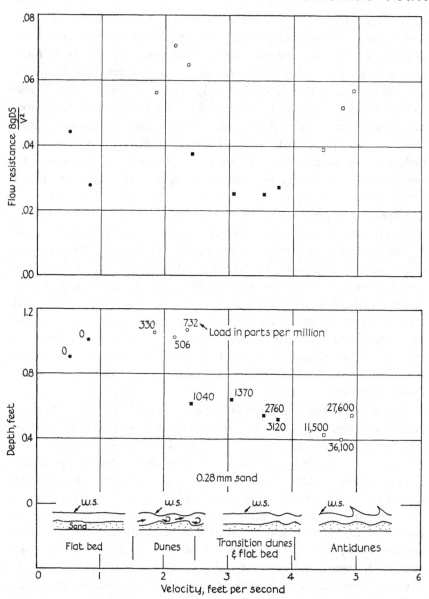

Figure 7-12.   *Bed configuration and changes in depth, flow resistance, and*
sediment transport, shown by selected flume data in which
size of sediment was not a variable.

an airplane—the whole surface seems alive with sand hugging the flat surface. The elimination of the dune protuberances is accompanied by a marked lowering of flow resistance, as the data in the upper part of Fig. 7-12 show. The rate of sediment transport also increases, as shown by the load figures in the lower diagram.

At a still higher velocity antidunes form, the resistance to flow again increases, and the sediment load continues to increase.

The flow resistance is not controlled directly or solely by the physical rugosity of the bed. The movement of debris along the bed in itself decreases the resistance to flow. This phenomenon was first reported by Buckley (1922) and later amply confirmed by others (for example, Vanoni, 1946, pp. 96 and 127). It appears to result from decreased turbulence due to damping by the sediment in motion, thus decreasing rate of energy loss.

Designed specifically to differentiate between the effect of bed configuration on flow resistance and the effect of damping of turbulence by sediment on flow resistance, the experiment by Vanoni and Nomicos (1960) illustrates well what we believe to be a real need in the field of river morphology—critical experiments designed to test a specific working hypothesis. These authors stabilized dunes that were formed by the moving sediment and duplicated the flow conditions, using clear water over the now immovable bed features. For the same bed form the resistance coefficient was 5 to 28% smaller when sediment was present in the flow than it was with clear water.

Vanoni and Nomicos noted that in the Mississippi River dune height generally increases with discharge. At Memphis dunes on the bed have been observed to attain a height of 34 feet. But the increase in dune height may be accompanied by decrease in flow resistance, concomitant with increase in suspended sediment load. Thus field observations and laboratory work lead to the conclusion that in a sediment-laden stream, flow resistance is altered in two ways: (a) by change of bed configuration, and (b) by damping of turbulence by the sediment load. In instance (a), the most marked increase in resistance occurs when dunes form on a previously flat or rippled bed. The most marked decrease occurs when at high stages and velocity the dunes are eliminated and the bed becomes sensibly flat. In natural channels the change of bed configuration has a much larger effect on resistance but the effect of the suspended load is not negligible, particularly where the bed is nearly flat and sediment concentrations are high.

Flow resistance of a movable bed may depend but little on the grain size exerted as skin resistance. The grain size may, however, help govern the nature, action, and form of the features built on the bed which exert the greatest influence on flow resistance.

The theoretical necessity for transformation from flat to rippled or dune-covered bed was shown by Bagnold (1956, p. 255). With increasing fluid stress on the bed a condition is reached in which there is a deficit in the resistance to movement which a plane grain bed can exert, and bed grains then will go on being eroded. But the amount eroded cannot be carried as bed load without an increase in applied stress; therefore, they must be redeposited. This results in a condition of bed instability, and a steady state can only be again attained if the grains are redeposited in a way which would create some new and additional tangential resistance. The formation of ripples or dunes satisfies this requirement.

To both the riverman and the geologist the depositional character of sand deposited under various conditions of bed configuration—and, indeed, in various parts of the dune—is of importance. Sand deposited on the steep face of a dune is unstable; it will not support any appreciable weight but acts quite like quicksand. But sand deposited on the back or gently inclined surface of the dune is closely packed and will support the weight of a man wading in the stream. These physical properties were first described and explained by Bagnold for desert dunes (1941, pp. 236–240), and a number of them have later been shown to apply to dunes formed under flowing water.

The importance of the changes in bed configuration is beyond the mere relation of the changing forms to flow resistance and sediment transport rate. The change of bed is a mechanism or process by which the interaction of hydraulic variables can readjust to promote and maintain a kind of equilibrium or steady-state condition in the open system represented by the water and sediment in the adjustable channel. As Bagnold's theoretical analysis showed, the changes in bed are automatic responses which block or restrain alterations that could, if unchecked, destroy the channel as a structure for carrying off water and sediment delivered from upstream. Such checks prevent or alter the sequence, rate, or type of erosion or deposition and, in doing so, regulate and adjust the channel system. Scour and fill, another regulating mechanism, is discussed in the following section.

But before ending this brief discussion of bed configuration it should be remarked that little or no detailed study either of the theory or the process of bed forms has been made. Nearly all the effort in laboratory work has been expended on measuring the simultaneous values of hydraulic factors and describing in most cursory fashion the concomitant forms on the bed. Few workers have attempted theoretical explanations of bed forms, and there is no detailed study of the forms themselves with measurements of their physical aspects.

Under given conditions of the independent factors—for example, discharge and introduced load—more than one combination of hydraulic conditions among the adjustable or dependent factors may exist. The combination of adjustable factors is not unique, and this fact should influence importantly the future direction of hydraulic experimentation. For example, sediment transport is not always a simple function of the external conditions in the moving fluid. The hydraulic variables become dependent variables as the quantity and character of the sediment inflow is altered (Brooks, 1956, p. 541). This view hardly comes as a surprise to the geologist, but it is only beginning to influence the type and direction of research in sediment transport.

The research in sediment hydraulics might well aim toward closer collaboration of hydraulic and geological experience and technique and toward simultaneous and close association of theoretical, laboratory, and field investigations.

River-bed Scour During Floods

With the rise in stage accompanying flood passage through a river reach, there is an increase in velocity and shear stress on the bed. As a result, the channel bed tends to scour during high flow. Because sediment is being contributed from upstream, as the shear decreases with the fall of stage the sediment tends to be deposited on the bed, or the bed fills. Channel scour and fill are words used to define bed cutting and sedimentation during relatively short periods of time, whereas the terms degradation and aggradation apply to similar processes that occur over a longer period of time. Scour and fill involve times measured in minutes, hours, days, perhaps even seasons, whereas aggradation and degradation apply to persistent mean changes over periods of time measured in years.

Some channels change their beds but little during a flood. Presumably this variation in bed stability is caused by differences in the type of material comprising the bed and its degree of consolidation, imbrication, packing, or cohesiveness. One indication of changes in channel is an ob-

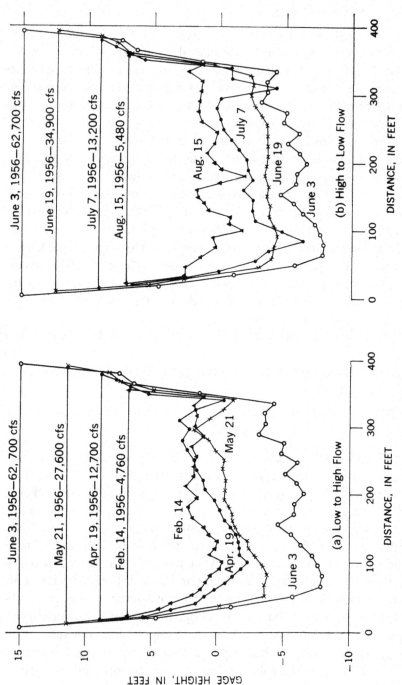

Figure 7-13. *Scour and subsequent fill during flood passage, Colorado River at Lees Ferry, Arizona, water year 1956.* **A.** *Low to high flow.* **B.** *High to low flow.*

served change with time of the relation of river stage to discharge. Such a change often seen in gaging-station records is, in stream-gaging parlance, a "shift in rating."

Figure 7-13 presents an example of bed scour, using data for the Colorado River at Lees Ferry, Arizona, showing cross sections at various discharges during the passage of a spring snowmelt. It can be seen that the whole width of the bed tends progressively to cut as the stage rises, and fills again during the falling stage. This phenomenon is characteristic of ephemeral streams and apparently of large rivers in semiarid climates. It is less typical of rivers in humid areas or those in high mountains, presumably because perennial flow tends to winnow away the fine material

Table 7-3. Selected data on amounts of scour observed in various rivers.

Maximum Depth of Scour Below Normal Bed Elevation (ft.)	Particle Size in River Bed, or Material Encountered	Flow Depth (ft.)	Location and Source of Data
10 to 15	silt, gravel	20	Pacolet River
22	sand, gravel	24 stage	Colorado River, U. S. Bureau of Reclamation, 1950
75	sand, gravel, cobbles	50 stage	Black Canyon, Colorado River (freq. = 1/50 yrs)
126	sand to gravel (cobbles)	35	Black Canyon
55	2″ × 6″ plank embedded in sand, gravel, in gorge 100–150 ft. wide		Black Canyon
32	cobbles moved, boulders smoothed to bedrock	? 12 to 20	Canadian River at Eufaula Dam
40	bank pilings in sand		Rio Grande
60	bridge pier in silt, sand		Lane and Borland (1954)
12 to 15	scoured to bedrock		Yellow River $w \approx 600$ ft. annual flood
0	fine sand	10 to 12	Colorado River, cable at Imperial Dam
20	very fine sand	10	Colorado River, Yuma, Lane and Borland (1954)
1.75 to 2 × regime depth	sand, silt	"regime" depth	Lacey in Blench (1957, p. 103)
$0.5y_1$	width constricted to $\frac{1}{2}$ that upstream	y_1 = upstream depth	bridge piers, Laursen (1960)

and the bed becomes armored with the coarsest segment of the bed material. Some observed depths of scour in different river sections are given in Table 7-3.

Study of the hydraulic geometry shows that the mean bed elevation at a river cross section depends not only on water discharge but is intimately related to changes in width, depth, velocity, and sediment load during the passage of the flood. During a flood passage, each of these parameters executes a hysteresis loop, usually in such a sense that at a given discharge the rising limb of the hydrograph is characterized by a larger sediment load, a higher velocity, and a smaller depth than the same discharge on the falling limb of the hydrograph. Similarly, mean bed elevation executes a hysteresis loop, the minimum bed elevation not necessarily being associated with maximum discharge during the flood passage.

Figure 7-14 shows the changes in cross section, and Fig. 7-15 the changes in hydraulic parameters, during passage of a snowmelt rise in the San Juan River near Bluff, Utah. In the cross sections (Fig. 7-14) it can be seen that as the flood passage began on September 9, when the discharge was 635 cfs, the bed was at an elevation on the gage datum of about 4 feet, and the bed rose about 2 feet as the discharge increased to 6,560 cfs on September 15. But between September 15 and October 14, when the peak discharge occurred, the bed scoured an average of about 6 feet. At the deepest part of the bed the scour was about 10 feet. From

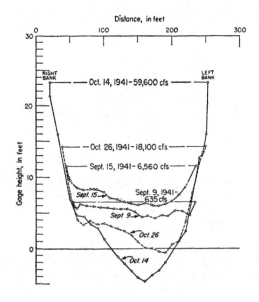

Figure 7-14.

Channel cross sections during progress of flood, September–December 1941, San Juan River near Bluff, Utah.

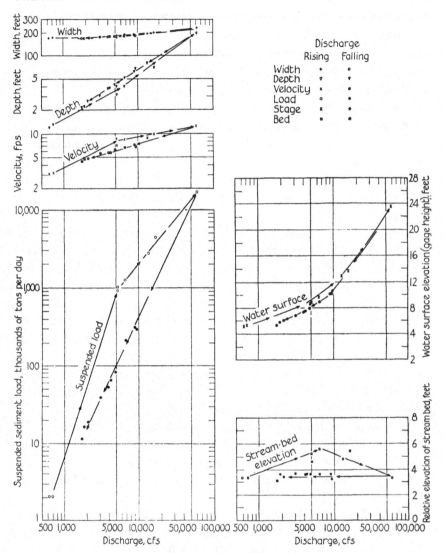

Figure 7-15. *Changes in width, depth, velocity, water-surface elevation, and stream-bed elevation with discharge during flood of September-December 1941, San Juan River near Bluff, Utah.*

October 14 to October 26 the discharge fell to 18,100 cfs and the bed filled about 5 feet.

Several points are of interest in this example. The bed first filled during the initial increase in discharge and then with further increase scoured.

The scour did not occur in proportion to discharge but, as will be shown, depended in part on the debris load coming into the reach. In this example the maximum scour did coincide with the highest discharge; however, this also may vary, depending upon the location of the measuring station along a given reach.

The streambed scoured to a depth about one-third of the amount the water surface rose. Expressed in another way, at the peak of the flood approximately one-quarter of the depth of water was accommodated by bed scour.

During the passage of the flood the width of the water surface increased progressively to the peak, but changed relatively little in terms of percentage.

Water-surface elevation (stage) increased with discharge in a curvilinear relation and did not retrace on the recession limb the same curve as on the rising flood. This change in stage for a given discharge on the rising as compared with the falling limb results in a shift of the rating curve at the gaging station, as previously mentioned.

The change in the rates of increase of depth and velocity with discharge is associated with the change in rate of increase of suspended sediment load with discharge. This change is also reflected in bed elevation as it varies with disharge. That these changes occur more or less simultaneously strongly indicates an interdependence but does not indicate which factors are independent and which dependent.

The interdependence of hydraulic and sediment factors is exemplified by the values of velocity, depth, suspended sediment load, and bed elevation at two equal discharges on rising and falling stages (Table 7-4).

It might appear at first glance that the larger suspended load on the

Table 7-4.   Values of hydraulic parameters at the same discharge (5,000 cfs) on rising and falling stage of flood of Sept.–Dec. 1941, San Juan River near Bluff, Utah.

Parameters	Units	Rising Stage	Falling Stage
Discharge	cfs	5,000	5.000
Width	feet	182	189
Velocity	feet/second	8.6	6.0
Depth	feet	3.2	4.4
Suspended load	tons/day	1,000,000	100,000
Elevation of bed above arbitrary datum	feet	5.3	3.3

rising stage was the result of the scouring effect of the high velocity. Under such a postulate, increasing velocities should be associated with scour and decreasing velocities with filling. On the contrary, the bed was filling during the conditions of rapidly increasing velocity as the discharge increased to 5,000 cfs. At a discharge of 5,000 cfs on the rising stage the velocity was 8.6 feet per second and the bed was filling. At the same discharge on the falling stage the velocity was only 6.0 feet per second, with neither filling nor scouring.

It seems likely that the debris load introduced from upstream into the reach in question was the independent factor and that the hydraulic factors changed in response to the changing load.

Assuming that energy slope was the same on rising and falling stage at the same discharge, note that the resistance to flow must have been lower when the sediment load was large. An increase of velocity and decrease of depth at constant slope is associated with a decrease of flow resistance. As mentioned earlier, an increase in sediment load is associated with a decrease in flow resistance. The stream reduces the flow resistance by modifying bed form, which results in a reduction of stage or depth. The variation in the velocity-depth relation at a given discharge then is related to the change in bed configuration that accompanies changes in hydraulic factors and load, and in part also to an alteration in the pattern of turbulence, which results in decreasing flow resistance with increasing suspended load. Unfortunately, only a few measurements on rivers relate the simultaneous change of the hydraulic factors to measured variations in the bed configuration.

Though scour followed by fill during a flood passage is a well-known phenomenon at particular cross sections where regular measurements are made, there is less evidence on the changes occurring simultaneously along any given river reach. Navigation experience on some rivers seems to show that when scour is occurring in a pool at a meander bend there is simultaneous filling on the bar or riffle at the crossover. Such evidence has led to the suggestion that while scour is proceeding at a measured cross section, fill is occurring simultaneously at some place immediately downstream. Most measuring sections associated with gaging stations are in pools, and there are few data simultaneous in time with the changes that occur on riffles or over bars in the vicinity.

In considering this problem, Lane and Borland (1954) computed the amount of sediment represented by the observed average depth of scour multiplied by the width and length of the alluvial reach of the middle

Rio Grande valley. They assumed that this amount of sediment would be moved during flood flow and that the same amount of sediment should be deposited in Elephant Butte Reservoir downstream—if, indeed, scour at one section were not balanced by fill in another. They found that the amount of sediment so computed would be very much larger than the measured amount actually deposited in the reservoir. They concluded that simultaneous fill must occur in some reaches of the channel in sections not measured or observed.

This computation of the expected rate of sedimentation involves the tacit assumption that the scoured sediment moved downstream at such a velocity that it went completely out of the reach during the time of a passage of the flood. This is tantamount to an assumption that the sediment was moving at a speed approximating that of the water. But the mean velocity of bed particles is much smaller than the mean velocity of the flowing water, especially when the particles are close together (large values of linear concentration).

In contrast to the reasoning used by Lane and Borland, recent measurements by the authors and our colleagues indicate that—at least in some reaches of river—at any one time scour takes place simultaneously through relatively long reaches, both on pools and over bars. Figure 7-16 shows longitudinal profiles of the channel bed and the amount of scour in a reach of the Rio Grande del Ranchos (near Talpa, New Mexico), a small perennial tributary of the Rio Grande. The planimetric map of this reach was presented on Fig. 6-7.

Thirty-two cross sections were measured in a channel distance of 900 feet. They were located both in pools and over bars in straight reaches and in bends. By comparing the cross sections at low and at high flow, the mean area of the bed which was scoured or filled was computed; these areas (in square feet) are plotted against distance in the lower part of Fig. 7-16. In this example only three cross sections out of 32 showed net fill; all the others showed a net scour. The data do not, however, show any systematic relationship between the relative amounts of scour in the pools as compared with that in the riffles. The stream studied averages 25 feet wide and the median of the bed material is 27 mm. Similar conclusions were reached in studies made in Wyoming over a period of 3 years in Baldwin Creek near Lander and in the Popo Agie River near Hudson. The evidence is not conclusive but it is evident that in some perennial streams carrying fine or moderate gravel, high flow is associated with channel scour over relatively long reaches. The amount varies along the stream but is not balanced locally by simultaneous fill in narrow places or on riffles.

In ephemeral streams flowing on sandy beds, typical of arroyos and sand channels of the semiarid climate, scour during flash floods also is characteristic. We have measured the amount of scour by means of chains placed vertically in the stream bed. The chain is weighted at its lower end by a rock or pin, lowered into a hole dug in the channel bed, and sand is packed around the chain while it is held in the vertical position. When scour occurs, the free upper end of the chain is bent from the

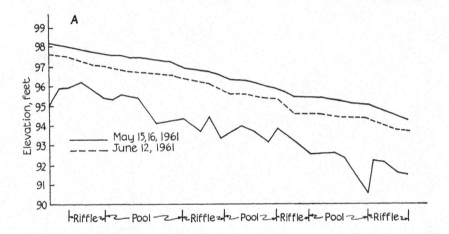

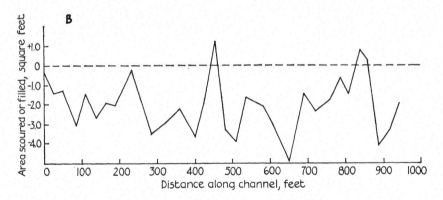

Figure 7-16.   *Profiles of a reach of Rio Grande del Ranchos near Talpa, New Mexico. A. Water surface at two stages of flow, and bed profile related to the higher stage. B. Area of scour or fill in cross sections along the stream; negative values connote scour in high flow relative to low flow.*

Figure 7-17.

Channel of Arroyo de los Frijoles below the Sand Plug, 10 miles NW of Santa Fe, New Mexico. In 1958 the stream passed on the other side of trees behind man. Fan deposited by tributary diverted channel to present position in 1961.

vertical and swept downstream; subsequent fill is deposited on top of the horizontal segment of the chain's length. After the flood the chain is relocated by survey and dug out. The position of the bend, and thus the maximum scour occurring at that point, is measured. Scour chains measure only the maximum amount of scour during the flood at each chain position. Chains do not measure the time relation of scour to flood peak.

Our chain measurements in Arroyo de los Frijoles near Santa Fe, New

Mexico, began in 1958 and are continuing. Sixty-three chains are arranged, one about every 1,000 feet, along a reach of channel about 6 miles in length; together with additional chains placed across the channel as a cross section, a total of 90 chains are measured. Each year of data usually includes two or three flash flows. The chains were surveyed after nearly every such flow.

The photograph in Fig. 7-17 is a general view of Arroyo de los Frijoles, and Fig. 7-18, **A,** contains a sketch map of the drainage area under observation. The location of the chains (by number) is shown along the 32,000 feet of the reach studied. Scour and fill data for a sample flow, for the year 1962, and for the period 1959–1962 are shown in Fig. 7-18, **B.** For each profile on this figure the lower dashed line represents the depth of scour. The upper dashed line represents the depth of fill. The heavy solid line represents the net change in bed elevation after scour and fill. By net

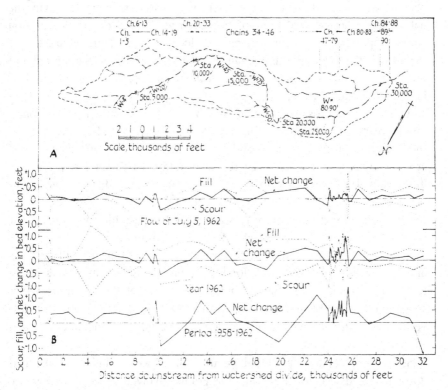

Figure 7-18. **A.** *Location sketch map of Arroyo de los Frijoles near Santa Fe, New Mexico.* **B.** *Net change of bed elevation during various periods.*

change is meant the rise or lowering of the channel bed during the year. The bottom graph of Fig. 7-18, **B,** is the net change in bed elevation during the 4 years of observation. In the reach from 0 to 25 thousand feet, where the data are best, two portions totaling 5,500 feet, or about one-fifth of the length, experienced net scour in the 3 full years of observation. The remaining four-fifths of the length experienced net fill. No characteristics of the channel provide indication of why some reaches scoured when the rest of the reach filled. Presumably the variation is due to local experience of discharge resulting from the spotty occurrence of rains that produce runoff.

Whereas the net result of the storms of 4 years was to make a bed rise over four-fifths of the observation reach—a rise that averaged about one-quarter foot—the scour that occurred and was later replaced by fill averaged about half a foot each year. In other words, the temporary scour, which was balanced by fill at a later time, was larger by far than the net fill that characterized most of the reach at the end of the observation period reported here. Considering the local and variable character of the runoff that caused these changes, the nearly compensating amount of scour and fill testifies to the remarkable tendency for the channel to maintain an equilibrium.

Transverse or cross sections of the channel behave much as the longitudinal profiles do. In addition to the 63 chains arranged along the channel to develop a profile of scour and fill, an additional 27 are placed in 8 cross sections in order to study the distribution of channel scour across the stream. In the example in Fig. 7-19 it can be seen that channel scour occurred in all 4 years of observation of this cross section. The net change for the whole section, however, was small because fill following scour returned the channel bed to near its former position.

These data indicate that scour may take place at nearly all sections along a reach of river during a flood. The time of maximum scour relative to the flood peak is not known, but the behavior of the Colorado River suggests that in most instances the two nearly coincide. Yet only a modest volume of sediment moves out of a river reach, as Lane and Borland showed.

Scour here apparently should be thought of as a dilation of the bed down to the depth of scour, and the grains involved in this dilation are in motion downstream but at a rate which on the average is much smaller than the rate of movement of the water. The ratio of grain speed to mean water speed is small when the concentration of particles is large, indicating

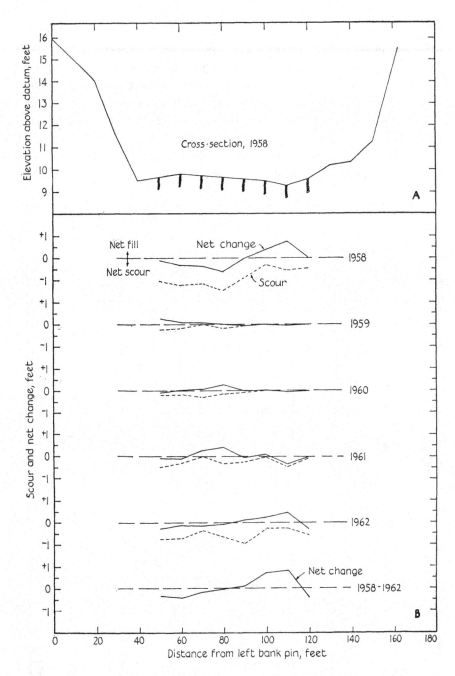

Figure 7-19.   *Scour and net change at Section 25,000 feet, Arroyo de los Frijoles, near Santa Fe, New Mexico, from 1958 to 1961. A. Original cross section. B. Scour and net change in bed elevation.*

that close proximity of one particle to another tends to increase the "slippage" between the water and the grains. When the grains are closely enough spaced, the ratio of grain velocity to water velocity decreases to zero. This effect, according to Langbein, is similar to the reduction in particle fall velocity with increasing sediment concentration reported by McNown and Lin (1952).

That the grains are in motion importantly affects the hydraulics, particularly bed configuration and flow resistance, and thence velocity and

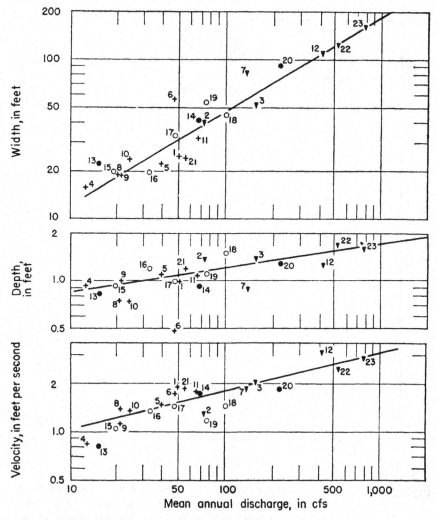

Figure 7-20.   *For legend see facing page.*

depth. Thus the work of scour and fill, in association with bed configuration, is an additional part of the mechanism governing the adjustment of size and shape in natural channels.

Variation of Hydraulic Characteristics in a Downstream Direction

Comparison of various cross sections along the length of a stream is most useful if the comparison is made for a flow of a given recurrence interval or frequency of flow. The most meaningful discharge for any discussion of channel morphology is that which forms or maintains the channel. Earlier it was brought out that this is a complex relationship but that the effective discharge can often be approximated by bankfull discharge. In many rivers the bankfull discharge is one that has a recurrence interval of about 1.5 years.

It has been shown that similar relations obtain for downstream changes in many rivers, whether one uses the mean discharge or bankfull discharge

---

Figure 7-20.

Width, depth, and velocity in relation to mean annual discharge as discharge increases downstream, Powder River and tributaries, Wyoming and Montana.

1. *Red Fork near Barnum, Wyo.*
2. *Middle Fork Powder River above Kaycee, Wyo.*
3. *Middle Fork Powder River near Kaycee, Wyo.*
4. *North Fork Powder River near Hazelton, Wyo.*
5. *North Fork Powder River near Mayoworth, Wyo.*
6. *South Fork Powder River near Kaycee, Wyo.*
7. *Powder River at Sussex, Wyo.*
8. *Middle Fork Crazy Woman Creek near Greub, Wyo.*
9. *North Fork Crazy Woman Creek near Buffalo, Wyo.*
10. *North Fork Crazy Woman Creek near Greub, Wyo.*
11. *Crazy Woman Creek near Arvada, Wyo.*
12. *Powder River at Arvada, Wyo.*
13. *North Fork Clear Creek near Buffalo, Wyo.*
14. *Clear Creek near Buffalo, Wyo.*
15. *South Fork Rock Creek near Buffalo, Wyo.*
16. *Rock Creek near Buffalo, Wyo.*
17. *South Piney Creek at Willow Park, Wyo.*
18. *Piney Creek at Kearney, Wyo.*
19. *Piney Creek at Ucross, Wyo.*
20. *Clear Creek near Arvada, Wyo.*
21. *Little Powder River near Broadus, Mont.*
22. *Powder River at Moorhead, Mont.*
23. *Powder River near Locate, Mont.*

as a basis for comparison (Leopold and Maddock, 1953, Fig. 11, p. 18). Because so many more data are available on mean annual discharge, it provides a convenient measure of flow for comparison, although it is important to recognize that in most instances the mean annual flow is not responsible for the observed conditions but simply reflects changes produced by the more effective flows. This distinction has already been mentioned in connection with the discussion of effective force in Chapter 3.

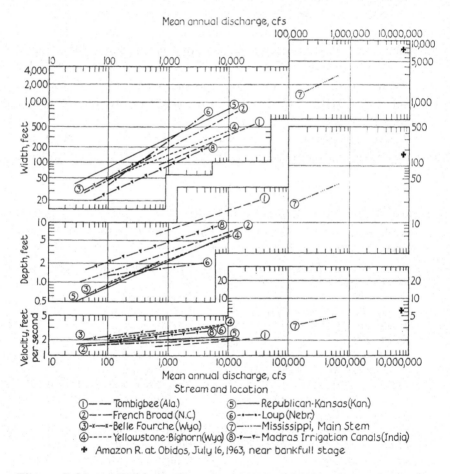

Figure 7-21.   *Width, depth, and velocity in relation to mean annual dis-*
charge as discharge increases downstream in various river
systems.

Mean annual discharge tends to have similar frequency of occurrence among many rivers of different types. This flow is equaled or exceeded on the average about 25% of the time—that is, about 25 days out of 100. Because mean annual discharge is published for nearly every gaging station in the United States, of which there are more than 7,000, it is a useful figure. Mean annual discharge usually fills a river channel to about one-third of its bankfull depth, as can be seen on the nondimensional rating curve, Fig. 7-10.

The width, mean depth, and mean velocity corresponding to flow at mean annual discharge, when plotted against discharge as discharge increases downstream, give curves that constitute further elements of the hydraulic geometry of stream channels. An example of such curves derived from measuring stations within a river basin is presented in Fig 7-20. Omitting the points showing values for individual gaging stations, downstream curves for various river systems are shown on a single graph for purposes of comparison in Fig. 7-21.

The only measurement to date of the Amazon River is that by Oltman et al.[1] of the U. S. Geological Survey in cooperation with the Hydrographic Department of the Brazil Navy. The measurement was made at Óbidos, about 400 miles above the river mouth. The data are plotted in Fig. 7-21. The other river data are for average annual flow, whereas the Amazon point is for near-bankfull stage. The graph indicates, nevertheless, that the general relations of the hydraulic geometry extend even to the world's largest river.

From these curves it can be seen that there is a considerable similarity in the slope of lines among various river systems, even though the intercepts vary. In addition, the stable irrigation canals in India—unlined canals which are allowed to scour and fill to achieve an equilibrium form—appear similar to the natural river channels. In Fig. 7-21 only the Loup River differs appreciably from the others. Data for this river were inserted to show the variance that may exist. Most rivers tend to increase downstream in width, in depth, and in mean velocity in quite a similar way. These lines may be described by the same equations that were used to describe changes at a station. The at-a-station and downstream values of exponents in the equations of the hydraulic geometry from a number of studies are summarized in Table 7-5.

[1] Oltman, R. E., Sternberg, H. O'R. (University of Brazil), Ames, F. C., and Davis, L. C., Jr., 1964, Amazon River Investigations reconnaissance measurements: *U. S. Geol. Survey Circular* 486, pp. 1–15.

Table 7-5.   Values of exponents in the equations for the hydraulic geometry of river channels;[1] $w = aQ^b$, $d = cQ^f$, $v = kQ^m$, $Gs = pQ^j$, $s = tQ^z$, $n' = rQ^y$.

	Average At-a-station Relations						Average Downstream Relations (bankfull or mean annual flow)					
	b	f	m	j	z	y	b	f	m	j	z	y
Average values, midwestern United States[2]	.26	.40	.34	2.5			.5	.4	.1	.8	−.49[5]	
Brandywine Creek, Pennsylvania[3]	.04	.41	.55	2.2	.05	−.20	.42	.45	.05		−1.07	−.28
Ephemeral streams in semiarid United States	.29	.36	.34				.5	.3	.2	1.3	−.95	−.3
Appalachian streams[4]							.55	.36	.09			
Average of 158 gaging stations in United States	.12	.45	.43									
10 gaging stations on Rhine River	.13	.41	.43									

[1] Symbols:
  Q, discharge.
  w, channel width.
  d, mean depth.
  v, mean velocity.
  Gs, suspended load transport rate.
  s, water-surface slope.
  n', roughness parameter of Manning type.

[2] Leopold and Maddock (1953, p. 26).
[3] Wolman (1955, pp. 23, 26).
[4] Brush (1961, p. 160).
[5] Leopold (1953, p. 619); using other data, Langbein obtained value of $Z = −.75$

Figure 7-22 summarizes a complete hydraulic geometry of a channel system for average values of the exponents. Cross section A represents a headwater station during low flow. Its position is given in the upper block diagram of a watershed in which low flow prevails. Cross section C represents the same headwater station at high discharge. Similarly, B is a downstream section at low flow, and D is the same section at high discharge.

The slopes of the lines in the graphs represent the average conditions observed in rivers. In the width-discharge graph, line A_0B_0 has a slope upward to the right equal to 0.5, or, in the expression $w = aQ^b$, $b = 0.5$, which is representative of the increase of width with discharge downstream.

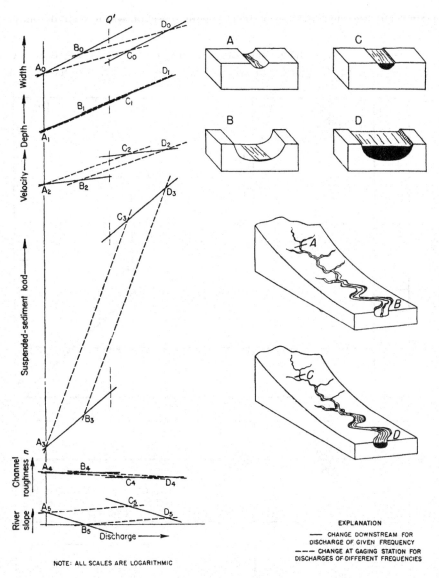

NOTE: ALL SCALES ARE LOGARITHMIC

EXPLANATION
—— CHANGE DOWNSTREAM FOR
DISCHARGE OF GIVEN FREQUENCY
––– CHANGE AT GAGING STATION FOR
DISCHARGES OF DIFFERENT FREQUENCIES

Figure 7-22.   *Average hydraulic geometry of river channels expressed by relations of width, depth, velocity, suspended-sediment load, roughness, and slope to discharge, at a station and downstream.*

Simliarly, the line A_0C_0 represents the increase of width with discharge at a station.

Lines in the suspended sediment load-discharge graphs are drawn to indicate the rapid manner in which load increases with discharge at a

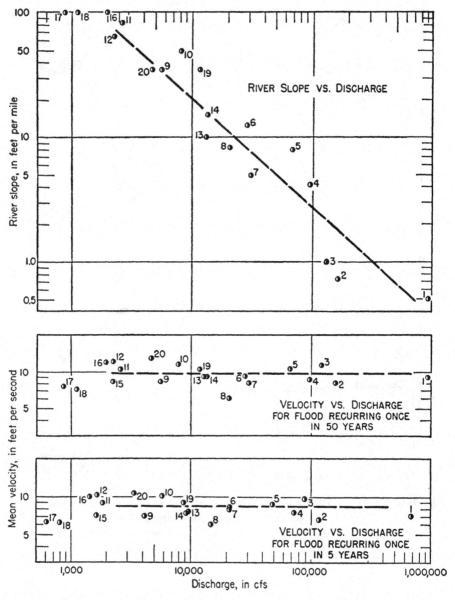

Figure 7-23.   *For legend see facing page.*

station (line A_3C_3, having a slope of $j = 2.5$), but in the downstream direction suspended load increases at about the same rate or slightly less rapidly than discharge. Line A_3B_3 has a slope of 0.8.

Channel roughness decreases with discharge at a station, and the downstream changes of velocity, depth, and slope indicate that, on the average, roughness decreases slightly downstream as well. On the whole the downstream reduction in flow resistance, resulting from decrease in particle size, is partly compensated by other forms of flow resistance, particularly that offered by bars and channel bends. The bends appear to counteract in part the downstream decrease in skin resistance associated with reduction in particle size.

In the graphs of energy slope against discharge the line A_5C_5 indicates a tendency for a very slight increase of slope with discharge at a station, which presumably is caused by a tendency for a slight shortening of the course taken by water as discharge increases. The line A_5B_5 sloping downward with increasing discharge is the downstream flattening of slope—that is, the longitudinal profile of the river.

Because velocity increases markedly at a station with increase in discharge it is important to note that the tendency for velocity to remain about constant or to increase downstream applies to flood flow as well as

---

Figure 7-23.

Change of mean velocity with increase in discharge downstream for flood flows, Yellowstone River Basin and downstream.

1. *Mississippi River at St. Louis, Mo.*
2. *Missouri River at Williston, N. D.*
3. *Yellowstone River near Sidney, Mont.*
4. *Yellowstone River at Miles City, Mont.*
5. *Yellowstone River at Billings, Mont.*
6. *Yellowstone River at Corwin Springs, Mont.*
7. *Bighorn River at Kane, Wyo.*
8. *Bighorn River at Thermopolis, Wyo.*
9. *Greybull River near Basin, Wyo.*
10. *Greybull River at Meeteetsee, Wyo.*
11. *Greybull River near Pitchfork, Wyo.*
12. *Wood River at Sunshine, Wyo.*
13. *Clarks Fork at Edgar, Mont.*
14. *Clarks Fork at Chance, Mont.*
15. *Rock Creek at Joliet, Mont.*
16. *Rock Creek near Red Lodge, Mont.*
17. *W. Fk. Rock Creek below Basin near Red Lodge, Mont.*
18. *Red Lodge Creek above Reservoir near Boyd, Mont.*
19. *Stillwater River near Absarokee Mont.*
20. *Rosebud Creek near Absarokee, Mont.*

to low flow. Mean velocity at a series of gaging stations along the length of the Yellowstone-Missouri-Mississippi River system was obtained from flood measurement data interpolated to represent floods of 50-year and 5-year recurrence intervals. Although variable from station to station in this particular river system, velocity remains essentially constant in the downstream direction at each flood frequency. In the same river system the river slope decreases from about 100 feet per mile to 0.5 feet per mile (Fig. 7-23). In general, river data indicate a slight downstream increase of velocity at bankfull stage.

It must be recognized that these are mean values. As noted in the discussion of the forces controlling channel shape, at constant slope and discharge the channel form will vary markedly with character of the bank material (Schumm, 1960). Such variations produce the kind of scatter observed in the downstream curve in Fig. 7-23.

Considering a downstream increase in discharge at a given frequency, the channel could conceivably accommodate the increasing flow in different ways. The channel could become increasingly wide, keeping its depth and velocity constant, or it could increase its velocity, keeping its depth and width constant. Each combination would require concomitant changes in channel slope and thus different longitudinal profiles, depending on the changing flow resistance downstream.

But in reality, for the river system as a whole, the values of the exponents are rather conservative. Width usually increases in a more consistent manner than any other factor, as the square root of the discharge, and mean velocity tends to increase slightly downstream in most rivers.

The consistency with which rivers of various sizes and in various physiographic settings make the adjustment to increasing discharge downstream suggests that there is a common general tendency or physical principle governing these adjustments. Because the channel adjustments are closely related to the profile of the river, the latter must be considered before seeking a more general statement of the physical principle governing the adjustments of both gradient and form.

Longitudinal Profile of the River Channel

Rivers increase in size downstream as tributaries increase the contributing drainage area and thus the discharge. Concomitant with the downstream increases in the channel's width and depth and the general tendency for bed-particle size to decrease, the gradient generally flattens. In general, the longitudinal profile is concave to the sky.

Figure 7-24 shows a number of longitudinal profiles of rivers located in different physiographic and climatic environments and varying in size. The profiles for rivers of roughly comparable order of magnitude are presented at the same scale, but the different sets of profiles are drawn to different scales. Nearly all are concave, but in some there are convex portions.

Several things should be noted in these profiles. The Nile does not gain any appreciable amount of water for a long distance downstream, owing to its long traverse through a desert area. The smaller Rio Grande also flows through an arid region in its downstream reaches, through which it gains no appreciable inflow. Both rivers have concave profiles, indicating that the decreasing gradient downstream depends on more than increasing discharge.

Even rills which develop in a year or two on man-made or new fills usually have similar profiles, so age alone is not the determinant of con-

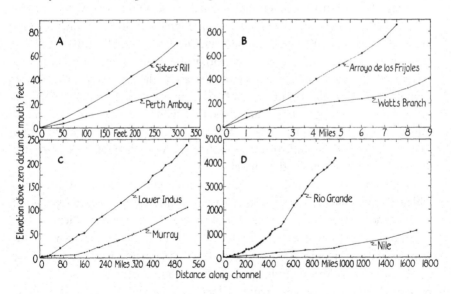

Figure 7-24. *Samples of longitudinal profiles of rills, creeks, and rivers of various sizes. All the examples show some tendency for concavity, even those in which discharge does not increase downstream, such as the lower portions of the Indus, Murray, Rio Grande, and Nile.*

cavity. Eliminating these simple possible explanations it is again necessary to consider the interdependence of several factors, focusing upon the downstream change in gradient.

Although implied in the previous discussions, thus far no explicit mention has been made of adjustments in slope in natural channels. Gradient, however, has been the primary concern of the geologist and geomorphologist inasmuch as it is the stream gradient or profile which is most indelibly impressed upon or revealed in the landscape. Although channel form may sometimes be preserved in ancient rocks or in alluvial terrace deposits, more often it is the gradient which can be inferred from terrace remnants, or perhaps from particle size in the rocks. Under some circumstances parts of the geologic record can be read from the remains of longitudinal profiles of past epochs. Through understanding of the factors related to the profile some idea of previous climate, river discharge, sediment load, drainage basin characteristics, and vegetation may be constructed, using stratigraphic and geomorphic evidence, of which the profile is an important element.

Along the length of a river the channel must adjust to several parameters which change independently of the channel itself. Other factors are related both to the environment and to the channel; still others are principally functions of action by or within the channel and only indirectly related to the regional environment. Certainly the lithology—the whole spatial pattern of the occurrence of different rocks or soil materials—is independent. Through these materials the channel is carved and maintained. Also, the tributaries bring water and debris into the channel, and thus the type and amount of debris entering along the length of a master stream is independent of the channel.

The tributaries and direct inflow from ground water swell the flow which the channel is required to carry. The increase of discharge along the channel, or in some regions the decrease, is also independent.

Factors affecting size and characteristics of the channel which are semidependent include: flow resistance, a function of debris size and of the form of transitory deposits such as dunes and bars; the mode of transport of debris; and channel pattern, including curves, bends, and islands. Debris size may be considered semidependent in that it is a function of the drainage basin but altered within the channel by abrasion and sorting.

Parameters which are most nearly dependent—that is, functions of the action by or within the channel—are river gradient, width, depth, and velocity. River gradient, however, particularly where it is partially controlled by lithologic and structural variations, may also be considered semidependent.

Thus the longitudinal profile can be considered to be a function of the following variables:

Discharge	Q
Load (delivered to the channel)	G
Size of debris	D
Flow resistance	n
Velocity	v
Width	w
Depth	d
Slope	s

The longitudinal profile is a graph of distance versus elevation. To include dynamic factors, distance must be related to one of the other variables. In many regions the geometry of the basin and of the drainage system is so orderly that the relation of distance along the channel to discharge can easily be specified because of certain regular factors. One is the stable and quite constant relation between length along the principal stream, L, and drainage area, A_d:

$$L \propto A_d^{0.64}.$$

Another regular factor is the relation between flood or bankfull discharge, Q_b, and drainage area:

$$Q_b \propto A_d^{0.75}.$$

The first equation was described in Chapter 5 and is applicable to many basins. The second varies somewhat among various climates but can be closely specified for many basins. The exponent of drainage area varies from 0.65 to 0.80, a general average of wide applicability being 0.70 or 0.75 in the second equation. For mean annual flow in humid regions the exponent is about 1.0. The lower exponent at higher flows is because of storage in the river valleys and because rains of high intensity rarely cover the entire basin but are instead widely and irregularly spaced. The values of these exponents need not be considered as wholly unexplained empiricism.

The variables listed above are interrelated, and from their interrelation is derived the relation of fall in elevation to distance along the channel, the longitudinal profile.

The derivation of values for any set of interdependent factors depends on the utilization of a set of equations equal in number to the variables.

But independent equations equal in number to the variables are not available. The addition of equations introduces new variables, and the numbers of equations and of variables continue to be disparate. For example, in relating sediment size to resistance it is found that a simple equation, such as the Shield's relation (see Rouse, 1950, p. 790), involving a tractive force as a function of a form of Reynolds number, introduces three new variables, but adds only two new equations. The relation of tractive force to size and form of bed features (dunes, ripples, bars) cannot be expressed in any simple form, so a new variable would be added but no new equation.

Therefore, to analyze the behavior of the river system it is necessary to simplify the actual conditions by making various assumptions and holding certain factors constant. The form of the resulting relations gives some insight into the problem.

An assumed increase in discharge downstream for a hypothetical small basin is shown in the upper right diagram of Fig. 7-25. It is desired to find what velocity and depths would be required if channel width is specified, roughness (flow resistance factor) is constant, and slope is given. The latter is defined in three arbitrary profiles differing in concavity as drawn in the lower right diagram of the Figure. Profile A is a straight line—that is, with no concavity. Two assumptions of width are made. In the one represented in diagrams $A1$, $B1$, $C1$, width increases downstream as a power function of discharge and at a rate similar to that observed in most natural channels. The second assumption is that width is constant downstream, results for which are diagramed in $A2$, $B2$, $C2$.

Having specified slope, roughness, width, and discharge, the two variables of depth and velocity can be calculated from the two equations available. The calculated changes of velocity and depth with increasing discharge downstream, based on the Manning equation, are shown by the six graphs in the left portion of Fig. 7-25. The calculations deal merely with the hydraulics of water flow and do not involve debris load. The assumption of constant channel roughness, as we have noted, would apply to some natural rivers.

Under the other assumptions made, a profile of no concavity would require a downstream rate of increase of velocity equal to that of depth— a result not observed in any natural channel. Depth always increases downstream faster than velocity. There are of course some channels having a straight profile and some with profile convex to the sky. But these profiles result, as will be shown, from factors other than a near equality of the downstream rates of increase of velocity and depth.

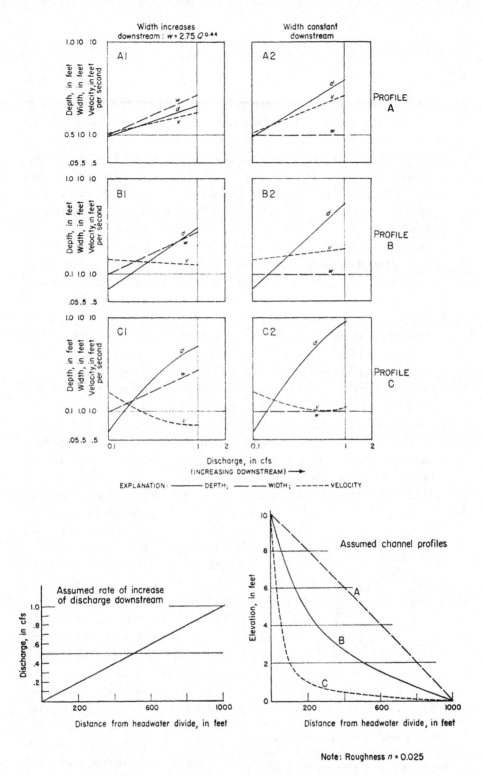

Figure 7-25. *Variation of velocity and depth downstream for three assumed longitudinal profiles; increase of discharge downstream is specified; two rates of change of width downstream are specified; channel roughness is constant.*

The downstream rates of change of width, depth, and velocity most like those observed in natural streams are the ones depicted in graph *B*1 of Fig. 7-25, and are associated with concave profile *B*.

The effect of the size and character of bed material influences the hydraulics of the channel through its effect on flow resistance. In his detailed study of the longitudinal profiles of streams of Maryland and Virginia, Hack (1957, p. 58) states that there is no direct correlation of bed-material size and channel slope if all localities in the region studied are considered. In streams whose bed material has a median size of 100 mm, for example, channel slopes range from 6 feet per mile to over 1000 feet per mile. But in individual streams or groups of streams, classified according to the geology of their basins, there is a systematic relation between the two variables.

The variations in the relation of bed-material size and channel slope are quite consistent when a third factor is taken into account. When localities of equal drainage area are compared, slope increases as a power function of median bed-material size, specifically as the 0.6 power of the median size. The specification of equal drainage area implies, for the Appalachian area studied, approximately equal discharge. With the addition of the restrictions of constant discharge and a given lithology this result is in qualitative agreement with general observations that coarse debris occurs in streams having steep slopes.

Hack's data also show that the change of slope downstream—concavity —is influenced by change in size of debris downstream. The more quickly the bed material decreases in size downstream, the more concave is the longitudinal profile. Conversely, where particle size increases downstream the profile has a small concavity. If the rate of increase of particle size downstream is large enough, the profile is a straight line or may even be convex to the sky.

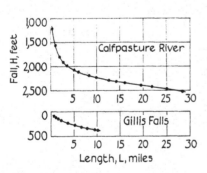

Figure 7-26.

Longitudinal profiles of two rivers; bed material size of Calfpasture River is constant downstream but increases downstream in Gillis Falls. [After Hack, 1957.]

Figure 7-26 shows longitudinal profiles of Calfpasture River in Virginia and of Gillis Falls in Maryland. Calfpasture River has essentially a constant size of bed material downstream, and in Gillis Falls particle size increases downstream. These differences are caused by differences in lithology of tributary streams and by wear and sorting in the channels themselves. Calfpasture River flows in a strike valley joined by tributaries draining sandstone areas. Along its course the tributaries provide a continuous source of uniform particles. In contrast, along Gillis Falls in the Piedmont of Maryland phyllite breaks into small fragments and the more resistant vein quartz becomes more abundant. As quartz becomes more important in the bed material, the median size increases.

It can be seen from the figure that the concavity of the profile of Gillis Falls is less than that of Calfpasture River, a difference associated with the different rates of change of particle size downstream.

Although each of these profiles describes an entire channel, the same principles would apply to a composite channel made up of segments dominated by distinctive lithologies. The steep straight profile of a segment in sandstone might be joined through a transition zone to a portion in limestone or shale on a flatter slope. Because sandstone usually breaks up more slowly than limestone, particles of sandstone might be carried well downstream into the transition reach. In shale, few if any large particles would be found save near a bedrock outcrop, inasmuch as shale disintegrates into small fragments. A composite profile formed of segments in different lithologies would reflect these changes. Each portion is in essence a separate curve, having a particular slope and concavity determined primarily by rock type. The steep slope in sandstone may be followed successively by flatter and more concave slopes in limestone and shale, as Hack has shown (1957, p. 75 et seq.).

Despite the many references in the literature to the downstream change of particle size which must result from abrasion and sorting, it is often difficult to relate these changes directly to changes in the longitudinal profile. The difficulty arises from the fact that particle size alters the flow in a complex manner, presumably having its effect through its influence on competence and on resistance. But flow resistance is determined not only by size of bed particles but also by form or configuration assumed by the particles on the channel bed.

Flow resistance unfortunately cannot be quantitatively specified from the type of channel boundary and the size and forms of materials making up the channel boundaries. The change of flow resistance along the river can be computed from measured values of velocity, depth, and slope for specific conditions along the river length where flow measurements have been made, but the relation of the computed resistance to the size and configuration of bed debris is so complex that at present no simple transition between the two can be made.

The observed downstream flattening of gradient, or concave profile, is generally attributed to an increase in discharge and to a decrease of particle size. Comparing different rivers at a constant discharge, steep gradients are associated with large bed material. When debris size decreases rapidly downstream, the channel slope also tends to flatten rapidly; that is, the profile is more concave.

It is perhaps simpler to see the necessity for steep slopes associated with coarse debris and flat slopes with fine debris, than to ascertain the effect on the profile of increasing discharge without any change in debris size. To carry an increasing discharge as tributaries enter, the channel may increase in width, depth, or velocity. An increase in velocity would be assisted by decrease in flow resistance. The diagram in Fig. 7-27 indicates how the hydraulic factors would tend to interact under certain specified conditions. If the usual downstream rates of increase of width, depth, and velocity were to characterize different rivers, the effect of different changes in channel roughness downstream on the profile would be shown by values along the vertical line in Fig. 7-27 at the value $f = 0.4$. Note that the assumed average value for perennial rivers of $y = -0.3$, or roughness decreasing downstream, is associated with $z = -0.95$, indicating a rather rapid decrease of channel slope downstream. If roughness were constant downstream, a condition which might be approximated by constant downstream size of bed material, then $y = 0$, and the associated ordinate value would be $z = -0.37$, a less concave profile than in the assumed perennial river.

One may see then that if it were possible to choose rivers for comparison in which certain factors were changed at specified rates, a stream which had constant particle size downstream would still have a profile concave to the sky associated with increased discharge downstream, if other hydraulic factors were changing downstream in the fashion usual for rivers. This case is exemplified by point 2 on Fig. 7-27, representing the Kansas-Republican River system.

If discharge is constant, it is still possible for a river profile to be concave. The principal point to keep in mind throughout any discussion of adjustment of river channels is that usually there are more dependent or adjustable factors than independent ones. It may not be possible to fore-

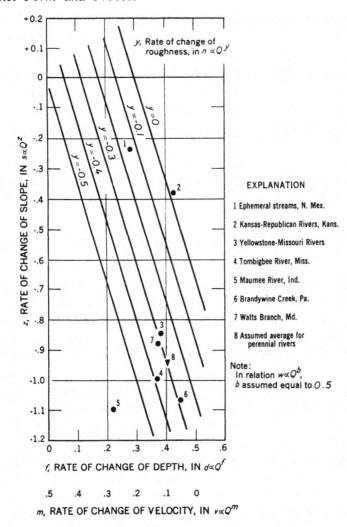

Figure 7-27. *Downstream relations of depth, velocity, slope, and rough-ness, expressed as values of exponents in equations relating these factors to discharge.*

cast the way in which the requirements are satisfied from the physical or hydraulic requirements alone.

Unless otherwise specified, the various elements of the channel may be assumed to be equally amenable to adjustment. The geologist has long referred to the "relative erodibility" of bed and banks, and this factor must again enter the analysis. A very resistant rock which crops out in a reach of channel will form a hump in the channel profile. If the banks of the channel are alluvium, then the width-depth ratio will probably increase as the gradient over the hump increases, assuming the discharge remains constant. Although much of the discussion of the longitudinal profiles of channels seems to assume that they are smooth curves or composites of smooth curves, this is rarely so. In addition to bars, pools, and riffles, variations in discharge and in lithologic and structural controls often produce irregularities of a larger scale. As a rule these dominant influences are associated with or compensated for by changes in channel form, pattern, and bed configuration.

Observations on Introduced Base Level

In addition to those controls of the channel profile exercised for the most part at a point or from the contributing watershed above a given reach, a channel system may be affected by changes of its base level, the level at which its distal portion joins a major body of water. The terms "local base level" and "temporary base level" are often used to refer to a level to which portions of a channel system flow, or to temporary falls or lakes which similarly may be the level to which rivers or streams may temporarily flow. Clearly these are not precise terms and on occasion are even used synonymously, or more often to distinguish local levels to which streams flow on their way to the oceans, to which by definition all exterior or unenclosed river systems ultimately flow.

In the abstract, the expected reaction of a river to a change in base level may seem rather obvious, but lack of ability to forecast the reaction indicates that the obvious guess is untrustworthy. Consider a river debouching into the sea. Base level for this stream is sea level. Should sea level be lowered for any reason, or the outlet dropped—for example, by normal faulting—the distal portion of the longitudinal profile will be steepened. The "oversteepened" reach, as it is sometimes called, may be expected to cause an increase in ability to transport sediment and erosion of the channel in the steeper reach. The change in the elevation of base level will be felt upstream for some distance, depending upon the flow and the characteristics of the underlying material. Details of this process are considered in Chapter 11 in connection with channel degradation. The change must be transmitted upstream as an alteration in gradient. If erosion flattens the slope or tough bedrock retards upstream migration of the steepened reach, the headward reaches of the basin will be unaffected by the change in base level. Similarly, if the rate of inflow of sediment into an upstream reach exceeds the capacity of even the steepened channel, the effect of the lowered base level will not be felt.

This reasoning, however, must be viewed in light of the preceding discussion of adjustment among hydraulic variables. There may well be some circumstances in which a wide variety of adjustments might equally well satisfy the requirements imposed by the change in base level. To the extent that a variety of alternatives are equally possible, the manner of the response to the change imposed may not be determinable from physical relations alone. Stochastic or probabilistic relations may also enter.

Man's time period of direct observation is too short to judge the final effect of some of his activities on geomorphic processes, whereas the effects are soon obvious in other instances. Unfortunately it is not easy to forecast which activities will have immediately observable effects and which will not. The effect of a change in base level imposed on a natural stream is one of those phenomena about which it is difficult to prognosticate. With this in mind, it is instructive to review some of the results following the imposition of new or artificial base levels on rivers by construction of dams.

Dams and the reservoirs they impound provide particularly good illustrations of the effect of rising base level on the longitudinal profile. Any obstruction of a stream which lessens its capacity or competence will promote deposition.

Following closure of a dam, water and sediment accumulate in the reservoir. Base level is raised from its former position, the bed of the channel, to the level at which the water surface of the reservoir intersects the original bed. The maximum water-surface elevation is the crest of the spillway for the dam. As illustrated by the Red River arm of Lake Texoma, shown in Fig. 7-28, the aggradational effect of the water is not felt upstream from the point of intersection of

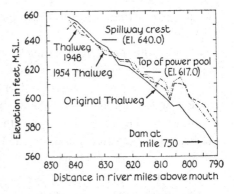

Figure 7-28.

Longitudinal profiles of the bed of the Red River above the reservoir of Lake Texoma, Texas, before and after construction of Denison Dam.

the sloping river channel and the backwater curve joining the river and water surface of the reservoir. Because this intersection is not abrupt but occurs in a transition reach along a backwater curve, the point of intersection is actually the upstream end of the backwater curve. On the exaggerated scale in the figure the backwater transition curve does not show.

Because vegetation develops on the deposited sediments in the upper reaches of a reservoir it is conceivable that the effect of the reservoir will be felt progressively upstream. As on the Rio Grande above Elephant Butte Dam, however, the profile of the Red River shows alternating fill and scour of the bed at different sections above the spillway crest elevation. Such alternations are shown by successive surveys of cross-sections. There is no consistent pattern of progressive aggradation. The Washita arm of Lake Texoma shows several feet of aggradation 3 to 5 miles upstream from the spillway crest elevation. There are numerous other examples, as on the Canadian River system and on the Missouri River. Completely filled reservoirs in the southeastern United States (Eakin and Brown, 1939, p. 139) indicate, too, that at most the effect of the reservoir is felt only at short distances upstream from the maximum pool elevation. The many sedimented mill dams in eastern and northeastern States have also but little affected the river profiles above them.

These reservoir studies indicate that a rising base level controls deposition in the river system only up to the level at which the backwater transition curve intersects the original stream bed profile. However, this is not the only type of effect noted. The situation seems to differ in gravel streams, for in some of these the effect of a barrier or dam is felt at some distance upstream.

In 1937 this problem was much discussed among engineers who were designing and building small barriers and checkdams across the channels of ephemeral arroyos, gullies, and gravelly washes in the semiarid west. Some maintained that as vegetation became established on the sediment deposited behind such a barrier, with time the wedge of sediment would gradually build upstream and ultimately completely fill the arroyo or gully. Others held, on the contrary, that the sediment wedge would cease building when it became essentially level at the elevation of the spillway.

The observations on a variety of reservoirs showed the general correctness of the latter rather than the former view. In small gullies over short distances sediment may accumulate in channels upstream from reservoirs for distances of several hundred feet (Miller, Woodburn, and Turner, 1962,

Fig. 5), but there is no evidence to suggest that a rising base level will affect deposition throughout a river system.

The difference of opinion led to a survey of the profiles of sediment deposits behind as many barriers as were known to have been in place over a period of years in channels of small to moderate size. The interesting report on these surveys by Kaetz and Rich (1939, unpublished) is well known, even though it is available only in a limited number of copies in a few libraries. The authors arranged to have the same profiles resurveyed in 1961. The principal results of both the 1939 and the 1961 surveys are shown in Table 7-6.

Kaetz and Rich concluded that the gradient of deposition varied between 30 and 60% of that which characterized the original bed of the channel. The steeper depositional slopes were on channels which carry coarse gravel. The resurvey 22 years later showed that in a few cases the wedge of deposition had extended upstream slightly but at a rate far slower than had been the average in the earlier period between date of construction of the barrier and the time of the first survey. Where there was an upstream growth of the depositional wedge in the last 22 years, it had been accompanied by a slight steepening of the depositional surface but the percentage increase was small.

Save for a complex case above Imperial Dam on the Colorado River, involving effects of a number of dams upstream (Lane, 1951), tentatively it appears that artificial barriers such as dams that raise base level of a channel affect only limited reaches upstream, and the gradient of deposition of the sediment wedge is appreciably less than that of the original channel.

The fact that deposition appears not to proceed upstream even though the slope is flatter below indicates that the channel has adjusted in such a way as to transmit the sediment across the depositional reach and into the reservoir. In addition, much of the time this transport is actually along a steeper gradient because the water levels in the reservoirs are rarely at the maximum elevations; hence the effect of the rise in base level is not felt in the depositional reach. Two alternatives then still remain. First, insufficient time has elapsed in which to have observed the "ultimate" changes. Second, channel changes accompanying deposition are such that the flattening of slope is compensated by increases in depth or by changes in other factors which maintain continuity of sediment and water transport. Such changes have been suggested by Rubey (1952, p. 135) to account for the narrow distributary channels on the flat gradients of the Mississippi River

Table 7-6.   Gradient of deposition behind diversion dams and barriers, and effect of passage of time on this gradient.

	KAETZ AND RICH, 1939 SURVEY[a]					
Dam and Location	Original Height, Bed to Spillway (ft.)	Drainage Area (sq. mi.)	Average Original Slope of Bed	Time[b] Period Since Construction (yrs.)	Average Slope of Deposition	Distance of Deposition Above Dam (ft.)
Muddy Creek Diversion, Mexican Springs, Arizona	20±[c]	5.6	.0105	5	.0025	2200
Deer Springs Diversion, Mexican Springs, Arizona	16[e]	5.2	.0071	4	.0019	3000
Norcross Wash Diversion, Mexican Springs, Arizona	17[e]	4.0	.0083	5	.0027	
Many Farms Diversion, Chinle Wash, 14 mi. from Chinle, Arizona	3 to sluice gate[e]		.0021	1	.0008	7500
Kayenta Diversion, 7 mi. NW Kayenta, Arizona	22?[e]	5.5	.0136	3	.0036	1400
Frazier Diversion, 12 mi. NW of Chinle, Arizona	14[e]		.0056	4	.0014	4300
Wepo Wash Barrier, 5 mi. SW of Polacca, Arizona	12[c] (same structure dams Wepo and Polacca Washes)		.0060	2	.0020	
Polacca Wash Barrier, 5 mi. SW of Polacca, Arizona			.0029		.0007	5800
Bell Rock Dam, 10 mi. abv. Laguna New Mexico	12[c]		.0064	3	.0030	3700
Laguna Cr. Diversion, 2 mi. NW of Kayenta, Arizona	8[e] (sluice 4 ft. lower)		.0022	5	.0008	8100
Oraibi Dam No. 1, Oraibi Wash, 30 mi. S of Oraibi, Arizona	7[e]		.0026	2	.0007	4000
Oraibi Diversion, Oraibi Wash, 14 mi. N of Oraibi, Arizona	3[e]		.0025	5	.0014	2200

Time Period Since Construction (yrs.)	Average Slope of Deposition	Distance of Deposition Above Dam (ft.)	Type of Material Deposited	Ratio of Slope of Deposition to Original Slope (%)	
				In 1939	In 1961
27 (Of two dams 1 mi. apart, applies to the one upstream)	.0052	3400?	Sand and clay; 1961 slope .0052, to 2400 ft., .0065, 2400 to 3500 ft.	22	50
26	.0019	3050	Sand, some fine gravel	27	27
27	.0055	5400	Clay and silt	33	66
23	.0021	>15000	Fine sand, becoming coarser upstream	38	100?[i]
			Fine sand, becoming coarser upstream	26	
			Sand, becoming coarser upstream	29	
			Fine to coarse sand	33	
			Fine sandy silt and clay	24	
			Fine sand, becoming coarser upstream	47	
			Fine sand, becoming coarser upstream	39	
			Fine sand with silt and clay	27	
27	.0014	2200	Sand with silt and clay	56	56

Table continued

Table 7-6 continued

Dam and Location	Original Height, Bed to Spillway (ft.)	Drainage Area (sq. mi.)	KAETZ AND RICH, 1939 SURVEY[a]			
			Average Original Slope of Bed	Time[b] Period Since Construction (yrs.)	Average Slope of Deposition	Distance of Deposition Above Dam (ft.)
Oraibi Dam No. 2, Oraibi Wash, at Oraibi, Arizona	6[e]		.0029	1	.0013	1100
Ramah Dam, above Ramah, New Mexico	12[e]	3±	.0135	34	.0076	2270
Cottonwood Creek Barrier, Farmington, Utah	23[d] (spillway 19 ft.)		.0464	15	.0206	950
Ford Creek Barrier, Centerville, Utah	8[d]		.0646	7	.0416	500
Willard Creek Barrier, Willard, Utah	30[d]		.0558	15	.0236	1025
Salt Creek Barrier, 7 mi. above Nephi, Utah	12[d] (spillway 8 ft.)		.0172	17	(.0071)[g]	1000
Manti Canyon Barrier, Manti, Utah	30[d]		.0276	48	.0147[h]	950
Salina Creek Barrier, 3 mi. W of Salina, Utah	12[d]		.0126	6	.0063	1050
Upper Cedar City Barrier, 4 mi. above Cedar City, Utah	27[d]	100	.0297	3	.0150	1650

[a] Kaetz, A. G., and Rich, L. R., Report of surveys made to determine grade of deposition above silt and gravel barriers: unpublished memorandum dated December 5, 1939, U. S. Soil Conservation Service, Albuquerque, N. M., on file in S.C.S. library, Albuquerque.

[b] Time elapsed between dam construction and 1939 survey; in that period and in period 1939–1961 some were washed out, others altered as noted.

[c] Myrick, R. M., and Leopold, L. B., Gradient of deposition behind dams: in preparation, U. S. Geological Survey, Washington, D. C., 1963; dams listed by 1939 survey were resurveyed if they could be found and were extant.

[d] From Kaetz and Rich, 1939; heights of dam listed as "to crest."

[e] From Myrick survey, 1961

| MYRICK SURVEY, 1961[e] | | | | Ratio of Slope of Deposition to Original Slope (%) | |
Time Period Since Construction (yrs.)	Average Slope of Deposition	Distance of Deposition Above Dam (ft.)	Type of Material Deposited	In 1939	In 1961
			Fine sand with silt and clay	65	
56	.0074	2900?	Sand with silt and clay	56	54
			Fine gravel and sand	44	
29	.040	400?	Fine gravel, increasing upstream to cobbles and boulders	64	62
37	.052	?[f]	Gravel and cobbles, increasing upstream to large gravel	42	
39	.016	5000	Silt and clay vary near dam; upstream is gravel 2–3 inches diam.		93?[i]
70	.029		Fine gravel, increasing upstream to cobbles and heavy gravel	53	100?[i]
28	(no change)		400 ft. upstream and beyond is gravel, increasing to boulders	50	
			Gravel to small boulders	51	

[f] Data in doubt; apparently 7 ft. of fill from 1000 to 2000 ft. above dam; gravel quarrying in bed.

[g] Bed slope .0071 seems not representative; .013 measured in 1939 from 400–1100 ft. more representative.

[h] 3 ft. of fill for 2000 ft. above barrier at same slope reported for original bed in 1939, probably due to raising spillway 4 ft. with logs.

[i] The high ratio applies only to zone relatively near dam; at some distance upstream it is presumed that the profile of sediment deposited behind the barrier would intersect original bed.

delta. The evidence to date suggests that the second alternative is probably correct: that the rise in base level is in fact accompanied by a set of adjustments. However, it also appears that rather than a unique combination of the variables there are a number of combinations which will satisfy the requisite demands of water and sediment transport.

Equilibrium, River Profiles, and Channel Geometry

It has been noted that equations expressing the average relations among hydraulic factors in stable or regime irrigation canals are similar in some respects to those describing the average hydraulic geometry of a river system. Graphs of width, depth, and velocity as functions of discharge for the canals, however, usually show very much less scatter than do similar graphs for river data. This lack of scatter probably can be attributed, at least in part, to the uniformity of the bank material throughout any given canal system. In India and Pakistan this bank material is a nearly constant clayey silt. The similarity in behavior of natural river channels and canals known to have adjusted themselves to achieve stable or equilibrium forms suggests that both have developed equilibrium channel sizes and shapes appropriate to the available discharge and character and quantity of sediment supplied. It is important to note that the equilibrium channel implied here by the hydraulic geometry or regime-type equations for canals is one in which sediment is moving, whereas the channel in cohesionless sand described earlier is in equilibrium at the threshold of movement. Both, however, imply that in an equilibrium channel a threshold of erosion on the banks at least limits the shape of the channel by controlling the width.

In the literature of geology and geomorphology, streams which exhibit evidences of this adjustability and stability have been called "graded." Davis used the word "grade" to mean the condition of "balance between erosion and deposition attained by mature rivers" (1902, p. 86).

Mackin (1948, p. 471) defined a graded stream as "one in which, over a period of years, slope is delicately adjusted to provide, with available discharge and with prevailing channel characteristics, just the velocity required for the transportation of the load supplied from the drainage basin. The graded stream is a system in equilibrium; its diagnostic characteristic is that any change in any of the controlling factors will cause a displacement of the equilibrium in a direction that will tend to absorb the effect of the change."

This definition, aside perhaps from some overemphasis on slope, well expresses the idea of equilibrium as it applies to rivers. Parallel to the longitudinal profiles of many rivers are series of terraces cut on bedrock or alluvium. During any single episode of stability represented by each terrace the river did not continually cut downward but maintained a kind of equilibrium as it swung laterally, widening the valley floor at the level now representing that episode.

Equilibrium does not demand a smooth longitudinal profile concave to the sky without undulations. Woodford (1951, p. 819) citing the Middle Rhine, points out that although the slight hump in its profile opposite the Kaiserstuhl is probably the result of a buried buttress of bedrock, ". . . no available evidence indicates that the equilibrium of the unregulated Rhine was less perfect here than in other parts of the Rhine graben."

Grade is generally considered synonymous with equilibrium. Despite difficulties of definition the concept of equilibrium is a useful one. It implies both an adjustability of the channel to changes in independent variables such as load and discharge and a stability in form and profile. The latter aspect, stability, is implied in the distinction between grade (equilibrium) and aggradation or degradation—the progressive building up or lowering of the channel bed. The unit of time here is significant; a channel in equilibrium may scour or fill. Those are short-lived changes. Of course an entire landscape is being reduced in elevation over geologic time, so the word stable does not strictly apply. Nevertheless, even while the channel is slowly eating away the land, its form and local gradient may remain constant and in quasi-equilibrium with available sediment and water. As a rule the condition of equilibrium has been observed, measured, or thought of in terms of some intermediate time scale. Thus Mackin (1948, p. 471), for example, refers to adjustments "over a period of years."

More recently Strahler (1957), Hack (1960), and other geomorphologist have used the term "dynamic equilibrium," referring to an open system in a steady state in which there is a continuous inflow of materials, but within which the form or character of the system remains unchanged. A biological cell is such a system. The river channel at a particular location over a period of time similarly receives an inflow of sediment and water, which is discharged downstream, while the channel itself remains essentially unchanged. Such systems are characterized as "open" as opposed to the "closed" systems more often described in studies of chemical equilibrium (Chorley, 1962).

From its headwaters to its mouth a natural river channel essentially represents a system in which potential energy provided by quantities of water at given elevations is converted to kinetic energy of the flowing water and dissipated in friction created at the boundaries. As the flow does not accelerate, all of the kinetic energy is dissipated by friction. In analyzing the behavior of the river channel system, however, primary interest lies not in the total energy in the system but rather in the way in which energy is distributed throughout the system. This emphasis upon the distribution of energy *within* the system is in a general way analogous to a consideration of the entropy of thermodynamic systems. From one point of view entropy may be said to be a measure of the energy in a system available for external work. The greater the entropy the more energy is "unavailable" for external work (King, 1962, p. 102). The natural process represented by the flow of water from the headwaters to the mouth of a river channel system is an irreversible process in which energy is transformed with an increase in entropy.

A complete description of the thermodynamics of the river system is exceedingly complex. Analogy with the thermodynamics of systems in a steady state, however, led Leopold and Langbein (1962) to consider the way in which energy might be distributed and dissipated in the river system.

There are eight interrelated variables involved in the downstream changes in river slope and channel form: width, depth, velocity, slope, sediment load, size of sediment debris, hydraulic roughness, and discharge. Inspection of the available equations involving these variables shows that there are more unknowns than equations. To arrive at a solution, two additional conditions concerning energy relations were postulated as reasonable. First, that along a given channel within the river system the power expended per unit area of the bed tends to remain constant—that is, equal at all positions along the river length. Second, that the power expended per unit length of the channel tends to be equal along the river length. The adjustment in the hydraulic variables necessary to accommodate an increase of discharge downstream takes place, it was assumed, in such a way that each variable changes as little as possible such that no single variable absorbs a disproportionate share of the required variation.

Consonant with the preceding descriptions of the river channel, the variables were expressed as functions of the discharge. As a test of the reasonableness of the assumptions made, the resultant solutions can be compared with the observed hydraulic geometry of rivers.

Though much more work needs to be done on the various sets of hydraulic equations that might be used, the principal elements in one of the derivations—the changes downstream in a river—are described here. The following three hydraulic relationships are assumed.

1. Continuity; discharge is the product of area of cross section times velocity.
2. Velocity is a function of depth, slope, and channel roughness.
3. Sediment transport is a function of stream power; this, combined with a relation of debris size to channel roughness, leads to a statement that sediment concentration is a function of velocity, depth, slope, and channel roughness.

For the theoretical derivation of the exponents in the hydraulic geometry it was assumed that discharge increases downstream, and sediment transport per unit discharge—that is, concentration—remains constant downstream in the river. In the factors considered, five variables are included in the above equations—width, depth, velocity, slope, and roughness. For the downstream case, roughness is eliminated by combining expressions for the velocity and sediment transport.

Assuming the proper units, the conditions involving the flow of water and sediment, as well as expressions for the distribution of power per unit area and power per unit length, can be given by the relations in Table 7-7.

Table 7-7. Hydraulic and energy statements used in deriving a theoretical hydraulic geometry, for the case of downstream changes.[a]

	Relation	Equation	Relation among Exponents
Hydraulic conditions	Continuity	$Q = wdv$	$b + f + m = 1$
	Hydraulic friction	$v \propto \dfrac{d^{2/3} s^{1/2}}{n}$	$m = \frac{2}{3}f + \frac{1}{2}z - y$
	Sediment transport	$C \propto \dfrac{(vd)^{0.5} s^{1.5}}{n^4}$	$0.5m + 0.5f + 1.5z - 4y = 0$
Energy statements[b]	Equal power per unit area	$\dfrac{Qs}{w} = vds$ is nearly constant	$m + f + z \longrightarrow 0$ (should be small)
	Equal power per unit length	Qs is nearly constant	$1 + z \longrightarrow 0$ (should be small)

[a] Symbols:
Q is discharge.   velocity, $v \propto Q^m$.   roughness, $n \propto Q^y$.
width, $w \propto Q^b$.   slope, $s \propto Q^z$. $\longrightarrow$ approaches or tends toward.
depth, $d \propto Q^f$.
[b] Omitting force units.

As for the units in the above statements, note that power expended per unit area is force times distance per unit time per unit area: (MLT^{-2}) $(L)(T^{-1})(L^{-2})$ or MT^{-3}. This is dimensionally equivalent to $\gamma Qs/w$, weight of water per unit volume times volume per unit time, per unit of bed width.

The power per unit stream length is force times distance per unit of time per unit of length: $(MLT^{-2})(L)(T^{-1})(L^{-1})$ or MLT^{-3}. This is dimensionally equivalent to pounds per cubic foot times cubic feet per second: γQs.

A profile in which Qs/w was constant would be only slightly concave, whereas a profile in which Qs was constant would be highly concave. This can be shown by a simple trial, assuming some values of Q and w representing average values in successive downstream reaches. Then various combinations of values of s, river slope, are chosen for the same reaches, each combination representing a specific assumed longitudinal profile. In these computations the summation of all values of s in any longitudinal profile must be constant (topographic relief is given or is a constant). It will be seen that when the product, Qs, is constant for each reach, the summation of those products gives a minimum value of the sum. Thus constant power expenditure per unit of stream length can be shown to be equivalent to minimum rate of work in the river system.

Such arithmetic trials will also demonstrate that the two energy statements can not be simultaneously satisfied. As one of the two is fulfilled, the fulfillment of the other is made impossible. The simplest resolution of this seeming conflict is to assume that the longitudinal profile occupies a position intermediate between the two. The mutual adjustment among the several variables—velocity, depth, width, slope, and power—can be achieved in such a way that the hydraulic statements are fulfilled and the increase in discharge downstream is most equally accommodated by each variable. This is accomplished in the solution by minimizing the sum of the squares of the exponents, m, f, b, z, and $(1 + z)$, for the hydraulic variables. The values so obtained are shown for the case of the downstream changes in river-channel factors as the top line in Table 7-8.

It is important to note here that the theoretical solution describes the behavior one would expect if rivers exhibited the tendencies postulated. One would not expect the mutual adjustment of all the variables in every river system, however, to be precisely the same. Thus the solution represents what might be termed the most probable distribution, or the one most likely to be observed in natural rivers satisfying the basic hydraulic conditions. In this sense it represents a modal value or central tendency.

Table 7-8. Adjustability characteristics of open channel systems in terms of independent and dependent variables, in quasi-equilibrium (not aggrading or degrading). [From Langbein, 1963.]

Type of Open Channel	Hydraulic Factors (I = Independent or Given; D = Dependent or Adjustable)							Values of Exponents in Hydraulic Geometry									
	Discharge	Sediment Load	Width	Velocity	Depth	Roughness	Slope	Width b Theory	Width b Data	Velocity m Theory	Velocity m Data	Depth f Theory	Depth f Data	Roughness y Theory	Roughness y Data	Slope z Theory	Slope z Data
River, in downstream direction	I[a]	I[b]	D	D	D	D	D	0.53	0.50	0.10	0.10	0.37	0.40	−0.22	−0.15	−0.73	−0.75
Tidal estuary, downstream	D[b]	D[b]	D	D	D	D	I[b]	.72	.72	.05	.06	.23	.22	.01	—	−.11	—
Meltwater stream on glacier	I	—	D	D	D	D	I	.50	—	.22	—	.28	—	−.04	—	—	—
River, at a station: cohesive	I	I	I[c]	D	D	D	I[c]	.25	.26	.32	.34	.43	.40	−.035	—	0	±0
noncohesive	I	I	D	D	D	D	I[e]	.50	—	.23	—	.27	—	−.04	—	−.12	−.11 to −.17
Canal system	I	I	D	D	D	D (minimum)	(minimum)	.47	.50	.17	.17	.36	.33	+.01	—	—	—
Flume recirculating: with slope specified	I	D	I	D	D	D	I	—	—	.45	.47	.55	.53	−.09	−.11	0	0
with depth specified	I	D	I	D	I[d]	D	D	—	—	1.0	1.0	0	0	−.50	−.50	1.0	1.0
Flume, nonrecirculating, sediment fed in, (G. K. Gilbert)	I	I	I	D	D	D	D	—	—	.42[e] / .45[f]	.42 / .45	.58[e] / .55[f]	.58 / .55	−.16[e] / +.06[f]	—	−.23[e] / .29[f]	−.25 / .29
Sediment fed in, erodible banks (Wolman and Brush)	I	I	D	D	D	D	I	.50	.50	.22	.22	.28	.28	−.04	—	0	0

[a] Imposed from upstream, or determined by what is introduced at the upstream head of the reach.
[b] Semidependent in that it depends on adjustments made in reach upstream.
[c] Not adjustable in any short periods of time, and thus controlled by "downstream" relations.
[d] Run at given depth but variable slope.
[e] $j = 0$ (sediment concentration constant).
[f] $j = 1$ (sediment concentration proportional to discharge).
Dash indicates no data or not available.

Langbein (1963) has applied the basic method to a variety of other types of channels and flumes. His results from the theoretical derivation are compared with values from field data in Table 7-8, and it can be seen that the agreement is close. The average relationships—that is, the hydraulic geometry of the "typical" river—agrees very closely with the results derived from his assumed energy conditions.

Of particular significance is Langbein's derivation of the Lacey equations, which describe the relations between velocity, depth, width, slope, and sediment load in stable irrigation canals. These are the "regime" canals known to be nonscouring and nonsilting, which are analogous to graded rivers or to rivers in quasi-equilibrium. The interrelations of the hydraulic parameters—that is, the hydraulic geometry—are not identical in the river and the canal, except that width increases downstream as the square root of discharge. In the canals the discharge decreases along the direction of flow as large canals are split up into distributaries. Also, the longitudinal profile of such canals is convex upward, in contrast to the concave profile of rivers, inasmuch as both small tributaries and distributaries have steeper gradients than main or trunk channels.

By a line of reasoning similar to that just described for computing the exponents in the hydraulic geometry in the downstream direction, Langbein (1963) introduced the specification that, in actual practice of canal design, slope is chosen by the engineer to be a minimum value compatible with the continuity of water and sediment transport. To maintain the canal at a high elevation the designer seeks to avoid increasing the slope. This means that the exponent z tends toward zero. This specification recognizes a characteristic inherent in the conditions of canal design in India but one which, though widely known, has received little emphasis in the large literature. Owing to the flat slopes characteristic of the immense alluvial valleys of the Indian subcontinent, canal designers have always been forced to design within the constraint imposed by the natural valley gradients, which are very flat. Although this specification of minimum change in slope is the only difference in the derivation of the exponents for canals and the derivation for the downstream changes in rivers, the computed exponents do differ for the river and the canal (Table 7-8).

The tidal channel, like the upland channels, is formed and maintained by the flow of water and sediment that it carries and is thus the creator of its own geometrical properties. It differs from the river in one important respect: whereas any segment of the upland receives an amount of water from its upstream drainage area that depends on that drainage area and not on the properties of the channel upstream, the tidal channel during the flood portion of the tidal cycle receives an amount of water not in proportion to a fixed drainage area but determined by the capacity of the channel supplying the water. In other words, the discharge at any section is itself a dependent variable, governed by the manner in which the flow shaped the channel in all channel segments between the point in question and the main bay or body of tidal water. In a terrestrial river, discharge is independent in that it is produced by the watershed, and to that discharge the channel accommodates itself.

For this reason the change in channel properties along the tidal channel is different from that of terrestrial rivers. Inspection of a map (Fig. 7-29) or aerial photograph is enough to reveal an impor-

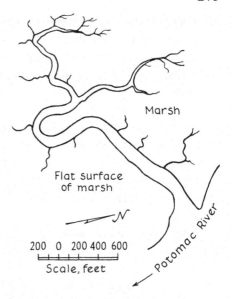

Figure 7-29.

Planimetric map of a small tidal estuary, Wrecked Recorder Creek, which enters the Potomac River on its west side about 1 mile south of Alexandria, Virginia. Note rapid downstream widening of the channel.

tant aspect of this difference. The width increases rapidly along the tidal stream and seems to flare outward toward the bay or ocean. Except for this rapid increase in width, the planimetric configuration resembles the upland river system in that the drainage net ramifies into many small tributary channels.

The comparison of increase of width with discharge downstream in rivers and estuaries is shown by the values of the exponent *b* in Table 7-8. The larger value for the estuary is a quantitative statement of the downstream widening, which represents the principal way in which the estuary accommodates itself to increased discharge. The discharge is not an independent but a dependent variable in the channel adjustment.

Table 7-8 shows not only the values of the exponents but also a notation on the dependence or adjustability of each of the principal variables in the various open-channel systems.

Because the hydraulic requirements do not completely specify how the dependent factors adjust among themselves in response to changes in discharge, it can be visualized that the manner of this adjustment will depend both on how many and which factors are dependent. The division among the dependent factors cannot be rationalized by an inspection of the numerical values of the exponents.

The results of these studies lead us to suggest a generalization that appears to have considerable importance in geomorphology. Landscape factors usually are relatively large in number because of the complexity of the physical processes involved and the number of such processes simultaneously acting. At the same time, the number of cases on which these processes operate is nearly infinite; that is, there are many hills, many channels, many rills.

When there are a large number of interrelated factors which must adjust among themselves in response to occurrences in the environment, such as storms or flows, it should be expected that there will generally be an indeterminacy in the manner of this mutual adjustment. The number of cases or examples in the landscape is so large that one may expect to see a spectrum of results of this interaction—a statistical distribution of the results. The central tendency in this distribution may, as in cases described here for channels, be possible of forecast, but it may not be possible to specify in any individual case the manner in which the adjustment took place. Where a particular factor dominates, such as a bedrock floor of a channel, the effects of this factor may be readily evident. However, where the alternatives become more nearly equivalent it becomes more difficult to specify the precise form in any given case. This is also true when one considers the spectrum of possible combinations from the range of natural occurrences.

This indeterminacy in a given case results from the fact that the physical conditions, being insufficient to specify uniquely the result of the interaction of the dependent variables, are controlled by a series of processes through which any slight adjustment to a change imposed from the environment feeds back into the system.

To take an example, a change of bed configuration feeds back its effect into the combination of velocity, depth, and slope if all are adjustable. Thus, change of bed configuration is one of the feedback mechanisms operating in the direction of quasi-equilibrium. But this is not the only kind of change which can occur and feed back into the hydraulic adjustment. Certain combinations of depth, slope, and sediment load result in increased capacity for transport of sediment of a particular size, and if such sediment is available locally on the bed it will be picked up. The local scour that results will alter the cross-sectional area and thus the depth-velocity relations. The response to this new adjustment will feed back but will always tend to check rather than to expand the initial minor change. Thus the system tends toward mutual adjustment and equilibrium rather than toward instability.

As another part of the feedback, the large frictional resistance caused by the formation of dunes tends to decrease velocity, requiring an increase in depth to maintain continuity. At a given slope the increased depth will tend to increase bed shear; this in turn may tend to increase sediment transport rate, which may result in washing off the tops of the dunes and possibly eliminating them. This decreases friction, increases velocity, and decreases depth, all in a direction opposite to that incurred by the growing dune. The system, therefore, is self-governing in that the initial change tends to set up a tendency to counteract the change.

Indeterminate systems in geomorphology, then, have certain characteristics. Such a system searches, as it were, for combinations that will meet all requirements of the physical conditions, but a balance of effects seems to be attained only in an average sense.

The type of indeterminacy here found in channels would appear to be applicable to many other geomorphic forms in which the number of interacting factors is large and the number of examples in nature very great. Hillslopes and drainage networks come to mind as possible examples in which stochastic or probabilistic tendencies may be influential. As a result, no two examples will be exactly the same, but there will be a central tendency recognizable in any set of samples.

The concept of a most probable behavior of the channel system provides a means of estimating the average rate of change of channel form in the downstream direction which is in close agreement with average measured values. The results leave little doubt of the validity of the two postulated tendencies—uniform distribution of energy dissipation and minimum total work in the system. The concept emphasizes the fact that the downstream changes among various hydraulic and load factors of a river cannot be determined from hydraulic considerations alone. The results provide a quantitative rationale for Rubey's suggestion that "graded slopes are not determined solely but only within rather broad limits by the imposed conditions of discharge, load, grain-size, and degree of sorting. . . . It is not improbable that the precise form taken by the adjustment is governed by something like the principle of least work" (1952, p. 135).

The considerations of the hydraulic geometry show that the concavity of the longitudinal profile results from the mutual adjustment of all channel factors as discharge increases downstream. The downstream increase of discharge associated with tributary entrance is typical of a large part of the land surface of the earth. In the arid zones, however, rivers whose headwaters drain a humid area may pursue a course of considerable length through which there is no tributary entrance and the discharge is essentially constant downstream. Exposition of some principles applicable to this case brings out certain additional points particularly with regard to effects of base level.

The higher the elevation above base level—the ocean, a lake, or another channel—the greater is the variability of slopes. This is reasonable on geometric grounds, for the range of probable slopes is reduced as base level is approached. This should not be taken to mean that the slope is physically controlled by base level. Rather, *on the average* it should be expected that slopes will decrease as base level is approached.

The longitudinal profile derived from considerations of energy distribution is one which represents a most probable distribution of variables. If sample situations are chosen by a random process in which the probabilities of alternative values of gradient are fixed, the average of a number of such samples normally distributed would reflect the most probable of the alternative possibilities. Random-walk models offer a method of deriving such samples which can then be averaged in some appropriate manner to obtain the most probable case.

It is possible to set up a model in which the probability of a decrement of elevation, ΔH, in a unit horizontal distance, Δx, decreases as the elevation, H, approaches H_0 or base level. In such a random-walk model of a longitudinal profile, with each forward step the probability of a simultaneous move downward decreases as the walk approaches base level (Leopold and Langbein, 1962, p. A8).

On a piece of cross-section paper let a point of origin for the random walk be at zero distance on the abscissa and 5 units on the ordinate. Each step of the walk will be determined by drawing a card from a specially prepared deck. On each random drawing of a card the walk proceeds one unit to the right; the card will also indicate whether simultaneously a move is made downward as well as to the right.

The cards to be drawn are made in such a way that the probability of drawing a card indicating a move downward depends upon the walk's height above the base level at the time of drawing. This may be accomplished by making up a pack of 5 white cards and 1 black one. The number

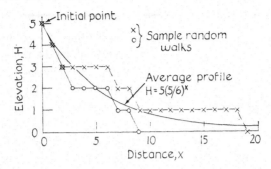

Figure 7-30. *Sample random walks used in the generation of an average stream profile.*

of white cards represents the total number of levels above base level, in this case initially 5. The cards are shuffled and one is selected at random. If a white card is drawn, the unit move to the right on the random walk is also accompanied by a unit move downward; after each draw the card drawn is always replaced by a black card. If a black card is drawn, the unit move to the right is not accompanied by a downward move. Thus the probability of choosing a white card and taking a downward move decreases in proportion as the number of white cards in the pack decreases. The initial probability of a downward step equals H 6.

The drawing of cards proceeds until the individual walk reaches base level. Two such walks, marked by crosses and circles, are shown in Fig. 7-30. The data pertaining to the derivation of the average random walk in Fig. 7-30 (solid line) are presented in Table 7-9. The table shows the

Table 7-9. Frequencies of random walks having a given elevation at various distances from origin, in percent, subject to the condition that probability of a downward step equals $H/6$.

Elevation (H)	Distance										
	0	1	2	3	4	5	6	7	8	9	10
5	100	17	3								
4		83	42	17	6	2					
3			55	56	39	24	14	7	3	2	1
2			27	46	50	45	37	29	20	14	
1				9	23	36	45	50	52	50	
0					1	5	11	18	26	35	
Average Elevation	5.0	4.17	3.48	2.90	2.42	2.02	1.68	1.40	1.17	0.98	0.81

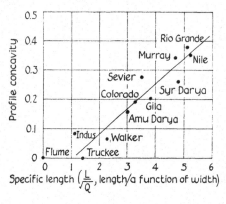

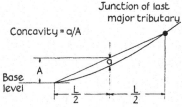

Figure 7-31.

Relation of profile concavity to specific length in rivers flowing through deserts where there is no change or decrease of discharge along the length considered. Channel width has been approximated by $\sqrt{Q}$*. [After Langbein, unpublished.]*

percentage of the walks that stood at various elevations at each distance along the horizontal. A number of such walks were constructed, and their ordinate values at selected positions on the abscissa were then averaged; these values define the average random walk (Fig. 7-30), which curves smoothly downward, becoming asymptotic to the horizontal axis.

These random-walk models demonstrate the derivation of the average condition resulting from a number of random choices governed by certain restrictions or rules. They also show how the operation of chance results in the deviation of individual examples from the most probable.

This discussion treats the theoretical longitudinal profile in the absence of any limit to stream length. But if length is made the principal restraint, then the probability of a downward step decreases with increasing distance downstream rather than with respect to elevation. Thus height and distance are reversed, so that with the length restraint the profile would have the form

$$x = L(p)^H,$$

where $p = L/(1 + L)$, L is total stream length, H is height above the elevation where x equals L, and x is horizontal distance from an upstream origin. The equation above is exponential with respect to distance rather than to elevation, but a more familiar way of expressing the relation would be to say that elevation varies as the logarithm of distance:

$$H \propto \log_p \frac{x}{L}.$$

This is a common form of longitudinal profile, as shown by Hack (1957, p. 70). As the available length is restricted, the profile is likely to be less concave and the angle at which the profile joins the base level will be steeper. This is true despite the fact that no physical control need be exercised by base level on the reaches of the river upstream.

This suggests that rivers having no increase in discharge over long reaches through desert or arid areas should have a concavity which is related to the length of the reach downstream from the last major tributary. Table 7-10 presents data on river reaches of approximately constant dis-

Table 7-10. Length, discharge, and concavity of profiles of rivers with uniform discharge. [From Langbein.]

River	Length (miles)	Discharge (cfs)	$L/\sqrt{Q}$	Concavity
Indus	500	200,000	1.1	.09
Nile	1,650	100,000	5.2	.35
Amu Darya	650	47,000	3.0	.16
Syr Darya	690	20,000	4.8	.26
Colorado	470	20,000	3.3	.19
Murray	515	12,000	4.7	.34
Rio Grande	450	8,000	5.1	.38
Gila	170	2,000	3.8	.20
Truckee	50	1,100	1.4	0
Walker	52	500	2.3	.07
Sevier	75	450	3.5	.28
Laboratory flume	.02	10	.006	0

charge extending through desert areas; from these data Fig. 7-31 is constructed. The concavity is defined as shown by the example included in the figure, the ratio of the maximum vertical distance between arc and chord to the elevation at that point. Length is expressed nondimensionally—as length of reach divided by river width—so that data for the several rivers may be compared. Width is proportional to the square root of discharge, which can thus be used instead of width, discharge usually being well known but width often unrecorded.

The data define roughly a straight line with scatter of points typical of river data, and this line crosses the abscissa at a finite value. That is, concavity is zero or the profile is straight when specific length equals some particular value, a finding in accord with the results of random-walk analysis in which length was increasingly constrained (Leopold and Langbein, 1962, Fig. 4).

Trials by random walks first suggested the relation of concavity to length for rivers in which discharge is constant. Further development of random-

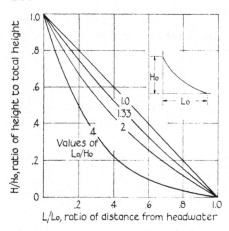

Figure 7-32.

Nondimensional diagram of profiles developed by random walks; it indicates that concavity is a function of the reciprocal of relief ratio, the ratio of basin relief to basin length. [From Langbein, unpublished.]

walk profiles showed that profile concavity is a function of relative length; this function can be generalized into a nondimensional diagram, which eliminates any reference to absolute relief or absolute length. This general relation, presented in Fig. 7-32, shows that concavity is an inverse function of the reciprocal of relief ratio, the quotient of height to length of a basin. The greater the relief for a given length of basin, the straighter is the profile. Conversely, the greater the length of a basin for a given relief, the more concave the profile. This has not been tested against field cases, but previous accordance of random-walk results with field data would suggest that this analogy might also yield reasonable agreement with data from the field.

The theory derived from probability considerations does not in any way contradict the fact that the usual concavity of the longitudinal profile results from the downstream increase in discharge. The length of channel between tributary junctions is usually too short for concavity to develop under conditions of constant discharge downstream.

Nor does the theory developed from probability considerations in any way minimize or render unimportant the effect of changes in bed-material size which result from varying rock types. Changes in bed-material size along the channel and their effect on slope or concavity have been discussed in a preceding section. The relation of grain size to hydraulic roughness provides a rationale for explaining how slope is affected.

Both channel forms and profiles derived theoretically from postulates about the distribution of energy in a stream-channel system appear to agree with experience in nature, which indicates that the postulates provide valuable insights into the behavior of natural streams. The energy relations provide a framework for describing the average behavior of river channels. Nevertheless, in association with these concepts the need remains for physical explanations of the processes by which complex channel adjustments actually take place.

Channel Pattern: Introduction

By channel pattern is meant the configuration of a river as it would appear from an airplane. The channel patterns that have been recognized are meandering, braided, and straight. Rivers are seldom straight through a distance greater than about ten channel widths, and so the designation straight may imply irregular, sinuous, or nonmeandering.

There is no sharp distinction between any of these patterns. Rather, river pattern is a continuum from one extreme to another. For purposes of definition we have used the ratio of channel length to downvalley distance as a criterion. This ratio, called sinuosity, varies in rivers from a value of unity to a value of 4 or more; rivers having a sinuosity of 1.5 or greater we have called meandering, and below 1.5 straight or sinuous. This definition is somewhat limited, and it might be well to require that meandering channels also have some degree of symmetry in their curvature. However, no measure of symmetry has yet been put on an objective basis, and in the present discussion the value of 1.5 for sinuosity will be used as a criterion for meandering.

A braided channel is one that is divided into several channels, which successively meet and redivide. Anastomosing, a term borrowed from medical usage, where it applies to dividing and rejoining blood vessels, is synonymous with braiding.

River patterns represent an additional mechanism of channel adjustment which is tied to channel gradient and cross section. The pattern itself affects the resistance to flow, and the existence of one or another pattern is closely related to the amount and character of the available sediment and to the quantity and variability of the discharge. Although separation of distinctive patterns is somewhat arbitrary, for simplicity each major pattern is discussed separately.

Straight Channels

It is easier to find examples of meandering or braided channels than to find long straight reaches. Even where the channel is straight it is usual for the thalweg, or line of maximum depth, to wander back and forth from near one bank to the other. Opposite the point of greatest depth there is usually a bar or an accumulation of mud along the bank, and these bars tend to alternate from one side of the channel to the other. In an idealized sense, flow and depositional pattern seen in the plan view is similar in straight and in meandering patterns. A detailed description of pools and riffles has been given earlier in this chapter.

One of the requirements for pools and riffles in nonmeandering streams appears to be some degree of heterogeneity of bed-material size. Channels that carry uniform sand or uniform silt appear to have little tendency to form pools and riffles. In part this might be attributed in sand channels to the concomitant high width-to-depth ratio associated with relatively non-cohesive bank materials, and this ratio in turn tends to promote a braided pattern. Pools and riffles are less well developed in braided than in straight but nonbraided channels.

Braided or anastomosing channels are often but not always associated with sandy or friable bank materials. Also, vegetation has similar effects; a change from nonbraided to braided character is sometimes associated with a change from dense vegetation along the channel banks to sparse or no vegetation. Whether these coincident changes are causally related cannot usually be ascertained, although the coincidence is suggestive.

A change in channel pattern is accompanied by a large change in flow resistance, but again the unraveling of cause from effect is usually difficult or impossible. There is no doubt that braided or meandering channels offer larger resistance to flow than otherwise comparable straight channels. A change in pattern along the stream probably reflects at least in part a channel adjustment in flow resistance—that is, in energy expenditure. That such a statement must be made tentatively and with qualifications suggests that an important area of needed research is in the nature and location of energy transformation in channels.

The pattern of flow in straight channels has been subject to considerable speculation. A tendency for lateral cutting of the stream bank results from the formation of alternating bars deposited along the channel banks. Several additional lines of evidence also suggest that filaments of water in straight rectangular channels do not move downstream in straight and parallel lines. Gibson (1909) observed two circulation cells in the plane perpendicular to the flow. Rotating in opposite directions, water converged at the surface toward the center of the stream. It was common in the early days of boating on the Mississippi River to attach a houseboat to a large and water-soaked log which floated mostly submerged, for such a log kept to a downstream course near the center of the river. This suggests that surface cross-currents move toward the center of the channel.

Vanoni (1946) observed longitudinal clouds or ribbons of sediment forming along the bed of a straight rectangular flume which carried sand sufficient only barely to cover the flume bed; others (Leliavsky, 1955, p. 185) have observed parallel streaks or ridges attributable to secondary circulation on the beds of straight channels transporting sediment at low rates.

Direct instrumental measurements of cross profiles of the water surface in straight reaches in natural channels are rare. Such profiles for selected rivers are plotted in Fig. 7-33 in terms of deviations above or below a straight line connecting water surface elevations on the opposite sides of the stream. These profiles show a marked central hump. The elevation of

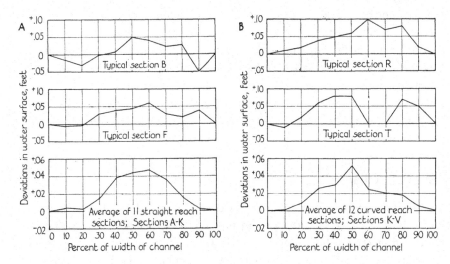

Figure 7-33.   *Cross-channel water-surface profiles for the Rio Grande del Ranchos near Taos, New Mexico. Channel width averages 23 feet in the measured reach, with extremes of 17 and 36 feet. Deviations shown are those which describe the differences in elevation between a straight line connecting the water-surface elevations observed at each bank and the actual water-surface elevations. This then sets the end-point (or bankside) deviations at zero and corrects for any superelevation. Plotted points are for given percentages of channel width and do not necessarily represent points of actual measurement. A. Straight reach. B. Curved reach. [Measurements by Leon A. Wiard et al., United States Geological Survey.]*

the water in the center of the stream is higher than the elevations at the two edges. We now have a number of such observations in straight reaches on streams of various widths up to several hundred feet, and, with expected minor variations, all tend to confirm the convexity of the surface profile.

Among practicing water engineers there has been disagreement concerning whether the cross profile also varied from rising to falling stage. As one example studied by us, cross profiles of water surface of the Wind River near Riverton, Wyoming, were measured at equal stages, when the river was rising and when it was falling; in both there was a dip or topographic trough in the cross profile, which otherwise was convex to the sky. R. M. Myrick, who made the measurements, noted that debris floating in the stream tended to be concentrated in this trough. Once caught in the topographic trough, such debris presumably could not escape and so tended to accumulate there. A trough was observed on both rising and falling stage, discounting the view sometimes heard that debris tends to collect in the center of the stream during rising stages and along the banks of the stream during falling stage. Others have suggested that a stream carries floating debris on the rising but not on the falling stage.

Despite their apparent simplicity, flow and sediment movements in straight natural channels are complex phenomena, requiring further detailed field observations.

Braided Channels

The separate channels of a braided stream are divided by islands or bars. Bars which divide the stream into separate channels at low flow are often submerged at high flow. Great areas of braided channels of proglacial streams (see Fig. 7-34) are in many ways similar to the braided channels that can be observed in the sandy detritus covering numerous short reaches of the concrete gutter at the edge of a city street. The building of central and lateral bars is an important part of the process of development and shifting of braided channels.

In a study of braided channels tributary to the Green River near Daniel, Wyoming, we noted the following characteristics. The several channels in a given reach were separated by vegetated islands whose upper surfaces tended to be at the same level as small unvegetated central bars obviously in the process of being built. The islands were composed of gravel essentially similar to that of the building bars, diameter 60–100 mm. The ages of vegetation taking hold on the new gravel bars indicated that the

Figure 7-34.

Braided channel of the Muddy River, at a point about 23 miles north of Mt. McKinley, Alaska. The river heads in the Peters Glacier, which flows off the north side of Mt. McKinley. [Photograph by Bradford Washburn.]

bars tend to build by additions at the downstream end and probably also on some parts of the lateral boundaries, but not on the upstream tip.

In flume experiments conducted in a channel molded in moist but uncemented sand, the introduction into the flowing water of poorly sorted debris at the upper end produced, with time, forms similar in many respects to those observed in the field. After 3 hours a small deposit of grains somewhat coarser than the average introduced load had accumulated on the bed in the center of the channel (Fig. 7-35). This represented a lag deposit of the coarser fraction which could not be carried by the flow. These same grains had already been carried downstream several feet, which suggests that when mixed with smaller material large grains may be moved under flow conditions incapable of transporting them when concentrated with particles of similar size.

Lag deposits of material slightly coarser than average can also be ob-

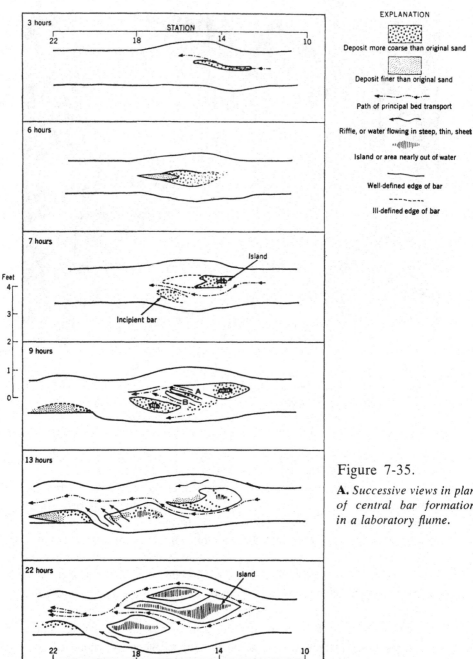

EXPLANATION

Deposit more coarse than original sand

Deposit finer than original sand

Path of principal bed transport

Riffle, or water flowing in steep, thin, sheet

Island or area nearly out of water

Well-defined edge of bar

Ill-defined edge of bar

Figure 7-35.

A. *Successive views in plan of central bar formation in a laboratory flume.*

PROGRESS IN DEVELOPMENT OF BRAID IN FLUME CHANNEL

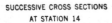

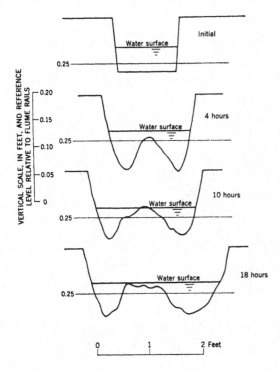

Figure 7-35.

B. *Successive views in cross section of central bar formation in a laboratory flume.*

served in ridges separating small anabranches on alluvial fans in desert areas. These appear to be similar depositional bars or ridges. A photograph of such a gravelly ridge on a fan in Death Valley is shown in Fig. 7-36.

As the instantaneous velocities in turbulent flow are subject to fluctuation, a brief decrease in intensity allows some particles to come to rest. Large particles may have rolled into the stream from eroding banks upstream. Velocities required to keep them moving are less than those required to reinitiate movement after they have come to rest. Once concentrated, the large particles form a locus for continued deposition.

As shown in the sketch at 3 hours of flow (Fig. 7-35), the band of principal bed transport lay on top of the submerged central bar, whereas grain movement in the deeper parts of the channel adjacent to the central bar was negligible at this stage. The central bar continued to build closer to the water surface. By the end of 7 hours of flow the central bar had been built so close to the water surface that individual grains rolling along its ridge actually broke the water.

Figure 7-36.

Ephemeral channels on surface of desert fan in Death Valley, California, which have central bars or gravel ridges comparable to those in gravelly channels of braided perennial rivers.

The growth surfaceward of a central bar tends to concentrate flow in the flanking channels, which then scour their beds or erode their banks (or both), as can be seen in the plan and in the cross section of Fig. 7-35. As the cross section is enlarged, the water surface elevation is lowered, and the bar, formerly just covered with water, emerges as an island. In a natural stream the emergent bar may be stabilized by vegetation, which prevents the island from being easily eroded and in addition tends to trap fine material during high flow. Thus the gravel tends to become veneered with silt.

Similar processes have been described in maps and by time-lapse photography on the White River, a braided stream emanating from the Emmons

Glacier on Mount Rainier in Washington State (Fahnestock, 1963). Flowing on an average slope of about 0.04, the White River changes with great rapidity in response both to random deposition of cobbles and boulders and to diurnal fluctuations in discharge accompanying glacial melting.

For example, within a period of 2 months during the summer melting season, in a single cross section carrying an estimated flow of about 200 cfs, at one time there were 5 distributary channels, four times there were 4 channels, three times there were 2 channels, and once there was only 1 channel (cross section No. 5).

As another example, Fahnestock noted that during a period of 8 days the water in a given reach shifted or switched a lateral distance of about 400 feet from one side of the valley to the other.

As in the laboratory, flow is often on elevations rather than depressions in the alluvial plain, and natural levees confine the channel in places (Fahnestock, 1963). Downstream from local chutes or channels of concentrated flow, deposition takes place, forming fanlike noses as the flow diverges in the area of deposition (Fig. 7-37). The levees and noses may also contribute to shifting the channel alignment by imposing barriers across active anabranches. In both field and laboratory, deposition on top of the

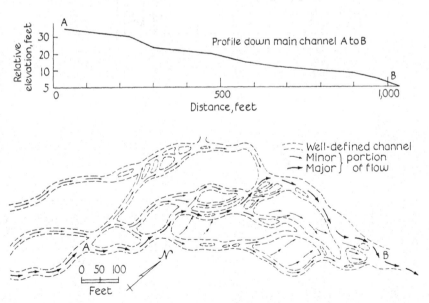

Figure 7-37. *Pattern of flow on gravelly valley train below Emmons Glacier, Washington. Note the dividing and joining of channels at obtuse angles. [From Fahnestock, 1963.]*

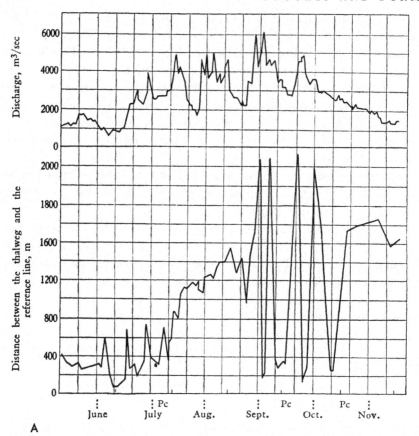

A

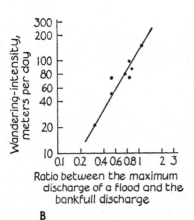

B

Figure 7-38.

Changes in braided river pattern, Yellow River. [From Chien, 1961.] **A.** *Shifting of the thalweg in a zone of convergence.* **B.** *Channel shifting as a function of the ratio of maximum discharge to bankfull discharge.*

bar often continues even as the depth of the water becomes exceedingly shallow.

Large braided rivers are also characterized by wide channels, rapid shifting of bed material, and continuous shifting of the position of the river course. An example for which observations are available over a considerable period is the Kosi River, a tributary to the Ganges in Bihar, India. The river rises in Nepal in the vicinity of Mt. Everest, and near Chatra emerges from the mountains at the apex of a flat cone built by the deposited sediments, with a gradient of .0009 near the apex, flattening to .0002 near Dhamaraghat. In the flood season the average discharge is about 160,000 cfs. Through this whole distance of 130 miles over the cone the river is braided. The last two centuries have witnessed a westward movement of the river over this cone, a lateral distance of some 70 miles. The movement has not taken place gradually but sporadically; the river is known to have shifted 12 miles laterally in a single year. The lateral movement of the Kosi River channel, on a line passing through Bhelhi and Puenea, is tabulated for various periods of specific observation.*

DATES	PERIOD (years)	DISTANCE MOVED (miles)
1736–1770	34	6.7
1770–1823	53	5.8
1823–1856	33	3.8
1856–1883	27	8.0
1883–1907	24	11.5
1907–1922	15	6.8
1922–1933	11	18.0
1933–1950	17	11.0

On the Yellow River, Chien (1961, p. 738) has shown that channel shifting varies with fluctuations in discharge (Fig. 7-38), and that the amount of shifting—that is, the lateral movement—is controlled by the spacing of constrictions or control points along the river. Likening the braided stream to a long elastic band along which oscillations will be propagated by a disturbance at any one point, Chien (p. 745) points out that by controlling the displacement at several points on the band the amplitude of the oscillations is reduced. Controls limited to bedrock or resistant strata on the bank only, as at the outside of a bend, may be ef-

* Unpublished report by M. P. Mithrani, "Note on the Kosi Problem," July 17, 1953, Patna, Bihar.

fective only at stages below overflow, but control points such as bedrock narrows, bridges, or revetments are effective at all stages.

If a reach of river possessing a single channel is compared with an otherwise similar reach in which the channel is divided by a bar or island, the braided portion will have a steeper slope (Fig. 7-39, **A**). This increase in slope is analogous to the variation of slope with discharge in a downstream direction as indicated in the hydraulic geometry. The undivided channel is comparable to the larger downstream channel, and each of the anabranch channels represents a stream with smaller discharge flowing on similar bed material. The ratio of slopes of divided to undivided channels was found to range from 1.4 to 2.3 in field situations and 1.3 to 1.9 in the flume (Leopold and Wolman, 1957, p. 51). The sum of the widths of the anabranches is also greater than the width of the undivided channel, the ratio varying from 1.6 to 2.0 in field examples and 1.05 to 1.70 in the flume.

There is a close relationship between braiding and meandering: a braided channel may exhibit curves that have a characteristic relation of radius to channel width, and the river has at least some reaches that would be called meandering. In other instances, as in Fig. 7-39, **A**, the anabranches of a braided stream definitely meander while the undivided channel does not. In overall plan, however, the channel course of the braided river is usually very much less sinuous than a meandering river of comparable size. Although the channels may meander at low stages, at overbank flow the braided river often moves nearly straight down its valley. Figure 7-39, **B,** shows that when two rivers of a given size of river (same discharge) are compared, braided channels occur on steeper slopes than meanders. Steeper slopes contribute to sediment transport and to bank erosion and are often associated with coarse heterogeneous materials. All these are conditions which contribute to braiding.

Where coarse material is available, braiding may result from the selective deposition of the coarser material, causing formation of a central bar and thus diverting the flow and increasing erosional attack on the banks. This was observed in the flume, in the gravelly channels studied by us in Wyoming, and in braided proglacial rivers described by Fahnestock (1963) and others. Even in fine material, however, irregular deposition of bars and bank erosion may produce a braided pattern. The shifting channel may move gradually during low flows, but during floods major changes in the position of the thalweg can be produced. Because deposition is essential to formation of the characteristic braided pattern, it is clear that sediment

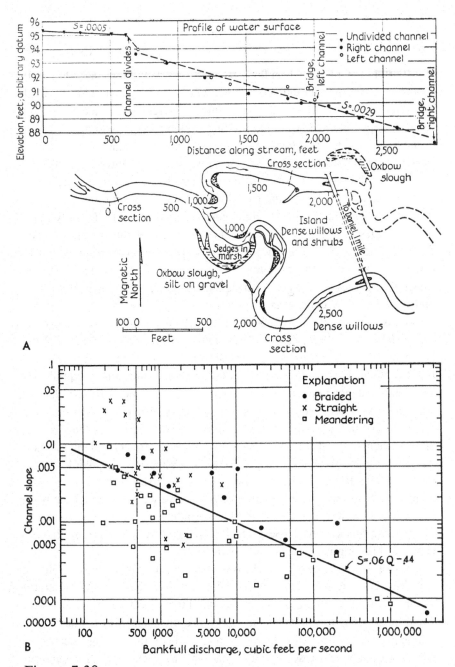

Figure 7-39.

*Relation of discharge to slope in braided and nonbraided rivers. **A.** Plan view
and profile of a channel which divides around an island, showing how the individual divided channels are steeper than the single one, Green River near Daniel,
Wyoming. **B.** Relation of discharge to slope and a line which separates data from
meandering and braided channels. [After Leopold and Wolman, 1957, and Balek
and Kolar, 1959.]*

transport is essential to braiding. It is also evident, however, that if the banks were unerodible and the channel width confined, the capacity of the reach for the transport of sediment would be increased, reducing the likelihood of deposition. In addition, any bars which formed would be removed as flow increased, since bank erosion could not take place. Thus, for the bars to become stable and divert the flow, the banks must be sufficiently erodible so that they rather than the incipient bar give way as the flow is diverted around the depositing bar. Sediment transport and a low threshold of bank erosion provide the essential conditions of braiding. Rapidly fluctuating changes in stage contribute to the instability of the transport regime and to erosion of the banks; hence they also provide a contributory but not essential element of the braiding environment. Heterogeneity of the bed material in the same way creates irregularities in the movement of sediment and thus also may contribute to braiding.

Because the braided reach is wide and shallow and the channel banks are unstable, the rate of sediment transport per unit width of channel may be relatively low. As we noted earlier (p. 201), anabranches of some braided channels appear to be near the threshold condition of equilibrium. In addition, deposition in such reaches is characteristic. Braided channels then may be associated with aggradation. It is also clear, however, that this need not be the case. Although deposition takes place it may be local and transient only. Deposits left at one moment are moved in the next. The timing may involve minutes or days, as Fahnestock so vividly showed on the White River, or months or seasons on rivers subject to periodic annual or seasonal floods. Transient as the individual channels may be, the reach as a whole may be stable, aggrading or degrading (Stricklin, 1961; Fahnestock, 1963). Several kinds of observations support the concept that braiding is a valid equilibrium form.

On Horse Creek near Daniel, Wyoming, for example, the stability of the channel pattern is demonstrated by the fact that islands separating anabranches have changed in outline but little in 60 years. Cottonwood Creek near Daniel, Wyoming, changes abruptly from a meandering to a braiding pattern. Through the short reach where this change occurs, no tributaries enter, and discharge and load are the same in both meandering and braided portions. There is no evidence of rapid aggradation or degradation. If the meandering reach may represent an equilibrium condition, then the braided must also do so.

When bedload in transport cannot all be carried through a particular reach for some reason—such as a local deficit in shear—it is logical that the largest particles are deposited. If this results in the growth of a central bar, the divided channels will be characterized by a larger width and greater slope but probably only a modest increase in mean velocity, if any. These are adjustments, then, which tend to increase the ability of the same discharge to carry a larger amount of bedload, a generalization supported by flume data and by inference from field experience which was summarized by Leopold and Maddock (1953, p. 29).

It appears, therefore, that braiding is a type of adjustment that may be made in a channel possessing a particular bank material in response to a debris load too large to be carried by a single channel. Braiding then represents a response or adjustment among the controlling variables which may provide an equilibrium condition over a period of time. The pattern in itself is not evidence that the channel is "overloaded." (This usually implies aggradation, which, it has been shown, need not be the case.)

Braiding is therefore considered to represent a particular combination of variables in a continuum of river shapes and patterns. Once established, the braided pattern may be maintained with only slow modification, and the braided river may be as close to equilibrium as are rivers possessing meandering or other patterns.

Geometry of Meanders

Nearly all natural channels exhibit some tendency to develop curves which seem to be proportional to the size of the channel. Because most streams are sinuous to some extent, it has often been suggested that the term meandering be restricted to channels exhibiting curves of considerable symmetry. This appears to be a reasonable suggestion. Again, however, because symmetry itself is a continuous function, the difficulty of defining a distinctive geometry remains. In Fig. 7-39, **B**, meanders were arbitrarily confined to sinuous channels having a ratio of channel length to valley length equal to or

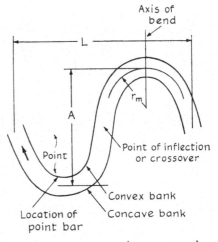

L = Meander length (wave length)
A = Amplitude
r_m = Mean radius of curvature

Figure 7-40.

Definition sketch for meanders.

exceeding the value of 1.5. But in a tabulation of 50 rivers ranging in size from models to large rivers, we found the sinuosity of the meandering rivers had a median value of 1.5; thus one-half had values less than that. Sinuosity is usually measured over a reach including several bends. Other terms describing a meander are defined by the sketch presented in Fig. 7-40.

Meander length is empirically related to the square root of effective or dominant discharge. Because channel width is also related to discharge it has been postulated that there is a fundamental relation between width and meander length. The best-fitting empirical relations, indeed, are those between channel width and meander length and between channel width and radius of curvature (Fig. 7-41).

Despite the scatter of points the relations among the factors appear to hold through a very large range of stream size, from laboratory streams a foot wide to the Mississippi River, a mile wide. Various authors have expressed the relations for rivers in alluvial materials by regression equations, a few of which are presented in Table 7-11.

The exponents in the regression equations are all so close to unity that

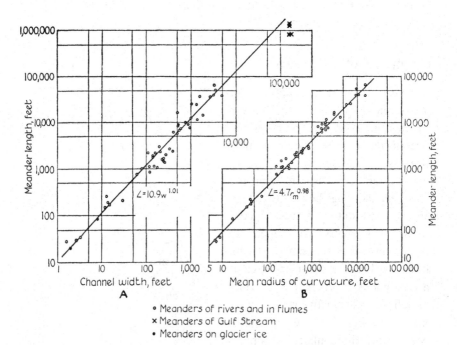

Figure 7-41.   *Relation of meander length to width* **(A)** *and to radius of curvature in channels* **(B).**

Table 7-11.   Empirical relations between size parameters for meanders in alluvial valleys.

Meander Length to Channel Width	Amplitude to Channel Width	Meander Length to Radius of Curvature	Source
$\lambda = 6.6w^{0.99}$	$A = 18.6w^{0.99}$	—	Inglis (1949, pt. 1, p. 144 Ferguson data)
—	$A = 10.9w^{1.04}$	—	Inglis (1949, pt. 1, p. 149, Bates data)
$\lambda = 10.9w^{1.01}$	$A = 2.7w^{1.1}$	$\lambda = 4.7r_m^{0.98}$	Leopold and Wolman (1960)

the relations between meander length (wavelength), amplitude, radius of curvature, and channel width may be considered linear. Among different groups of data the values obtained for the coefficient relating meander length to channel width are of the same magnitude; meander length ranges from 7 to 10 times channel width. Measured along the channel itself, distances between analogous points of the wave are larger, varying from 11 to 16 times the channel width. It was pointed out earlier that the spacing of successive riffles in a straight channel averages 5 to 7 times the channel width. Successive points of inflection or crossovers of a meander are thus spaced also at about 5 to 7 channel widths measured along the channel; since these inflection points are relatively shallow, they are quite comparable to riffles in a straight reach. On the basis of this similarity of bed profile and the spacing of shallow riffles we conclude that the processes which may lead to meanders are operative in straight channels.

In contrast to meander length, amplitude correlates only poorly with meander length. Tentatively, it may be inferred that amplitude of meander loops is determined more by erosion characteristics of stream banks and by other local factors than by any hydrodynamic principle. In uniform material, amplitude does not progressively increase nor do meander loops cut off during downstream migration of bends (Friedkin, 1945, p. 15).

Study of a sample of 47 alluvial streams in the Great Plains area showed that a higher sinuosity is associated with small width relative to depth, and with greater cohesiveness in channel boundaries; cohesiveness in this study was measured by the percentage of silt and clay in the soil material (Schumm, 1960). Laboratory studies (Friedkin, 1945, p. 14) also show that substitution of inhomogeneous for homogeneous sediments in the bank materials of a meander can alone reduce the sinuosity from 1.48 to 1.22, all other factors remaining constant.

This effect of inhomogeneities in sediment was well documented by Fisk (1952), who showed that local alignment and configuration of the channel of the Mississippi River is governed to a considerable extent by local variations in bank materials. Abandoned meander loops or oxbow lakes become filled over a period of time with fine-grained "backswamp" materials, which form clay plugs highly resistant to erosion by direct shear but susceptible to slumping in large blocks when undermined. Because these clay plugs are of limited depth and overlie previously deposited sandy alluvium, the substratum of friable materials may be scoured with relative ease, causing the bank of fine-grained material to collapse and slump.

In a sample of 50 rivers differing in size as well as in physiographic province a large proportion of individual bends were found to have a value for the ratio radius of curvature to width in the range 2 to 3, with a median value of 2.7. Two-thirds of the values were in the range 1.5 to 4.3. It is this constancy that makes planimetric maps of river bends look alike regardless of scale. When a map of a reach of the Mississippi River is laid next to one for a small creek, each to a scale that makes the meander lengths equal on the printed page, one cannot tell by inspection which is a map of a large river and which of a small river.

Flow in Meanders

Regarding the tendency toward a nearly constant value of the radius-width ratio, some studies suggest that control is exercised through the relation of geometry of the bends to flow resistance. Studies of pipe bends have shown that when the ratio of curvature to width has a value near 2, a minimum resistance due to curvature is found. Similar results have been found in open channels in some laboratory experiments.

Bagnold (1960) suggested that at this value there is incipient breakaway of water filaments from the convex boundary of the bend. As radius of bend curvature is reduced in a channel of uniform width, the velocity distribution becomes progressively more asymmetric and the main flow tends to recede from the inside region of great curvature toward the outer concave boundary. Consequently, in the region near the inner convex boundary, shear rate is reduced. Just prior to the breakdown of flow into large eddies—that is, just prior to breakaway from the convex wall—there is a reduction in boundary resistance near the convex wall and a restriction of the flow to a portion of the channel width. In this effective width the effective ratio of radius to width is somewhat greater than that of the channel itself. Bagnold shows that this leads to a minimum flow resistance at a radius-to-width value corresponding to incipient breakaway; this value is approximately 2. Additional field data are required to confirm the significance of the unique value of the ratio of curvature to channel width, and both theoretical and laboratory studies are needed to explain fully the mechanics of the system.

Measurements in meandering streams and in curved flumes allow the construction of a generalized picture of the flow pattern in a meander, presented in Fig. 7-42. This isometric view of the two principal components of velocity at various positions in the bend shows the main features. Superelevation of the water surface near the concave bank does not show at the scale of the diagram but is implied by the velocity distribution. The amount of superelevation is proportional to

$$\frac{v^2 w}{g r_m},$$

where v is the mean velocity, w is channel width, r_m the mean radius of curvature, and g the acceleration of gravity (Leliavsky, 1955, p. 124).

At the crossover or point of inflection of the bend (section 5, Fig. 7-42) the cross-sectional shape is not completely symmetrical but is slightly deeper near the bank which was concave in the bend immediately upstream. Downstream from the crossover the section (indicated at 2) becomes approximately symmetrical but some cross-channel component of flow is present because this section is in the bend.

The velocity in a meander crossover is not symmetrically distributed. In general, the high velocity that hugged the concave bank in the bend upstream is still present to some degree when the curvature becomes zero at the crossover, as is suggested by the velocity vectors in sections 1 and 5.

The highest velocity at any point tends to be located near the concave bank just downstream from the axis of the bend. Individual filaments of water accelerate and decelerate along a streamline, and thus the maximum velocity is not always associated with the same parcels of water.

Measured distributions of velocity also show that the maximum point velocity in a bend occurs not at but below the water surface. These observations agree with measurements made in a curved flume of rectangular cross section.

The slight lack of congruence of streamline curvature with bank curvature leads to the tendency for the locus of point-bar deposition to occur downstream from the axis of the bend. Similarly, Parsons (1960) found

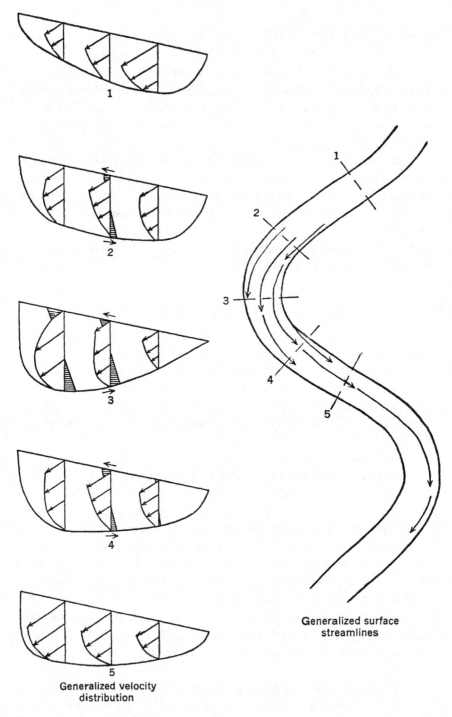

1

2

3

4

5

Generalized surface
streamlines

Generalized velocity
distribution

Figure 7-42.   *Isometric view of generalized diagram of flow distribution
in a meander.*

that the place where engineering revetments most often fail and where bank erosion is most frequent is just downstream from the axis of the bend. With both bank-caving and point-bar deposition concentrated slightly downstream from the bend axis, river curves tend to move downvalley.

Material slumping into the bed by bank caving is caught in the transverse component and carried toward the middle of the channel near the bed. If the location from which it was derived is far enough downstream in the bend, such material is not carried across the bed to the other side of the channel but moves into the crossover without having crossed the channel. Once into the reversed curve, it is drawn toward the same bank from which it started. This seems to be the explanation of the observation of Friedkin (1945, Plate 6) that sand eroded from a concave bank in the laboratory river did not cross the channel but was deposited on a point bar downstream and on the same side of the channel.

The vigorous crosscurrents near the bed in a bend can transport considerable quantities of bed material toward the convex bank. This must be at least in part the mechanism of point-bar building. Measurements of suspended load in movement in a crossover show a nearly uniform concentration of material across the section, but concentration increases markedly near the convex bank in a curved reach (Eakin, 1935, p. 471; Leopold and Wolman, 1960, p. 781).

Initiation and Development of Meanders

The vagaries of nature provide endless opportunities for perturbations in the flow—local bank erosion, chance emplacement of a boulder, fallen trees—any one of which would alter the path of a straight channel. Thus one need hardly inquire why a channel is not straight. But a random succession of chance perturbations might be expected to result in random bends of a variety of shapes. Though this would describe many channels, the existence of beautifully symmetrical meander bends and the remarkable similarity of bends in rivers of different sizes and in different physiographic settings still need explanation. There is not yet any theory or dynamic principle that explains quantitatively the characteristic geometric relations common to meandering channels. In the absence of such a general principle, however, attempts have been made to explain at least qualitatively how symmetrical successive bends develop and grow and how the size of bends is related to the size of the river.

Reviewing the experience of European engineers, Leliavsky (1955, see especially pp. 111–141) concluded that it is the consensus of these workers that helical flow is the dominant factor. Prus-Chacinski (1954) also credited helical flow as the basic mechanism. He showed that by the introduction of an artificial secondary circulation at the entry of a given bend it is possible to produce various kinds of secondary circulations in the next successive bend, which, in turn, affects the circulation in the next bend, and so forth.

It is clear that helical flow plays an important role in the process of erosion and deposition, but it has not yet been shown how helical flow results in the observed relations of width, curvature, and meander length.

Although denying that helical flow exists in wide rivers, Matthes (1941, p. 633) postulated a requirement closely allied to it. He indicated that bank cutting and the orderly transfer of sediment to its place of deposition on point bars were principal requirements for meandering. Friedkin's (1945, p. 4) concept is essentially identical: "The only requirement for meandering is bank erosion. . . . When all outside disturbing influences were eliminated [in the laboratory] . . . , the development of a series of uniform bends from an initiating bend was positive and capable of duplication."

These statements are probably true but they do not appear to tell the whole story; nor are they necessarily clues to a still more basic principle. A meander could not exist, of course, if the banks were unerodible or if they were completely unstable. The meander pattern of meltwater channels on the surface of glaciers have nearly identical geometry to the meander bends in rivers, yet these meandering channels on ice are formed without any sediment load or point-bar construction by sediment deposition. A fact of unknown significance, but interesting in itself, is that the geometry in plan view of meanders in the Gulf Stream is also similar to that of rivers (see Fig. 7-41), another example of meanders without debris load and, in this instance, without confining banks.

Several workers have attempted to express the oscillatory motion of a meander in terms of a gravity wave. An example of such a wave is a seiche in a lake, the period of which is much too short, however, to account for the observed wavelength of river meanders. To overcome this difficulty, Hjulstrom (1949, p. 84) expressed the gravity wave in terms of a seiche having a fetch equal to the width of the meander belt and a water depth equal to that of the river. Wave celerity he considered to be the mean downvalley velocity of the river. Exner's (1919) derivation was based on the identical principle, but he found the calculated velocities to be higher than those observed in rivers. Presumably to overcome this difficulty, Hjulstrom made the wavelength also dependent on a coefficient of turbulent friction. This approach has the difficulty that it makes meander length dependent on wave amplitude. Measurements do not seem to confirm this relation.

The forces determined by velocity distribution, including helical circulation, can account for the shape of the cross section in a meander, the depositional and erosional pattern, and the downvalley migration, but they do not seem to be sufficient to explain the characteristic dimensions and proportions of meandering channels. The existence of meanders on glacier ice also implies that erosion and deposition are collateral rather than governing principles of meander development and movement.

Consideration of various rates of energy dissipation in shallow and deep sections suggests that a change from meandering to nonmeandering pattern has a high degree of improbability. This implies that a closer approximation to equilibrium is achieved by meandering than by nonmeandering pattern, but it does not provide a basic cause for the observed geometry of the meander. Through a long reach consisting of many alternating pools and riffles one can draw a line representing the average energy gradient. This average grade line is shown as the dashed line in Fig. 7-43. Relative to this line, energy is lost rapidly over the riffle, where the actual hydraulic grade line falls below the average. This is the hatched zone *D*, below the average grade line. In the pool, however, energy is being dissipated per unit length at a less rapid rate than the average. This zone in the lower end of the pool and in the upper portion of the riffle is *A* in the figure.

It has been shown that the most probable distribution of energy and that represented by the steady

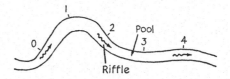
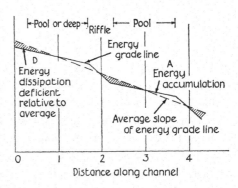

Figure 7-43.

Diagram of the plan view of a hypothetical reach of channel showing location of pools and riffles, and longitudinal profile of total energy (energy grade line) for same reach.

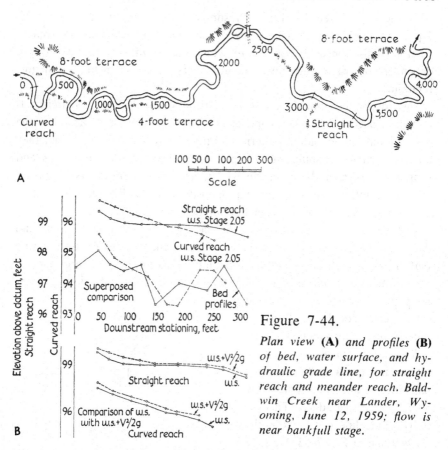

Figure 7-44.

*Plan view **(A)** and profiles **(B)** of bed, water surface, and hydraulic grade line, for straight reach and meander reach. Baldwin Creek near Lander, Wyoming, June 12, 1959; flow is near bankfull stage.*

state in an open system is one in which the rate of energy "loss" in each unit of distance along the stream is equal. As the change in energy represents a conversion of energy within a given reach of river, a steady-state condition would be one in which the energy grade line should be uniform.

A plot of data from a field case is presented in Fig. 7-44. The graph includes the energy grade line and actual water surface elevation and bed elevation for two nearby reaches of the same stream, Baldwin Creek near Lander, Wyoming; the two reaches are a meander bend and a straight reach respectively. Data were taken at high flow, near bankfull stage. Discharge, bed-material size, and other factors were the same in the two reaches. The energy grade line is the water surface profile adjusted for differences in velocity distribution in each cross section; the kinetic energy head is $v^2/2g$ multiplied by a correction factor.

Note that the bed profiles of the two reaches are quite comparable. The energy gradient is somewhat steeper in the meander reach than in the straight reach. Other conditions being equal, a meandering stream is steeper than a nonmeandering one because of the necessity of overcoming the additional friction loss due to channel curvature. The main difference between the energy profiles of the straight and meandering channels is that in the straight reach even at bankfull stage the alternation of steep water surface slope over the riffle and flat gradient over the pool is not eliminated, but in the meander reach the slope of the energy grade line is uniform. Thus it appears that the meandering reach more closely approximates the most probable condition of uniform energy loss in each unit of the channel length.

For the Popo Agie River near Hudson, Wyoming, a larger river, similar results were obtained. The change in regularity with stage may be shown more clearly by computing the mean deviations of the actual profiles from a mean straight line. For example, the deviations from a straight-line profile of bed elevations and energy grade line at bankfull show the following for Popo Agie River:

> Meander reach
>> Bed   0.76 feet
>> Bankfull energy grade line   0.16 feet
>> Ratio, bankfull/bed   0.21
>
> Straight reach
>> Bed   0.67 feet
>> Bankfull energy grade line   0.25 feet
>> Ratio, bankfull/bed   0.37

Thus, although the bed elevations of the meander reach are less regular than those of the straight reach, the reverse is true of the bankfull energy profile; the profile of the meander reach is more regular than the profile of the straight reach.

Similar relations exist among the mean deviations of the profiles of Baldwin Creek:

> Meander reach
>> Bed   0.43 feet
>> Bankfull energy grade line   0.04 feet
>> Ratio, bankfull/bed   0.09
>
> Straight reach
>> Bed   0.24 feet
>> Bankfull energy grade line   0.12 feet
>> Ratio, bankfull/bed   0.50

More data are needed to confirm this observation. Also the contrast between straight and meandering reaches may be more appropriately indicated by differences in relative variability between bed and energy profile rather than by absolute differences. The difference between a curved reach of the meander and a comparable straight reach appears to reflect the added flow resistance produced by the curve at just that portion of the stream where, in the absence of curvature, energy is being expended at less than the average rate, which is in the pool. The pool in the meandering stream

Figure 7-45.

Valley meanders. Left: *River Leach, Gloucestershire, near Aldsworth, looking southeast, a misfit stream on the backslope of the Cotswold Hills, England. The present channel is of insignificant size.* Right: *River Evenlode, SW of Combe, Oxfordshire, England. Valley meander bend shows clearly with steep concave bank and gently sloping convex bank. Present stream meanders within valley meander, manifestly underfit.* [*Photograph from Cambridge University Collection.*]

coincides approximately with the zone of maximum curvature. The effect of the meander, then, is to introduce flow resistance due to curvature in just such a way that the stepped profile of the straight reach is replaced by a uniform utilization of energy throughout the whole length of the meander reach. In this way the meandering pattern approaches more closely the condition of equilibrium as defined by the entropy concept than does the nonmeandering one. It would then be expected that a meandering reach, in all probability, would not change into a nonmeandering one, since the former is a closer approximation to equilibrium than is the latter. This explanation applies only to channels characterized by pools and riffles. It appears to provide an explanation of why channels are commonly sinuous and long straight reaches are rare.

One aspect of the emphasis upon energy distribution as applied to river channels is that the channel tends to adjust to a condition in which the rate of work expended in the system is minimum. The river channel has the possibility of internal adjustment among hydraulic variables. Given a tendency to form pools and riffles, the most probable channel pattern would appear to be a sinuous one. If a channel is sinuous, then, all bends tend to have the same ratio of radius of curvature to channel width, and this value of the ratio provides a minimum flow resistance. Both are compatible with and support a concept of the tendency toward equilibrium, which is the condition of steady state in an open system.

Meandering Valleys

Even if the explanations of the meandering patterns of alluvial streams continue to be incomplete, it is relatively easy to imagine that there might be mechanisms in streams effective enough to cause meanders in alluvium. Meanders in bedrock, however, are also common, and these too possess regular geometric proportions. One often sees a valley which winds between adjacent hills in a course so symmetrically sinuous that, quite apart from the pattern of the river flowing within the valley, it may be called a meandering valley. Figure 7-45, **A** and **B**, shows meandering valleys in which the bends are amazingly similar, and clearly the valley meanders were not cut by the present stream.

Where the bounding hills are bedrock, then the valley must have been developed in its sinuous pattern by river processes effective even in rock. Figure 7-46 and Table 7-12, for example, show the relation of meander length to valley width in valley meanders in rock compared to those in alluvium. There is more scatter

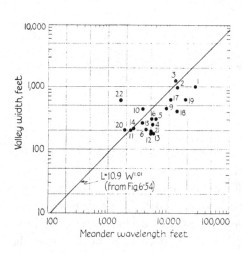

Figure 7-46.

Relation of meander length of valley meanders to valley width. Data from Table 7-11.

Table 7-12. Wavelength and valley width in rockbound channels of the Susquehanna River (North and West Branches and Tributaries) and other streams.

River and Reach	Valley Width at Channel Level (ft.)	Wavelength (ft.)
1. West Branch, Susquehanna River, near Williamsport, Pennsylvania	1,000	27,000
2. West Branch, Susquehanna River, Avis to Great Island, Pennsylvania	984	14,000
3. West Branch, Susquehanna River, between Glen Union and Hyner	1,260	13,000
4. Kettle Creek, 5 miles from Butler Hollow	250	5,400
5. West Branch, Susquehanna River, between Little Bougher and Karthams	300	6,000
6. Mushannon Creek, Peale to West Branch, Susquehanna River	200	4,200
7. Mushannon Creek, Peale to Bench Mark 955	200	4,000
8. Mushannon Creek, Bench Mark 955 to West Branch, Susquehanna River	200	5,000
9. West Branch, Susquehanna River, Shawville to Frenchville Station	450	9,000
10. West Branch, Susquehanna River, Clearfield Creek to Lick Run	450	3,700
11. Clearfield Creek, 1 mile below Lost Run	200	2,400
12. West Branch, Susquehanna River, Curwensville, Pennsylvania	178	5,000
13. West Branch, Susquehanna River, above Good, Pennsylvania	178	5,600
14. Pembina River, Alberta, Canada	210	2,580
15. Conodoguinet Creek, near Harrisburg, Pennsylvania	270	3,600
16. Raystown Branch, above junction with Susquehanna	300	5,100
17. Juniata River, at Yuling School Bench Mark 590	615	10,600
18. Juniata River, above Lewistown, Pennsylvania	400	14,000
19. North Branch, Susquehanna River, north of Scranton, Pennsylvania	615	19,000
20. Gunpowder River, below Pretty Boy Dam, Maryland	200	1,900
21. North Fork, Shenandoah River, Virginia	190	5,000 (Hack) 1,700
22. South Fork, Shenandoah River, above Front Royal, Virginia	700	1,700

in the relationship, but it is apparent that the length is directly proportional to the channel width in meanders in rock as well as those in alluvium. In these rock meanders wavelength is 15 to 20 times valley width, a relation close to the value given by Inglis (1949, p. 144) for what he termed "incised" rivers.

Detailed descriptions of meandering valleys have been made by Dury (1953). He has studied the phenomenon of the meandering valley with a view to ascertaining whether it was cut by a single large stream that filled it from one valley side to the other. In many of these sinuous valleys, the present river meanders on a flood plain within the valley, executing bends unrelated to the locus of the valley bends and proportional in size to the present river width. These exemplify the classic "misfit" stream, apparently too small for the valley within which it flows. Dury made borings and detailed stratigraphic sections in flood-plain materials of meandering valleys, particularly in southern England. His studies provide many instances showing that the cross-sectional form of a meandering valley tends to be asymmetric at the curves of the valley, with a deeper cut into underlying bedrock near the concave valley side than on the convex bank. This is similar to the cross section of a meandering river channel, as shown diagrammatically in Fig. 7-42. An example of the bedrock outline of a cross section across a meandering valley is shown in Fig. 7-47. Note that the present stream (shown in solid black in the upper portion of the plan view) wanders back and forth within the confines of the sinuous valley. The valley cross sections resemble cross sections of a meandering river.

From such valley cross sections it is not possible to ascertain to what depth the river flowed—if indeed a single river once filled the valley from one side to the other. Dury approached the problem, however, by exploiting the tendency for a relationship between drainage area—which has a functional relation to bankfull discharge—and meander length. This relation for a variety of rivers and for meandering valleys, using in the latter case the meander length of the valley, is shown in Fig. 7-48. It can be seen that these data on valleys (set of data in upper part of Fig. 7-48) define as good a relation as do data on rivers (lower group of data). For a given drainage area, meander length in valleys is about 10 times as large as for rivers flowing in these valleys.

From many examples Dury argues that the meandering valleys generally were cut by rivers larger than the present streams flowing in them, and these larger discharges in past times were associated with Pleistocene climate. The relation of meander length and discharge requires that to

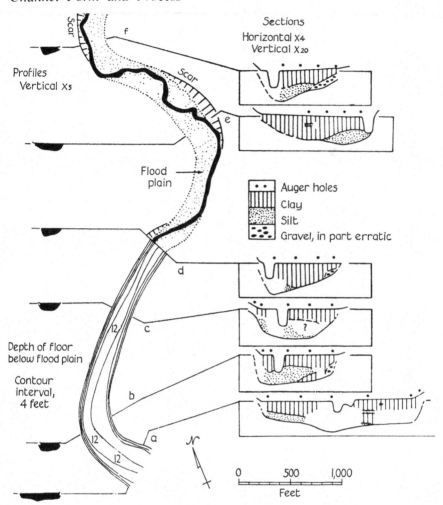

Figure 7-47. *Plan and cross sections of valley of the River Itchen, near Southam, Warwickshire, England. [From Dury, 1953.]*

produce a meander length larger by a factor of 10 requires a bankfull discharge larger than the present by a factor of 80 to 100. that is $(10)^{1.8}$ where b = 0.5. This certainly need not mean that average precipitation was larger than present by that ratio, for in a different climatic regimen a variety of factors would be different from the present, including probably vegetal association, infiltration rates, temperature, rainfall intensity, and others.

Dury's work presents an imposing argument that channel-forming discharges in many locations in Europe, England, and the United States were

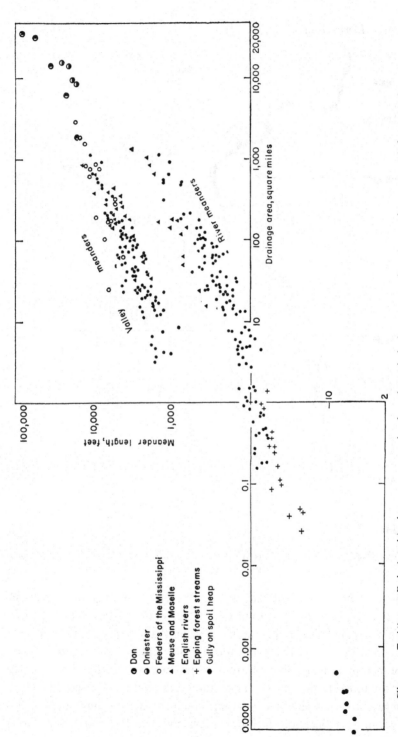

Figure 7-48.   *Relationship between meander length and drainage area for rivers (lower set of data) and mean-dering valleys (upper set of data). [From Dury, 1960.]*

far larger during the Pleistocene than at present. A problem remains, however, of reconstructing a picture of rainfall-runoff relations that would produce this large change in discharge with acceptable climatologic assumptions. In addition, some recent work by C. W. Carlston (personal communication) suggests that the meandering valleys may have a hydraulic geometry somewhat different from alluvial rivers; that is, the downstream exponents differ from those of rivers, as is also true of the canals and flumes shown in Table 7-7. Detailed study of individual bends of valley meanders also suggests that differences in geologic structure and lithology lead to differences in wavelength of meanders in rock. With accumulation of additional measurements the apparent distinctive grouping of valley versus river meanders shown in Fig. 7-48 may disappear in favor of a family of curves. If the two groupings remain distinctive it may well be that the different wavelengths of channels in rock, as compared with those in alluvium, result in part from differences in control or independence of parameters in the hydraulic geometry.

Understanding the processes by which a river cuts symmetrical bends in hard rock still remains a major problem in research. Rivers such as the North Fork of the Shenandoah in Virginia and the North Branch of the Susquehanna between Tunkhannock and Scranton, Pennsylvania, or the San Juan in Utah (Fig. 7-49) traverse bedrock, but in most reaches the

Figure 7-49.

Aerial photograph of valley meanders in bedrock cut by the San Juan River, Arizona. [Courtesy University of Illinois Committee on Aerial Photography.]

A

B

Figure 7-50.

Meanders related to glacial ice. **A.** *Meander of glacial meltwater channel cut in ice.* **B.** *Meander in rock superimposed from meander on ice.*

present rivers occupy the entire cross sections of the valleys. In many rivers meander length is proportional to channel width, but the ratio is not readily explained by changes in discharge, inasmuch as many of the meandering rivers in bedrock have probably had quite different climatic histories. Variations in bank material alter sinuosity, and similar variations are to be expected in alluvium and different types of rock. Hack (in preparation) has observed that the meandering pattern of a single river in Michigan undergoes major changes as the river traverses different kinds of rock. He suggests that the pattern is associated with variations in channel width and bank material. When a channel confined by bedrock occupies the entire valley width, bar building on the convex side of bends is inhibited. In soft rock and alluvium a smaller, narrower channel is constructed by deposition on the convex sides of the bends, and the meander pattern has a shorter wavelength and smaller radius of curvature.

It might also be supposed, assuming that most rivers in bedrock have less frequent cutoffs, that the meander length of valley meanders in bedrock could be longer. This is probably true, but the Pembina River in Alberta, Canada, has cut off in the recent past (Crickmay, 1960, p. 377), and its wavelength, though somewhat less than that of many of the other meanders in bedrock, is still closer to the average for bedrock than that for alluvial channels.

Few studies have been made of flow and geometry of meanders in bedrock. Because of the difficulty of visualizing how a channel could maintain a regular sinuous pattern while cutting across hard-rock strata, it has often been assumed that the meandering pattern was initiated in an overlying sedimentary cover and superimposed on the tougher rock below as the river intrenched itself into the strata below. Very often these presumed overlying strata have been eroded away, and hence the hypothesis is difficult to verify. In most cases there appears to be no need for such a two-cycle hypothesis. Comparative photographs in Fig. 7-50 suggest, however, that the mechanism may be valid and applicable in some cases. Figure 7-50, **A**, shows a meander on the steep ice front of the Dinwoody Glacier; **B** shows a meander pattern cut in rock once overlain by a glacier. The remarkable similarity between the two in form, scour marks, and slip-off slope suggests that the meander pattern in the rock may have been superimposed from a meander in the overlying ice.

Other meanders in bedrock—such as those on the Green River where it passes athwart the Uinta Mountains in Utah (Powell, 1875, p. 153)— suggest that the river was antecedent to the uplift. That is, the river appears

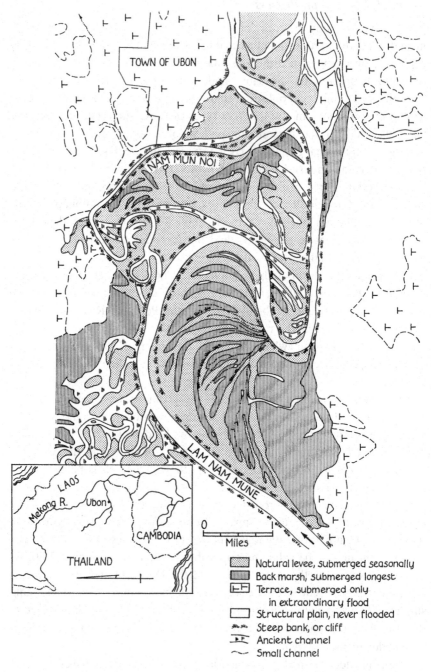

Figure 7-51.   *Flood plain features of lower Nam Mune, a tributary of the Mekong, Thailand. [After Ohya.]*

to have maintained its course, trenching the structure as the latter formed. In the absence of stratigraphic evidence it is impossible to distinguish between an antecedent river and one which was superimposed from an overlying cover. Evidence of antecedence has been presented in a number of studies, but the mechanism by which an elegant meandering pattern is maintained as a river cuts through thousands of feet of diverse lithologies in various structural attitudes remains unclear.

The River Flood Plain: Introduction

The flood plain is a strip of relatively smooth land bordering a stream and overflowed at time of high water. Major features of river flood plains have frequently been described in the literature. There is general agreement on the designation of these features, although the origin of some of them remains in doubt. A typical flood plain will include the following features, which can be seen in Figs. 7-51 and 7-52.

1. The river channel.
2. Oxbows or oxbow lakes, representing the cutoff portion of meander bends.
3. Point bars, loci of deposition on the convex side of river curves.
4. Meander scrolls, depressions and rises on the convex side of bends formed as the channel migrated laterally downvalley and toward the concave bank.
5. Sloughs, areas of dead water, formed both in meander-scroll depressions and along the valley walls as flood flows move directly downvalley, scouring adjacent to the valley walls.
6. Natural levees, raised berms or crests above the flood plain surface adjacent to the channel, usually containing coarser materials deposited as flood flows over the top of the channel banks. These are most frequently found at the concave banks. Where most of the load in transit is fine-grained, natural levees may be absent or nearly imperceptible.
7. Backswamp deposits, overbank deposits of finer sediments deposited in slack water ponded between the natural levees and the valley wall or terrace riser.
8. Sand splays, deposits of flood debris usually of coarser sand particles in the form of splays or scattered debris.

On large rivers, such as the Mississippi, the Sacramento, or the Mekong (Fig. 7-51), the majority of these features are distinctively displayed in

Figure 7-52.

Meander bend of Buffalo Creek, near Gardenville, New York. In foreground
house stands on a terrace but floodplain borders both sides of stream in left and
right middle distance. In the center the vegetation shows concentric bands as-
sociated with flood-plain scrolls, and plants are stratified by age, decreasing in
size and age toward the river. The band of unvegetated gravel and sand at the
convex bank is typical of many meanders. [Photograph by G. S. Smith, U. S
Soil Conservation Service.]

what might be called the classic model. On smaller streams, however, many
of them may be hard to distinguish as the flood-plain deposits are subject
to rapid removal and alteration.

Valley flats which would usually be considered "flood plains" may include those formed by different processes, such as landslides, low-angle
fans, and perhaps others. Hack and Goodlett (1960, p. 347) studied examples in the Appalachian Mountains of Virginia that were caused by
torrential movement of regolith from steep mountain slopes during intense
storms. The valley deposits caused by this process would fall into any
usual definition of flood plain. The features occurring on valley flats formed

by different processes may not be apparent and, indeed, detailed work may be necessary to determine the origin of a given feature. This is likely to be particularly true where the downvalley gradient is large and flash floods of great force are particularly effective. The flood plain may be highly uneven and dissected by irregular scour channels randomly located. Braided channels frequently display a wide variety of island forms. Some islands in the Amazon are encircled by natural levees forming closed depressions or basins (Sternberg, 1960). More often the islands are transient features and the flood plain is traversed by numerous abandoned channels.

The distribution of vegetation on many flood plains reflects the distinctive depositional and erosional environments (Drury, 1956). These differences can often be clearly seen on aerial photographs.

Floods and the Flood Plain

Looking at a river channel with well-defined banks, particularly on a rising stage when water just begins to overflow the banks, one is tempted to attach some special significance to the bankfull stage. The channel form itself, however, provides no obvious clue as to the possible significance. Hydrologic studies suggest at least a reasonable working hypothesis.

In a study of the mechanics of flood-plain formation, we analyzed the available data on the frequency of bankfull flow. There is a remarkable similarity in the frequency of bankfull stage on a variety of rivers in diverse physiographic settings and differing greatly in size. The recurrence interval of the bankfull stage appears to be in the range of 1 to 2 years, although some localities studied diverge greatly from this value. At stations where the flood plain is clearly defined and its elevation accurately known, the recurrence interval is closer to 1 than to 2 years. In general, a value of 1.5 years seems a good average. This means that the discharge in a river will equal or exceed bankfull 2 out of 3 years on the average.

An independent study of rivers in Britain was made by Nixon (1959) without knowledge of the American work. He concluded that the bankfull stage represents on the average the 0.6% point on the duration curve; hence the frequency of bankfull stage is .006 × 365, or 2.2 times per year, a recurrence interval of about $\frac{1}{2}$ year.

The authors have initiated a program of observation, enlisting the help of engineers of the U. S. Geological Survey to observe the discharge when flood flows have actually begun to move out onto the flood plain. By determining the corresponding discharge and stage at a nearby gaging

Table 7-13. Observed bankfull flows, measured discharge, and computed recurrence interval.

River and Location	Drainage Area (sq. mi.)	Mean Annual Discharge (cfs)	Maximum Flow of Record		Bankfull Flow					
			Discharge (cfs)	Gage Height (ft.)	Discharge (cfs)	Gage Height (ft.)	Recurrence Interval (yrs.)	Date of Observation	Observer	Remarks
Wabash River near New Corydon, Indiana	258	202	4,690	17.59	1,240	14.14	1.1	6-21-58	Swing	Slight flow regulation and diversion
Eel River at North Manchester, Indiana	416	372	7,500	14.00	2,160	7.23	1.2	2-17-59	Swing	
Wabash River at Delphi, Indiana	4,032	3,490	145,000	28.4	9,550	8.92	1.07	1-18-60	Lipscomb	
Wildcat Creek at Owasco, Indiana	390	366	10,200	13.3	3,390	7.58	1.7	4-25-61	Swing	
Wabash River at Covington, Indiana	8,208	6,979	200,000	35.1	20,900	15.84	1.1	4-4-61	Carrico	
Wabash River at Riverton, Indiana	13,100	10,880	250,000	26.4	31,900	16.02	1.24	4-26-59	Swing	
Fall Creek at Millersville, Indiana	313	264	22,000	16.3	3,400	8.73	1.48	4-23-61	Swing	Flow regulated
Bean Blossom Creek at Bean Blossom, Indiana	14.6	16.3	1,720	11.42	1,090	8.40	1.9	2-21-59	Tate	
White River at Newberry, Indiana	4,696	4,475	130,000	27.5	14,700	12.77	1.1	5-9-58	Swing	Slight flow regulation
Youngs Creek near Edinburg, Indiana	109	104	10,700	13.4	1,270	7.08	1.1	3-6-61	Swing	Right bank leveed, but broken
Driftwood Creek near Edinburg, Indiana	1,054	1,142	34,500	16.55	10,100	13.21	1.3	3-7-61	Swing	Right bank leveed at gage
Trempealeau River at Arcadia, Wisconsin	552	1,320	26,000	12.07	2,230	5.21	1.4	3-27-61	Kaupanger	
Wapsipinicon River near Dewitt, Iowa	2,330		18,900	22.34	5,080	9.36	1.15	3-15-61	Plantz	
Wahoo Creek at Ithaca, Nebraska	272	53.4			±2,500	±20.00	±1.0	8-6,7-58	Brice	
Silver Creek at Ithaca, Nebraska	72	9.14	2,450	12.22	450	9.5	2.0	8-6,7-58	Brice	
Nemaha River at Falls City, Nebraska	1,340	659	51,400	27.44	32,000	26.2	4.0	7-10,11,12-58	Brice	Dredged and straightened; probable reason for high recurrence interval
Powder River near Baker, Oregon	219	111	1,820	7.05	946	5.26	3.0	5-26-58	Oster	
Watts Branch near Rockville, Maryland	4.3				430	4.05	2.2	5-1-58	Leopold	
Bogue Chitto near Tylertown, Mississippi	502	875	45,700	33.50	7,000	16.0	1.2	4-12-55	Trotter	
Clear Creek at Bovina, Mississippi		36			2,770	21.73	1.25	5-1-58	Wilson	

station the recurrence interval of the incipient flood stage could be accurately fixed. Of 19 such observations collected to date the recurrence interval of the flow considered in the field to represent bankfull varies from 1.07 to 4.0 years, and 15 of the 19 lie in the range 1.07 to 1.9 years (Table 7-13).

At time of low flow the determination of bankfull stage or flood stage is not in itself a simple matter. As everyone who has tried to make this determination has discovered, the difficulty lies not so much in finding the recurrence interval of a particular discharge but in deciding in the field exactly what stage to call bankfull. Slight differences in elevation make large differences in discharge and thus in recurrence interval. Averaging a large number of channel cross sections has proved both laborious and inefficient in resolving variations in size and shape. A method that seems to give the most reproducible results consists of two parallel surveys, one a longitudinal profile of the bed along the center of the channel and the other along the flood plain. An average line is drawn through the points on each profile. The vertical difference between these two average lines when applied to the average bed elevation in the vicinity of the gaging station gives an elevation on the gage datum which may be called the bankfull stage. This procedure, however, has as yet had limited application and needs further trials to test the reproducibility of the results.

Unfortunately the data usually available in gaging-station records are not adequate for determining bankfull stage. Since 1960 all new gaging stations established by the U. S. Geological Survey will include a determination of bankfull stage, which will be recorded in the station description.

Once the gage height (relative to the datum of a gaging station) of bankfull flow is determined, the rating curve of the station may be entered to read the bankfull discharge, which in turn may be used in a flood frequency curve to determine the recurrence intervals.

For various stages within the channel it is relatively simple to define average values for the relation between stage and frequency of discharge. The relationship between overbank stage, discharge, and frequency, however, has received less attention and is more complex. One tentative relation is shown in Fig. 7-10. This needs better definition by further study, but presently available data suggest that the flood plain is overflowed to a depth equal to $\frac{8}{10}$ of the mean height of channel banks with a recurrence interval of about 50 years. This relation is obviously dependent on the width of the flood plain, and it is presumably the fact that width of flood plain often bears some rough relation to width of channel (size of river) that an average relationship is derivable at all.

The properties of the channels and the drainage net of tidal marshes have been investigated only in a few cases. One analysis of a tidal channel system tributary to the Potomac River near Alexandria, Virginia (Myrick and Leopold, 1963) shows that the logarithmic relation between average stream length, numbers of streams, and stream order follows the pattern which Horton first demonstrated for terrestrial rivers. A sketch of the channel and drainage net of this estuarine system is presented in Fig. 7-29.

Bankfull stage in an estuary occurs when overflow of the tidal flat begins. This is a matter of some interest because the bankfull condition occurs on nearly every tidal cycle, or twice a day, but the bankfull condition varies in discharge and thus effectiveness for channel formation, depending on the range of stage and maximum stage reached. The bankfull stage in the Potomac tributary tended to coincide with the maximum discharge and maximum velocity during a tidal cycle with a particular range of tide and a high tide equal to the average value. The frequency with which this average high-tide level is reached coincident with a range of stage that produces maximum velocity at bankfull stage is much less than twice a day. This one example suggests that the frequency of that occurrence is of the same order as the bankfull stage in terrestrial rivers, about once a year, but here again the finding is in need of further test.

Flood Plain Formation

Two related processes are probably responsible for the formation of most of the flood plains of the great rivers of the world. These are deposition on the inside of river curves and deposition from overbank flows. As a river moves laterally, sediment is deposited within or below the level of the bankfull stage on the point bar, while at overflow stages the sediment is deposited on both the point bar and over the adjacent flood plain.

Mackin's description (1937, p. 826) of the origin of flood plain deposits of the Shoshone Valley, Wyoming, is, with only minor modifications, applicable to nearly all flood plains formed by lateral swinging of a stream. During times of low flow (Fig. 7-53, **A**) the slack-water parts of the channel are slowly filled and material (f) accumulates, particularly near the convex bank. During high-water periods (Fig. 7-53, **B**) the channel section is enlarged, both by scouring out loose bottom fill and by erosion of the concave banks of curves, although deposition takes place where flow sepa-

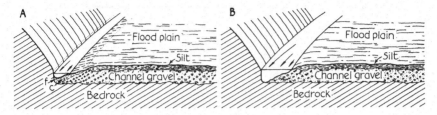

Figure 7-53. *Formation of flood plain by lateral swinging of the river, as exemplified by the Shoshone River, Wyoming. [From Mackin, 1937.] (**A**) Low-water stage; f, fine detritus laid down in the stream channel during periods of normal flow; c, coarse detritus laid down in the channel at the end of a period of high water. (**B**) High-water stage.*

rates from the bank even at high flow. As the flood waters subside, deposition occurs on point bars and channel floor.

The flood plain gravel typical of the Shoshone River and many other streams in which gravel is an important part of the load is "formed as a single sheet by lateral accretion; it is not really a floodwater [overbank] deposit at all, but a channel deposit; its maximum possible thickness, in a stream that is not aggrading, approaches the maximum depth of effective flood scour" (Mackin, 1937, p. 326).

Water in excess of channel capacity must flow overbank, inundating parts of all of the flood plain. Fine materials are thus deposited from suspension over the flooded area, and these deposits are labeled "silt" in Mackin's drawing.

Mackin concludes: "The flood plain is composed of two parts, distinctly different in origin. The gravel represents the coarsest material moved along the stream channel, and is characteristically unstratified. The overlying silts and clay are deposits from suspension . . ." (p. 826).

Because the flood plain deposits of many streams in subhumid areas are comprised mostly of silt with only a thin and irregular basal gravel, it is believed by some that the flood plain is comprised principally of overbank deposits in the same manner as the thin covering of silt observed by Mackin in the Shoshone Valley. Two lines of evidence indicate that in a stable or equilibrium channel most of these fine-grained flood plain silts are deposited exactly as is the gravel comprising the flood plain of the Shoshone River. The first is the nature of observed materials deposited by lateral accretion within the channel. The second is the relatively constant value of the recurrence interval of the bankfull stage.

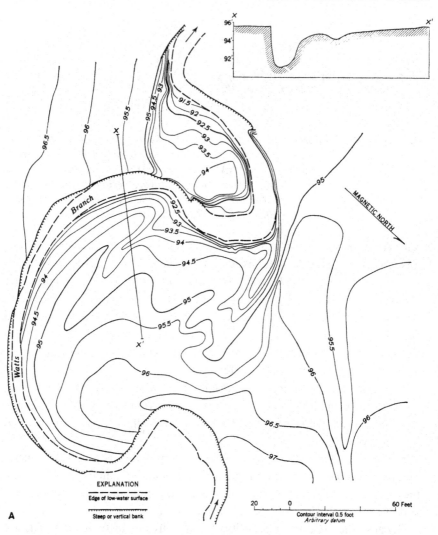

A

EXPLANATION

––– – ––– – ––– Edge of low-water surface

•••••••••••••••• Steep or vertical bank

Contour interval 0.5 foot
Arbitrary datum

The nature of the progressive deposition on a point bar is shown by profiles of cross sections we have measured annually over a period of years on several bends of Watts Branch near Rockville, Maryland. Sections show concomitant erosion on the concave bank and deposition on the point bar.

At the cross section pictured (Fig. 7-54), the left bank rises to a low terrace that can be traced in some detail over the valley. There are extensive areas near the present channel comparable in elevation with the level depicted on the right side of the drawing, and we believe this level is the flood plain related to the present stream regimen. The cross sections show

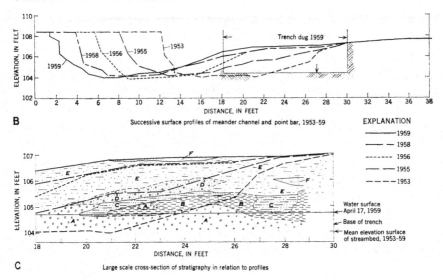

B Successive surface profiles of meander channel and point bar, 1953-59

C Large scale cross-section of stratigraphy in relation to profiles

Figure 7-54.

*Successive cross sections measured over a period of years at Watts Branch near Rockville, Maryland. **A.** Map of locality. **B.** Cross sections. **C.** Stratigraphy of point bar with cross sections superimposed so that successive positions of surface may be related to the deposited materials. Key to stratigraphic units: A. Gravel, mostly 3–8 mm with considerable 8–20 mm and a few 64 mm. B. Olive-gray clayey silt, with organic matter, small mica flakes. C. Orange-brown, mottled sandy silt with some clay, lenses of leaflike organic material, mica. D. Coarse sand, brown-stained, with pebbles up to 8 mm, some fine roots in situ, lenses of silt. E. Brown sandy silt gradually changing upward in places into fine sand. F. Fine sand with some silt.*

that during the time the channel moved laterally a distance equal to more than one channel width, the concurrent deposition on the point bar kept the channel width about constant. Various sections measured over the 8-year period showed that the net volume of deposition was about equal to erosion. At the point of maximum curvature, erosion slightly exceeded deposition; just downstream from this point deposition exceeds erosion.

The sections also show that no appreciable deposition occurred on the flood plain surface away from the sloping nose of the point bar; that is, overbank deposition was either absent or too small to measure.

Our observations of the material deposited in successive years showed that most of the fine-grained materials, which generally characterize the upper portion of the flood plain of Watts Branch, were deposited as point bars or within the channel, rather than by water spreading over the flood plain across the valley in what would be called overbank flow.

The stratigraphy of a trench dug in the flood plain of Watts Branch at one of the measured cross sections is plotted in Fig. 7-54, **C,** together with the successive positions of the surface during the eight years of observation. Contacts between materials of different textures are more or less horizontal, with some tendency in places to be parallel to past surface profiles. Gravel characteristic of a portion of the stream-bed material is observed in a position well above the channel bed, indicating that some of the coarse bed material is carried up the ramp presented by the sloping point bar and there deposited. But point-bar materials, as can be seen, are variable, generally grading to finer texture near the top of the section. The bulk of the point-bar deposit pictured is, indeed, the silt so typical of the upper part of flood plains throughout the area.

The constancy of the recurrence interval of the bankfull stage has already been noted. That the frequency of flooding is nearly the same in regions of very diverse runoff—from tropics to semiarid regions—implies that the size of the river channel is appropriate to the quantity of flow provided by the watershed. It is also apparent, however, that if overbank flooding by sediment-laden waters does occur, some deposition will in all likelihood be associated with it. If there were continuous deposition the channel would gradually appear to become depressed within its own alluvium. The regular frequency of flooding indicates that this is not the case, and hence some mechanisms must counteract this depositional tendency. Several mechanisms can be suggested.

First, progressive lateral migration of the river channel removes portions of the flood plain and hence limits the elevation of the surface. Second, where overbank flows are deficient in sediment, relatively little material is available for deposition. This may occur in many rivers where the sediment derived from the drainage basin is removed during the initial runoff period and the peak sediment load precedes the peak stage. Third, flow over the surface of the flood plain is often irregularly distributed, and velocities may be high enough to produce scour rather than deposition. Figure 7-55, for example, shows the irregular flow directions of flood waters and associated deposits in the Kano River basin in Japan (Aramaki and Takayama, 1960, p. 141). Here deposition increased downstream, then decreased in a reach in which the channel gradient flattened appreciably, while the particle size of the sediment decreased. A similar association of velocity, scour, thickness, and particle size of deposits in a single cross section of the Quinebaug

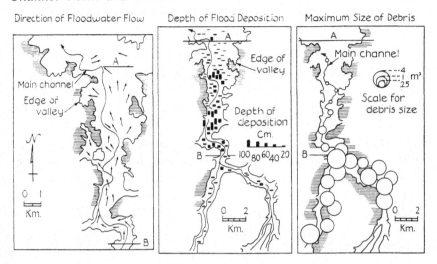

Figure 7-55. *Action of the 1958 flood in the Kano River, Japan, including flow direction. Thickness of sediment deposition and maximum size of deposited material is indicated. [After Aramaki and Takayama, 1960.]*

River, Connecticut, produced by a flood of August 1955 was given by Wolman and Eiler (1958).

The relative amount of sediment in the flood plain resulting from lateral and overbank deposits varies, depending upon the flood characteristics of the basin and the availability and size distribution of sediment. We have suggested that 60 to 80% of the sediment in many channels is deposited by lateral accretion. Lattman (1960), in a detailed study of a small stream in Pennsylvania, stated that an unmeasured but large percentage of the flood plain deposits were provided by lateral accretion and one-fifth of the total was made up of colluvium.

By definition, progressive deposition both within the channel and on the flood plain is aggradation. Under these conditions overbank deposition may comprise the most significant portion of the flood-plain deposits.

In the absence of aggradation or degradation the flood plain in essence provides storage for sediment in its movement from the surface of the drainage basin through the valleys to the mouth of the river. Sediment is stored for varying amounts of time, depending upon its size, location with the drainage system, and location within any given valley. Movement of sediment is discontinuous within the channel itself, and temporary storage within the flood plain lengthens the periods of dormancy.

The episodic character of discharge, of course, produces discontinuities in transport. To the extent, for example, that catastrophic events are localized, large quantities of debris may be contributed to the drainage system at selected localities. These sediments are not transported continuously through the system. Rather, as Tricart (1959, p. 298) points out, pebbles and boulders may be transported a few feet or perhaps a few miles during a single flood. The intermittent character of the flows at successive localities in the drainage system thus produces a discontinuous movement of sediment. To the extent that equilibrium is maintained in the system as a whole, it consists of fluctuations around a mean set of conditions over a period of time.

Under less flashy or more uniform regimens, as on Seneca Creek, Maryland, the time during which materials in a streambed are in transport constitutes only 0.3% of the time these materials are available for transport. A rough computation based on the age of trees on the flood plain of Seneca Creek suggests that the river swings laterally across its flood plain no faster than half its width in 80 years. If a sediment particle were trapped in flood-plain deposits, on the average it would not again be exposed to transport until the channel swung past that position again. The flood plain of Seneca Creek is about 7 times as wide as the channel, so a trapped particle may not be again exposed to transport for more than a thousand years. This is only one example of course, and the estimates are crude.

Rough as this estimate may be, it gives an idea of the order of magnitude of the time for which particles may be in storage—that is, temporarily deposited within the river valley. It appears that the storage time during which a particle may be subject to weathering is very much longer than is the period during which the particle would be exposed to abrasion during transport. The weathering during these long periods of storage probably is an important influence in the reduction of particle size by abrasion during the short periods of actual transport.

According to the view presented here, the flood plain is a depositional feature of the river valley associated with a particular climate or hydrologic regimen of the drainage basin. Sediment is temporarily stored in the flood plain en route through the valley and, under equilibrium condition averaged over a period of years, net inflow of sediment equals net outflow. An alteration of the conditions of equilibrium through tectonic changes or by changes in the hydrologic regimen, including changes in sediment and water yield, will result in altering the flood plain and lead to degradation and terrace formation, or to aggradation.

REFERENCES

Aramaki, M., and Takayama, S., 1960, Flood deposits and their physico-chemical characters caused by the flood of the Kano River basin in 1958: *Geog. Review of Japan*, v. 33, no. 3, pp. 137–150.

Bagnold, R. A., 1941, The physics of blown sands and desert dunes: Methuen, London, reprinted 1954, 265 pp.

Bagnold, R. A., 1954, Experiments on a gravity-free dispersion of large solid spheres in a Newtonian fluid under shear: *Roy. Soc. London Proc.*, ser. A, v. 225, pp. 49–63.

Bagnold, R. A., 1956, The flow of cohesionless grains in fluids: *Roy. Soc. London Phil. Trans.*, ser. A, no. 964, v. 249, pp. 235–297.

Bagnold, R. A., 1960, Some aspects of river meanders: *U. S. Geol. Survey Prof. Paper 282-E.*

Balek, J., and Kolar, V., 1959, Statistical parameters of river bends: *Vodohospodarsky Casopis* (Slovenska Akademia Vied, Bratislava, Czechoslovakia) v. 7, pp. 237–246 (with English summary).

Blench, T., 1957, Regime behavior of canals and rivers: Butterworths, London, 138 pp.

Brooks, N. H., 1956, Mechanics of streams with movable beds of fine sand: *Am. Soc. Civil Engrs. Proc.*, v. 83, no. HY2, paper 668.

Brush, L. M., Jr., 1961, Drainage basins, channels and flow characteristics of selected streams in central Pennsylvania: *U. S. Geol. Survey Prof. Paper 282-F*, pp. 145–181.

Buckley, A. B., 1922, The influence of silt on the velocity of flowing water in open channels: *Inst. Civil Engr. Proc.*, v. 216, pp. 183–211.

Chien, N., 1961, The braided stream of the lower Yellow River: *Scientia Sinica*, v. X, no. 6, pp. 734–754.

Chorley, R. J., 1962, Geomorphology and general systems theory: *U. S. Geol. Survey Prof. Paper 500-B.*

Crickmay, C. H., 1960, Lateral activity in a river of northwestern Canada: *Journ. Geol.*, v. 68, pp. 377–391.

Davis, W. M., 1902, Base-level, grade, and peneplain: Geographical essays, XVIII, pp. 381–412, Ginn, Boston.

Drury, W. H., Jr., 1956, Bog flats and physiographic processes in the Upper Kuskokwim River region, Alaska: Gray Herbarium Contribution No. 178, 130 pp.

Dury, G. H., 1953, The shrinkage of the Warwickshire Itchen: *Coventry Nat. Hist. and Sci. Soc. Proc.*, v. 2, pp. 208–214.

Eakin, H. M., 1935, Diversity of current-direction and load-distribution on steam-bends: *Am. Geophys. Union Trans.*, pt. II, pp. 467–472.

Eakin, H. M., and Brown, C. B., 1939, Silting of reservoirs: *U. S. Dept. Agric. Tech. Bull. No. 524*, 168 pp.

Exner, F. M., 1919, Zur theorie der flusmäander: *Sitz. Akad. Wissenschaften Wien*, Math.-Natur w. Klasse, Ab. IIa, 128 Band, 10 Heft, pp. 1–21.

Fahnestock, R. K., 1963, Morphology and hydrology of a glacial stream: *U. S. Geol. Survey Prof. Paper 422-A.*

Fisk, H. N., 1952, Mississippi River Valley geology relation to river regime: *Am. Soc. Civil Engr. Trans.*, v. 117, pp. 667–689.

Friedkin, J. F., 1945, A laboratory study of the meandering of alluvial rivers: *U. S. Waterways Engr. Exper. Sta.*, 40 pp.

Gibson, A. H., 1909, On the depression of the filament of maximum velocity in a stream flowing through a natural channel: *Roy. Soc. London Proc.*, ser. A., v. 82, pp. 149–159.

Gierloff-Emden, H. G., 1953, Flussbett-veränderungen: Rezenter Zeit, Erdkunde, Bd. VII, pp. 298–306.

Hack, J. T., 1957, Studies of longitudinal stream profiles in Virginia and Maryland: *U. S. Geol. Survey Prof. Paper 294-B*, 97 pp.

Hack, J. T., 1960, Interpretation of erosional topography in humid temperate regions: *Am. Journ. Sci.*, v. 258A, pp. 80–97.

Hack, J. T., and Goodlett, J. C., 1960, Geomorphology and frost ecology of a mountain region in the Central Appalachians: *U. S. Geol. Survey Prof. Paper 347*.

Hack, J. T., 1963, Glacial geology and post-glacial drainage evolution in the Ontonagon area,, Michigan: in preparation.

Henderson, F. M., 1961, Stability of alluvial channels: *Am. Soc. Civil Engrs. Proc.*, v. 87, *Journ. Hydr. Div.*, pp. 109–138.

Highway Capacity Manual, 1950, Committee on Highway Capacity, Highway Research Board, Bur. Public Roads, U. S. Dept. Commerce, Govt. Print. Off., Washington, D. C., 147 pp.

Hjulstrom, F., 1949, Climatic changes and river patterns, in Glaciers and climate: *Geog. Annaler*, H. 1–2, pp. 83–89.

Hjulstrom, F., 1952, The geomorphology of the alluvial outwash plains (sandurs) of Iceland and the mechanics of braided rivers: *8th Gen. Assem. Proc.*, 17th Intl. Congr., Internatl. Geogr. Union, pp. 337–342.

Horton, R. E., 1945, Erosional development of streams and their drainage basins; hydrophysical approach to quantitative morphology: *Geol. Soc. Am. Bull.*, v. 56, pp. 275–370.

Inglis, C. C., 1949, The behavior and control of rivers and canals: *Res. Publ.*, Poona (India), no. 13, 2 vols.

Kaetz, A. G., and Rich, L. R., 1939, Report of surveys made to determine grade of deposition above silt and gravel barriers: U. S. Soil Conservation Service, Albuquerque, N. M., unpublished.

Keulegan, G. H., 1938, Laws of turbulent flow in open channels: *Journ. Res. Natl. Bur. Standards*, Research Paper 1151, v. 21, pp. 707–741.

King, A. L., 1962, Thermophysics: Freeman, San Francisco, 369 pp.

Koechlin, René, 1924, Mecanisme de l'eau: Librairie Polytechnique, Béranger, Paris.

Lane, E. W., 1951, discussion of Stanley, J. W., 1950, Retrogression on the lower Colorado River after 1935: *Am. Soc. Civil Engrs. Proc.*, v. 76, pp. 1–12, Separate D-28, pp. 1–3.

Lane, E. W., and Borland, W. M., 1954, River-bed scour during floods: *Am. Soc. Civil Engrs. Trans.*, v. 119, pp. 1069–1080.

Lane, E. W., 1955, Design of stable channels: *Am. Soc. Civil Engrs. Trans.*, v. 120, pp. 1234–1279.

Langbein, W. B.: 1942, Hydraulic criteria for sand waves: *Am. Geophys. Union Trans.*, pt. 2, pp. 615–618.

Langbein, W. B., 1963, A theory for river channel adjustment: *Am. Soc. Civil Engr. Trans.*, in press.

Lattman, L. H., 1960, Cross section of a flood plain in a moist region of moderate relief: *Journ. Sed. Pet.*, v. 30, no. 2, pp. 275–282.

Laursen, E. M., 1960, Scour at bridge crossings: *Am. Soc. Civil Engrs. Proc.*, v. 86, no. HY2, sep. no. 2369, pp. 39–53.

Leliavsky, S., 1955, An introduction to fluvial hydraulics: Constable, London, 257 pp.

Leopold, L. B., 1953, Downstream change of velocity in rivers: *Am. Journ. Sci.*, v. 251, pp. 606–624.

Leopold, L. B., and Maddock, T., Jr., 1953, The hydraulic geometry of stream channels and some physiographic implications: *U. S. Geol. Survey Prof. Paper* 252.

Leopold, L. B., and Miller, J. P., 1956, Ephemeral streams—hydraulic factors and their relation to the drainage net: *U. S. Geol. Survey Prof. Paper* 282-A.

Leopold, L. B., and Wolman, M. G., 1957, River channel patterns; braided meandering and straight: *U. S. Geol. Survey Prof. Paper* 282-B.

Leopold, L. B., and Wolman, M. G., 1960, River meanders: *Geol. Soc. Am. Bull.*, v. 71, pp. 769–794.

Leopold, L. B., Bagnold, R. A., Wolman, M. G., and Brush, L. M., Jr., 1960, Flow resistance in sinuous or irregular channels: *U. S. Geol. Survey Prof. Paper* 282-D.

Leopold, L. B., and Langbein, W. B., 1962, The concept of entropy in landscape evolution: *U. S. Geol. Survey Prof. Paper* 500-A.

Lighthill, M. J., and Whitham, G. B., 1955, On kinematic waves. I. Flood movement in long rivers, pp. 281–316. II. A theory of traffic flow on long crowded roads, pp. 317–345: *Roy. Soc. London Proc.*, ser. A, v. 229.

Mackin, J. H., 1937, Erosional history of the Big Horn Basin, Wyo.: *Geol. Soc. Am. Bull.*, v. 48, pp. 813–894.

Mackin, J. H., 1948, Concept of the graded river: *Geol. Soc. Am. Bull.*, v. 59, pp. 463–512.

Matthes, G. H., 1941, Basic aspects of stream meanders: *Am. Geophys. Union Trans.*, v. 22, pp. 632–636.

McNown. J. S., and Lin, P. S., 1952, Sediment concentration and fall velocity: *Proc. Second Midwestern Conf. on Fluid Mechanics, Engr. Exp. Sta., Ohio State Univ. Bull. No. 149*, Sept.

Miller, C. R., Woodburn, R., and Turner, H. K., 1962, Upland gully sediment production: *Internatl. Assoc. Scientific Hydrology. Commission of Land Erosion*, Publ. No. 59, pp. 83–104.

Miller, J. P., 1958, High mountain streams; effects of geology on channel characteristics and bed material: *New Mex. State Bur. Mines and Min. Resources, Memoir 4*, 51 pp.

Myrick, R. M., and Leopold, L. B., 1963, Hydraulic geometry of a small tidal estuary: *U. S. Geol. Survey Prof. Paper 422-B*.

Nixon, M., 1959, A study of the bankfull discharges of rivers in England and Wales: *Inst. of Civil Engr. Proc.*, Paper No. 6322, pp. 157–174.

Ohya, M., 1961, Geographical study of flood in the Basin of the Mekong Tributaries: Report of Reconnaissance Team organized by the Japanese Government, Sept. 1961.

Parsons, D. A., 1960, Effects of flood flow on channel boundaries: *Am. Soc. Civil Engr. Proc.*, v. 86, no. HY4, pp. 21–34.

Powell, J. W., 1875, Exploration of the Colorado River of the West and its tributaries: Smithsonian Institution, Washington, D. C., 285 pages.

Prus-Chacinski, T. M., 1954, Patterns of motion in open-channel bends: *Assoc. Internat. d'Hydrologie*, Pub. 38, v. 3, pp. 311–318.

Quirashy, M. S., 1944, The origin of curves in rivers: *Current Sci.*, v. 13, pp. 36–39 (London).

Rouse, H., ed., 1950, Engineering hydraulics: Wiley, New York, 1033 pp.

Rubey, W. W., 1933, Equilibrium-conditions in debris-laden streams: *Am. Geophys. Union Trans.*, pp. 497–505.

Rubey, W. W., 1952, Geology and mineral resources of the Hardin and Brussels quadrangles (in Illinois): *U. S. Geol. Survey Prof. Paper* 218.

Scheidegger, A. E., 1961, Theoretical geomorphology: Springer-Verlag, Berlin, 327 pp.

Schoklitsch, A., 1933, About the decrease of the size of bed material in river beds: translated by B. B. Eissler from *Akad. Wiss., Wien*, v. 142, pp. 343–366.

Schumm, S. A., 1956, Evolution of drainage systems and slopes in badlands at Perth Amboy, New Jersey: *Geol. Soc. Am. Bull.*, v. 67, pp. 597–646.

Schumm, S. A., 1960, The shape of alluvial channels in relation to sediment type: *U. S. Geol. Survey Prof. Paper* 352-B, pp. 17–30.

Smith, K. G., 1950, Standards for grading texture of erosional topography: *Am. Journ. Sci.*, v. 248, pp. 655–668.

Sternberg, H. O., 1960, Radiocarbon dating as applied to a problem of Amazonian morphology: *Comp. Rendus du XVIII Congr. Intern. de Geog.*, Tome 2, Travaux des Sect. 1, II, et III, pp. 400–424.

Strahler, A. N., 1952, Dynamic basis for geomorphology: *Geol. Soc. Am. Bull.*, v. 63, pp. 923–938.

Stricklin, F. L., Jr., 1961, Degradational stream deposits of the Brazos River, central Texas: *Geol. Soc. Am. Bull.*, v. 72, pp. 19–36.

Stuart, T. A., 1953, Spawning migration, reproduction, and young stages of loch trout: Freshwater and Salmon Fisheries Res. Sta., H. M. Stationery Off., Edinburgh, 39 pp.

Tricart, J., 1959, Évolution du lit du quil au cours de la crue de Juin, 1957: *Bull. de la Section de Géographie*, v. 72, 403 pp.

U. S. Bur. of Reclamation, 1950, Boulder Canyon Project, Final Reports, Pt. III, Bull. 1, Geographical Investigations, 232 pp.

Vanoni, V. A., 1946, Transportation of suspended sediment by water: *Am. Soc. Civil Engrs. Trans.*, no. III, pp. 67–133.

Vanoni, V. A., and Nomicos, G. N., 1960, Resistant properties of sediment-laden streams: *Am. Soc. Civil Engrs. Trans.*, v. 125, pt. I, pp. 1140–1175.

Wolman, M. G., 1954, A method of sampling coarse river-bed material: *Am. Geophys. Union Trans.*, v. 35, no. 6, pp. 951–956.

Wolman, M. G., 1955, The natural channel of Brandywine Creek, Pennsylvania: *U. S. Geol. Survey Prof. Paper* 271, 56 pp.

Wolman, M. G., and Leopold, L. B., 1957, River flood plains: some observations on their formation: *U. S. Geol. Survey Prof. Paper* 282-C.

Wolman, M. G., and Eiler, J. P., 1958, Reconnaissance study of erosion and deposition produced by the flood of August 1955 in Connecticut: *Am. Geophys. Union Trans.*, v. 39, no. 1, pp. 1–14.

Wolman, M. G., and Brush, L. M., Jr., 1961, Factors controlling the size and shape of stream channels in coarse noncohesive sands: *U. S. Geol. Survey Prof. Paper* 282-G, pp. 183–210.

Woodford, A. O., 1951, Stream gradients and Monterey sea valley: *Geol. Soc. Am. Bull.*, v. 62, pp. 799–852.

Chapter 8   Hillslope Characteristics and Processes

Features of Slopes

Hillslopes are the part of the landscape included between the crest of hills and their drainage lines. Slopes are interdependent with stream channels and the geometry of drainage basins, but if initially one looks only at the characteristics and processes which are operative on the slopes themselves it is possible to describe the hills and begin to isolate some of the factors which control their form and development. Unfortunately, neither the forms nor processes of hillslopes are as well known as are those of rivers, and hence the description will be more topical and less comprehensive. Many processes rather than a single one are responsible for the forms of hillslopes, and in addition the evidence suggests that changes occur on slopes much more slowly than in stream channels. For this reason less is known about the combinations of processes which are occurring on hillslopes in various environments. The primitive state of knowledge of the development of hillslopes makes necessary a more elaborate description in the absence of unifying theoretical or physical models or comprehensive measurements.

To discuss specific characteristics of slopes, several additional definitions are necessary. There is as yet no terminology generally agreed upon, but the following definitions are based in part upon Savigear's concepts. The crest-slope is the upper, the mid-slope the central, and the foot-slope the basal part of the hillslope (Fig. 8-1). On any given slope, distinct segments representing any single part may well be missing.

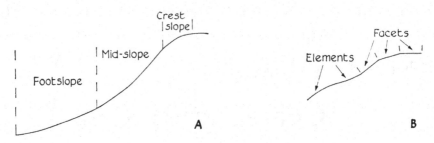

Figure 8-1.   *Definitions of terms applicable to the description of slopes.*
A. Parts of a slope. B. Segments of a slope. Facets are
straight; elements are convex or concave.

A slope facet is a rectilinear (straight) segment of a slope or slope profile. The slope may consist of a number of such rectilinear parts.

A slope element is either a smooth concave or a smooth convex area of a slope or is a smooth concave or smooth convex portion of the profile of the slope.

The total length of a slope is the distance from the divide to the channel line or valley surface. This distance is measured on the horizontal and normal to the contours of the surface of the slope. It is the horizontal projection of the slope distance which is measured along the sloping surface.

The slope as a whole may have a single angle of inclination if the entire surface is rectilinear. But every facet or element will have an angle of inclination on slopes that are not uniform.

A segment is simply a part of a slope, regardless of its form or shape.

It is customary to describe the forms of hillslopes by means of simple profiles, like those in Fig. 8-1. These can be compared by fitting smooth mathematical curves to them and comparing the constants in the equations. As the slope forms become more complex, however, the fitting of mathematical curves also becomes more complex and with it the comparison of features of hillslopes. For the Middle Atlantic region of the United States, Hack and Goodlett (1960) have shown that many convex slope profiles can often by fitted by straight lines on logarithmic paper if the quantities plotted are (a) the fall, measured as the vertical distance from the ridgecrest to a point on the slope; (b) the horizontal distance from the ridgecrest to the point, with the origin at the crest or top of the slope.

Such a plot and the accompanying equation for a hillslope in Martinsburg shale is shown in Fig. 8-2, **A.** The equation is of the form

$$H = CL^j,$$

where H is the fall, L the horizontal distance, and C and f are constants; C is a measure of the steepness and f is a measure of the curvature or the rate of change of slope. The higher the value of C, the steeper is the slope at a given horizontal distance. Similarly, the higher the value of f, the greater the curvature (Fig. 8-2, **B**). The use of such curves in comparing hillslope forms on different rock types is discussed in connection with the relation of rock type to slope form.

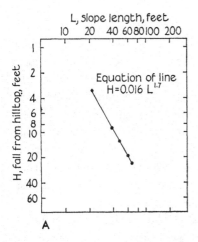

A

Figure 8-2.

A. *Convex hillslope on Martinsburg shale in Virginia. Slope length is horizontal distance from crestline to point on slope. Note logarithmic scale.* [*From Hack and Goodlett, 1960.*] **B.** *Hillslopes in Maryland and Virginia, showing variation in curvature and steepness of crest slopes. Sandstone is steepest and has least curvature.* [*From Hack and Goodlett, 1960.*]

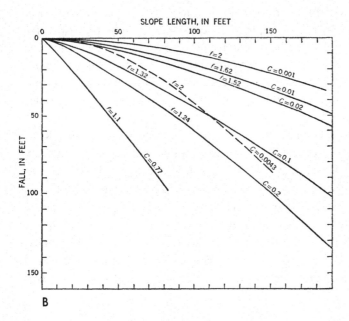

B

Controls of the Form of Hillslopes

In a large frame of reference, the characteristic controls of hillslopes are of course the same as the major forces controlling the development of all landforms. Thus their form and development are functions of that hardy triumvirate: structure, process, and time. These broad parameters, however, can be broken down into more specific terms when applied to the development of slopes.

The form of any slope is ultimately determined by the relationship between the rate of disintegration of the underlying rock and the rate of removal of this rock debris from the surface and base of each segment and from the slope as a whole. The rate of preparation of material is, by definition, the rate of weathering. Weathering in turn includes the physical and chemical breakdown of the rock into smaller fragments and different materials. The distribution of particle sizes and their physical behavior are in turn controlled by composition and structure of the rocks. Structure includes the attitude of the beds, the degree of jointing, type of deformation, and the geometry of such features as foliation and lineation. Composition includes mineralogy, cementation, porosity, and texture.

Although there is available some generalized knowledge of the different behavior (including erodibility) of various types of rocks, it is most often indicated by the degree of relief or dissection on different types of rocks. A good deal more work is required to clarify the differential aspects of weathering and of removal in different types of rocks. Vegetation may exercise both direct and indirect control on slope form through its mechanical action in holding the soil, by its influence on the force and availability of water at the surface, and through its effect on chemical and physical weathering.

Combined with these properties are the relatively few processes which act upon the surface of the land—running water, moving ice, wind, and mass movement of complexes of soil, water, and ice. These erosional processes may work on the slope itself or at the base or crest of the slope in molding its form. The angle of a rectilinear slope, for example, appears to be governed by the rate of removal of material at the base. Thus stream channels, even though occurring only at the base and not on the surface of the slope itself, may be a dominant control of the actual form of the surface. Wave and current action at the coast and glacial scour can work in the same way.

There are innumerable possible geometric forms which are produced by the interaction of these processes upon diverse lithologies and structures. Although many observers have stated that convex-concave forms are "most common," there is no statistical evidence to support the contention, nor is the statement particularly useful in understanding the characteristics of the landscape. Cliffs occur where undercutting produces free faces and competent cap rock is available which does not crumble in place. The causes of undercutting may be different in each place but the resulting forms may be the same. Where debris piles up at the foot of a slope it means simply that production exceeds removal, and such a slope will be less steep than the one from which material has been removed. More will be said about these developmental aspects in the chapter on evolution of hillslopes.

Despite the burden of detail which it fosters, we consider below a number of specific processes in an attempt to relate process, structure, and form in relatively simple cases where a single process can be isolated much as in a laboratory experiment.

Relation to Lithology and Environmental Controls

In analyzing the characteristics and forms of slopes in various environments one comes immediately to the deceptively simple task of associating various processes with various types of rock. To the extent that these processes are climatically controlled, the forms of the hillslopes will reflect either the past or present climate of an area. Unfortunately, this seemingly easy matter is complicated by lack of detailed knowledge of the processes operative in specific climates and on specific types of rocks. Although one can find specific examples of landforms which are the result of the exclusive operation of a single process, much more commonly a landform is the result of innumerable processes acting on diverse lithologies. Despite this oft-repeated qualification, these "controlled" examples provide our only real insight into the operation of specific mechanisms.

Mass Movements: Introduction

Mass movement is a term which means the movement of materials on slopes under the influence of gravity without benefit of the contributing force of independent agencies such as flowing water or wind. It includes movements ranging from free rockfalls to slow creep of material on very low gradients. Although forming a continuous series from the standpoint of slope formation, one can separate slides and falls consisting of isolated movements of discrete units from flows in which movement involves internal deformation of all elements. A classification of mass movements is given in Table 8-1.

Table 8-1. Classification of mass movements.[a]

Type of Movement		Type of Material			
Falls		Bedrock		Soils	
		Rockfall		Soilfall	
Slides — Few Units	Rotational / Slump	Planar	Planar	Rotational	
		Block glide	Block glide	Block slump	
Slides — Many Units		Rockslide	Debris slide	Failure by lateral spreading	

		All Unconsolidated			
		Rock Fragments	Sand or Silt	Mixed	Mostly Plastic
Flows — Dry	Rock fragment flow	Sand run Loess flow			
Flows			Rapid earth-flow	Debris avalanche	Slow earth-flow Creep
					Solifluction Mudflow
Flows — Wet			Debris flow: Sand or silt flow	Rubble flow: Dirty snow avalanche	

Complex	Combinations of materials or type of movement

[a] After Highway Research Board Landslide Committee.

Landslides and Rockfalls

There are a number of characteristic landforms associated with various types of slides and flows. A variety of these are shown in Fig. 8-3. Rockfalls occur where steep faces are maintained by erosion through headward sapping and by mechanical or chemical processes, and where weathering, jointing, freezing and thawing, earthquakes, differential settling, or other

MATERIAL

BEDROCK   **SOILS**

TYPE OF MOVEMENT

FALLS

Rockfall   *Soilfall*

SLIDES

Rotational   Planar

Slightly
deformed

Slump   *Block glide*

Deformed

Debris slide

FLOWS

Dry

|

Increasingly wet

↓

Wet

Debris avalanche

Debris flow

COMPLEX LANDSLIDES

Figure 8-3.   *Simplified diagram of types of slides and rockfalls [After Highway Research Board Landslide Committee, Special Report 29.]*

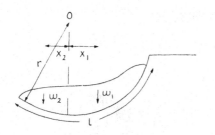

Driving moment = resisting moment

$$W_1 x_1 = W_2 x_2 + srl$$

Figure 8-4.

Equilibrium of forces maintaining a slope subject to sliding.

agencies produce conditions and trigger actions releasing sections of rock. In nival regions frost bursting is the most significant force in producing rockfalls. Rockfalls leave pockmarked slope facets at their points of origin and generally produce large talus forms (screes), the surface slope of which depends upon the type of rock, the height of free fall, and the weathering products.

More distinctive scars and depositional features are produced by landslides and rapid flows. Perhaps the most common is the spoon-shaped scar associated with shear failures along arcuate planes. Hummocky topography at the base of a hill is characteristic of landslide topography. Where a hillside is subject to sliding, the surface may consist of many concave scars and convex depositional elements produced by multiple slides. Where slides are controlled by bedding planes or other planes or by geologic contacts (including artificial fills), the shape of the scar will conform to the shape of the contact or plane. Thus, permeable lithologies such as sandstone underlain by impermeable shales may slide along planes which are lubricated by moisture. Similarly, certain clay soils containing large amounts of sodium may absorb moisture and hence become subject to movement on moderate slopes.

The condition of movement of any particular landslide can be extremely complex. Simply stated, however, a slide is a shear failure which will be set in motion when the stress along the potential surface of rupture exceeds the resistance to shear along that surface. This unbalanced condition may be brought about by (a) an increase in the weight of the overlying material due to the absorption of water; (b) a decrease in the resisting mass due, for example, to undercutting a slope; (c) a decrease in shear strength of the material itself, which may be caused by the absorption of moisture.

An equation describing the equilibrium of forces (see Fig. 8-4) has the form

Driving moment = resisting moment,

$$W_1 x_1 = W_2 x_2 + srl,$$

where r is the radius of the arc along the plane of sliding, and s is the shearing resistance cohesive material; shearing resistance $s = c + \rho \tan \phi$, where c is the cohesion and ϕ is the internal angle of friction; W_1 is the weight of the portion of the soil tending to produce failure, and W_2 is the weight of soil tending to resist it. Any action tending to reduce the right side of the equation will contribute to sliding. Raising the water table would tend to reduce s and thus lead to slippage or failure. A similar result is produced by reducing x_2 or W_2 by removing material from the base of the slope.

Applying these physical principles to the evaluation of the regional distribution or significance of landsliding as a geomorphic agent, pertinent areas may possess the following attributes: (1) deep weathering; (2) in the absence of deep weathering, sedimentary or planar structures of variable lithology; (3) swelling clays; (4) large quantities of moisture; (5) perennial, seasonal, or diurnal ice formations; (6) earthquakes, (7) areas of undercutting by wave action or rivers. Clearly, there are few areas with all these characteristics developed to a maximum. The tropics and the snow and ice regions of the high altitudes appear to be two regions in which climate and topography combine to promote landsliding as a major geomorphologic agent. Locally, of course, any single factor may make landsliding the dominant erosive force on a given slope or slopes.

Precipitation and thawing in cold snowy regions trigger slides from the slopes. The composition of the slide varies according to the source but, in any case, the rapidity of movement and the movement of large amounts of debris as single units do not allow for sorting.

Steep slopes are maintained by scars resulting from slides in both tropical and nival regions. To the extent that the slide debris is not removed from the foot-slopes, the hillslopes will consist of scarred segments in crest-slope and mid-slope and of hummocky and lobate forms on the foot-slope.

Rapid Flows of Wet Debris

Slides and rapid detachment of large masses of rock and debris are often associated with or are the precursors of flows of earth and debris. When large amounts of water saturate the soil, mudflows and debris avalanches occur. The photographs in Fig. 8-5 show comparable forms produced by mudflows consisting of poorly sorted debris in semiarid and in arctic environments. Although mudflow may connote finer materials than are found in debris flows, the latter term is quite often used in a

Figure 8-5.

Above: *Mudflow levees along a steep desert channel on slopes of Picacho Moun-tains, near Phoenix, Arizona.* [*Photograph by Howard T. Chapman.*] Right: *Mudflow levees associated with slope processes in an arctic climate, northern Sweden.* [*Photograph by Anders Rapp.*]

generic sense to describe all such mass movements. In many cases winnow-ing may have removed the fine material from the matrix, leaving primarily the coarser fraction.

If the quantity of water in a mudflow is not large, the viscous mass moves downhill, depositing a lobate tongue at the foot of the slope. However, Sharp (1942) has observed that when the flow is more fluid boulders ac-cumulate near the snout, obstructing the flow. When the moving fluid behind attains sufficient force, the boulders are shoved aside, producing mudflow levees. Mudflows will often follow weak channel courses previously established by similar flows or by small watercourses on the surface. Veloci-ties of flow are highly variable. On newly deposited ash at Paricutín, Mexico, Segerstrom (1950, p. 67) observed movement of the front of a flow only 3 inches wide at a velocity of 2 feet per second on a 25% slope. The same flow halted a minute later on a slope of 15%. A larger flow 2

meters wide had a velocity of about 0.5 foot per second on a gradient of 12 to 15%. Much higher velocities have been observed in large mudflows in the mountains of southern California. Here velocities of the larger surges averaged 8 to 10 feet per second with a maximum of 14 feet per second on slopes of about 6°. Surges are common as flows move over obstructions and along channels which vary in shape and gradient.

The surface form of mudflows that come to rest on open foot-slopes will in general be lobate, with the thickness of the deposit being inversely proportional to its fluidity and to the gradient of the slope. In narrow channel and valley bottoms the poorly sorted debris forms a very irregular surface, which may have local relief of 10 to 20 feet. The deposits are hard, broken, and consist of all kinds of debris, including boulders, silt, trees, and other vegetation.

A close cousin to the mudflow, perhaps best described as transitional between mudflow and landslide, is the snow avalanche. Rapp (1960) and others have pointed out that avalanching snow can be an effective agent in both eroding and transporting rock debris. Such avalanches most often accompany thawing of snow and ice. The avalanching snow carries debris of sizes ranging from boulders to silt and sand. Individual avalanches may contain many cubic yards of debris. The larger the avalanche, of course, the greater its energy, and large avalanches from side hills may transport debris well out into flat valley bottoms. Boulders measuring 12 to 15 feet in diameter have been so moved by sliding, rolling, and jumping. After deposition following melting of the snow, isolated boulders and cobbles may remain as protective covers on top of ice pinnacles.

Once the snow of an avalanche has melted, the deposit will have a distinctive character, but aside from the scattering of boulders and gravel in a localized area the deposit has no distinctive form of its own. But avalanches, mudflows, and debris flows are all effective in removing accumulating debris from the base of the steeper slopes on which the flows originate. Coursing over talus, for example, they remove the upper portion of the talus slope, contributing to the maintenance of a steeper crest slope or rockwall and thus to the erosional form of hillslopes. Aside from erosive channels on alluvial fans and talus slopes, however, the principal forms are depositional lobes, mudflow levees, and hummocky surfaces on foot slopes and valley bottoms.

Creep

In geomorphology the term creep is applied to the slow movement of rock and soil downslope solely under the influence of gravity. Because this mode of movement is set off from other kinds of mass movement in terms of relative velocity only, no clear-cut distinction really exists between this and the more rapid rates of flow. It will also be noted that the term solifluc-tion appears in Table 8-1 beneath creep. Solifluction is essentially a form of creep, although it is usually used to describe creep on slopes subject to frost action either in areas of permafrost or in nival areas subject to freez-ing and thawing. Creep is probably common to virtually all climatic en-vironments, but the present discussion will concern primarily slopes that are dominated by creep.

In areas of perennially frozen ground (permafrost), the contact between the overlying mantle which thaws during warm periods and the underlying permanently frozen ground provides a zone along which movement may readily take place. Solifluction lobes such as those shown in Fig. 8-6 are characteristic of hillslopes in such regions. Rates of movement by solifluction measured in a number of localities are on the order of several centimeters per year. Over a period of four or five years there may be no movement on some slopes, others may show rates of one to two feet per year. As in glacier ice or in a stream of water, flow is more rapid at the surface and in plan view may also be more rapid in the center of a moving unit rather than at the edges. This distribution tends to produce the ob-

Figure 8-6.

Slope in arctic region, showing solifluction lobes on hillslope. [Photograph by Anders Rapp.]

served lobate form, although customarily, on slopes of 15 to 35°, solifluction can operate at very moderate gradients—as low as 2 or 3°. Theoretically, this permits denudation on very flat slopes in regions of high frost
activity. The postulated result has been termed cryoplanation, meaning
the planation of the landscape through the action of frost.

There is a difference between the action of perennial ground frost and
that created by alternate freezing and thawing of needle ice. In the latter
case the growth of frost crystals forces the particles of soil upward or
perpendicular to the surface of the slope. Upon thawing these particles drop
again to the surface, having been displaced by an amount equal to the
component of the downhill slope parallel to the surface. This motion,
combined with the moisture and lubricating effect of melting water, provides
a very strong mechanism for movement on hillslopes. It is clear that such
movement will take place in all environments in which freezing and thawing is experienced. Presumably the net result of the process of creep in
frost regions would be a topography of hillslopes whose average gradient
on various types of rocks is gentler than would be the case in areas less
subject to creep.

In addition to these features associated particularly with the process of
creep or solifluction, in areas dominated by frost activity there are a vast
number of microsurficial forms produced by frost action, coupled with
creep and other processes of transportation. The specific mechanisms by

Table 8-2.   Some features and processes associated with frozen
                 ground.

PROCESSES	
Frost wedging	Produces rubble by mechanical breakage of rocks (sometimes called frost bursting)
Solifluction	Downslope movement comparable to soil creep but generally more rapid
Stone streams	Linear concentrations of blocky boulders
Rock glaciers	Lobate form of angular boulders without fines, comparable in overall shape to a glacier caused by frost inclusion in boulder mass
FEATURES	
Involutions	Irregular contorted structures such as waves or folds in soil
Patterned ground	Includes polygons, circles, stripes, and steps formed of coarse particles
Frost cracks	Nearly vertical cracks formed by ice wedges. Polygonal pattern on the ground surface. Filling of abandoned cracks by silt creates pseudomorph wedges

which these many forms are produced are beyond the scope of this book, but inasmuch as frost features often provide the most significant clue to past climatic conditions in a region, a partial listing is given in Table 8-2. Most are produced by freezing and thawing and by frost heaving, accompanied by various rates of downslope movement. It is interesting to note that many of the forms are not unique to frost regions but do occur wherever comparable mechanisms are available. Thus, stone polygons and stripes are found in desert regions where salt formation and decay with drying and wetting of the soil itself provide mechanisms of alternate heaving and settlement analogous to those of the ice regions. Phenomena such as landslides may be common to and produce similar landforms in such diverse regions as the humid tropics and the high latitudes of the north and south. Nevertheless, some features are unique to frost regions, and a preponderance of relict forms provide valuable evidence of previous climatic conditions.

Creep, however, is by no means confined to regions of frost. Many years ago Gilbert (1909) described rounded or convex crests of hillslopes formed on granite weathered to kaolin in the gold mine areas of California. He ascribed the convexity to creep of surficial material, reasoning that the rate of creep would be proportional to the slope if the depth of erosion was the same at each point on the convex segment of the slope. At points distant from the crest the slope would have to be greater to transmit the material eroded from above. The convex crest of the hillslope thus contrasted with the concave profile formed by running water, along which increasing volume of flow was compensated for by decreasing slope.

Recently more detailed observations of the process of creep have been made by Schumm (1956) on badland topography in South Dakota. Rain falling on the permeable Chadron formation is readily absorbed. As it is, the clays at the surface become sticky and the loose mass of clay aggregates which was initially dry begins to slump, each aggregate sliding as a unit on the underlying surface. On similar terrain in even more arid regions, having rainfalls concentrated in short periods of time, particles may form into larger sheets of aggregates, which move gradually downslope. The characteristic rounded forms of such a clay badland are shown in Fig. 8-7; Schumm's measurements of slope retreat are shown in Fig. 8-8. Gilbert's hypothesis that the depth of erosion on the convex form would be everywhere equal is not supported by these data for individual years; erosion increased from the base of the slope to the crest. If no aggradation occurs, the rate of creep on the steep part of the slope must be great in order to

Figure 8-7.

Characteristic surface textures and hillslope forms in clay badlands, South Dakota and Nebraska.

A. *Rounded forms on White River Formation near Crawford, Nebraska.*

B. *Nearly straight slopes and slightly rounded divides, near* **A.** *Rounded forms in* **A** *associated with small drainage density, wide and shallow channels compared with those in* **B.** *The material in* **A** *appears and feels harder, more indurated, and less erodible than that in* **B.**

C. *Pediment pass between two miniature badland residuals; note sharp break between hillslope and pediment, as well as the concavity and smoothness of the pediment. [Photograph by S. A. Schumm.]*

D. *Frost action during winter months loosens surface of hillslope, and some slope materials roll downslope onto the miniature pediment. [Photograph by S. A. Schumm.]*

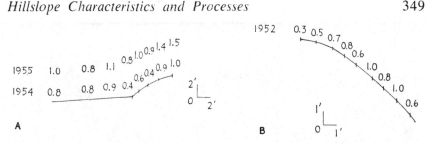

Figure 8-8. **A.** *Erosion by creep on Chadron formation in the badlands, South Dakota. [After Schumm.] The numbers represent depth of erosion in inches. Note that erosion is greatest at the crest.* **B.** *Erosion by rainwash on fill at Perth Amboy, New Jersey. Erosion is roughly proportional to the gradient of each segment. [After Schumm.]*

transmit the large amount of material derived from above. In the badlands, creep is restricted to a particular clay lithology. Thus, as is also the case of running water as an agent, both the process and resultant form of the land are functions of the underlying lithology rather than of the climate.

Schumm* points out the similarity of rates of creep that he measured on Mancos Shale slopes in Colorado in a semiarid environment with those reported for Greenland and Scandinavia. On a 34% slope in Greenland the creep observed by Washburn occurred at a rate of 0.2 feet per year; on a slope of the same inclination in Colorado the rate observed was 0.05 feet per year. On an 82% slope in Sweden, Rapp reported a rate of 0.33 feet per year, and on the same inclination in Colorado, Schumm observed 0.31 feet per year. These rates are much higher than that on grassed slopes in England measured by Young—0.0006 feet per year on a 66% gradient—or the barely perceptible movements measured by pressure recording devices on slopes in Ohio (Everett, 1962).

Every student of geomorphology is familiar with the photographs and drawings showing tombstones and telephone poles leaning upslope as the soil in which they are lodged creeps downslope. Quantitative observations of either the rate of such movement or of its regional distribution are usually lacking, but the rounded forms, displaced headstones, and leaning fences throughout much of the world suggest that creep is prevalent in virtually all regions. Penck (1953 translation, p. 119) stated that creep occurs in all climates "so long as the mean slopes of the land exceed that minimum gradient," which he estimates to be 5°. He believed that even in

* Personal communication, June 21, 1962.

forest regions slow movement by creep takes place through the network of tree roots (p. 81). The root system, he stated, acts like a sieve; the colloidal substances and small pieces of rock move downhill; the larger fragments either remain behind or may move on at a relatively slower pace according to circumstances. Observations are rare but some effects noted by Schumm are shown in Fig. 8-9.

Measurements of creep under various climatic and physiographic conditions are much needed. Rates of movement, differential velocities at various depths, relation to vegetation, effects of differing degree of inclination—all should be measured. A few investigators are beginning to make such observations, and some new tools are becoming available. One such tool is a metal strip to which strain gages are attached, the electrical wires coming

Figure 8-9.

*Effects of soil creep, Mancos Shale, Badger Wash, Colorado. [Photographs by S. A. Schumm.] **A, B, C.** Effects of creep on vegetation. **D, E, F, G.** Seasonal contrast of surface; **D, E,** show loosening of surface due to frost action in winter, and **F, G,** compacting of surface due to beat of rain.*

out to the surface (Williams, 1959). Successive measurements with some type of Wheatstone bridge also provide a measure of strain, although the amount of downslope movement has to be determined by some calibration.

We placed four strain gages on a 3-foot metal strip, which was set vertically in the ground on a forested hill near Bethesda, Maryland. The regolith developed on schist at this location is 20 to 40 feet deep. The local gradient of the hillslope at the installation averages 33%, and annual rainfall is about 40 inches. The strip was installed in 1961, and readings of the electrical circuit have been made at 1-week intervals. The record shows some indication of discrete and discontinuous movement during summer periods of high-intensity rain, but the period of measurement is still too short to indicate whether these indications are actual movement.

On the same forested slope a line of rods was installed in 1958 to measure mass movement directly. Approximately on a contour, 20 iron pins—rods ½ inch in diameter and 0.7 foot in length—were driven at 5-foot centers on a straight line. Two reference markers were placed at each end, one a 6-foot iron bar 1 inch in diameter, the other a lag screw placed in the base of a large tree. Each pin and reference marker was notched at the top with a file to show the exact position of the line of sight at time of original placement. At each resurvey the marks on the four reference pins were sighted in and experience indicated that four such markers satisfactorily stayed in alignment. The position of each of the 20 pins was then recorded relative to the line of sight.

After 4 years the results are equivocal. In the 6 resurveys made, the average distance of the pins from the line of sight varied from 0.23 inch uphill to 0.01 inch downhill. The latest resurvey (1962) showed that the tops of the pins are on the average slightly uphill from the line of sight. On this latest survey all pins were recorded as uphill. But the magnitude of the observed movement is close to the limit of precision of the survey.

Tentatively the data are interpreted to mean that no measurable movement has been recorded. But there is the possibility that the pins are rotating—owing to greater downhill movement below the surface than at the surface, the top of a pin appears to have moved uphill from a net rotation about a point near the midlength of the pin. These details are mentioned to indicate the order of precision of survey needed under field conditions where downhill movement is probably quite slow.

Similar lines of pins were established in a semiarid environment, and the initial results show measurable movement. Data presented below are for a line established at the authors' Arroyo de los Frijoles project area near Santa Fe, New Mexico, where the annual rainfall is about 14 inches. The line is established parallel to the axis of a small ephemeral gully whose bed width is about 2 feet, side slopes 45°, and depth 8 to 10 feet. The pins are on these side slopes bordering the channel. The slant length of the slopes is thus 12 to 13 feet. The material is stream-deposited alluvium, sandy silt in texture and probably Pleistocene in age. The horizontal component of downhill movement by creep for two resurveys is presented in Table 8-3. The indicated average movement per year in the two resurveys is 0.25 inch (8.4 mm).

Young (1960) measured creep on hillslopes in England by horizontal pins placed in the undisturbed soil of the wall of an excavated trench or pit. The pit is refilled after placement of the pins and excavated again for

Table 8-3.   Mass movement of pins on 45° slope in semiarid environment, Arroyo de los Frijoles Project near Santa Fe, New
Mexico.

| Pin No. | Horizontal Component of Yearly Downslope Movement (inches) | |
	1959–1961 (Average)	1961–1962
1	.19	.45
2	.03	.00
3	.53	−.07[1]
4	.50	.26
5	.39	.26
6	.56	.30
7	.36	.05
8	.58	.14
9	.09	−.10[1]
10	.09	−.18[1]
11	.36	.14
12	.01	.08
13	.05	.11
14	.67	.44
15	—	.21

[1] Minus indicates uphill movement.

resurvey. Measurements on grassed and wooded slopes of 20 to 30° gradient
in the Upper Derwent Basin, Southern Pennines (England), underlain by
shale and sandstone, showed downslope movement of the upper 10 cm,
excluding the topmost organic horizon, averaging 0.25 mm per year.

Overland Flow: General

Perhaps the most ubiquitous process occurring on hillslopes is erosion
and transport by running water. The flow itself may occur either in small
rills or as sheets of water of moderate depth over larger surfaces. To
produce erosion two requirements must be met. First, the rate of rainfall
must be sufficient to produce runoff; second, as in rivers, the force provided
by the moving runoff must exceed the resistance of the soil material to
erosion. A picture of the behavior of rainwater and runoff on slopes will
be concerned first, with the characteristics of the flow itself, and second,
with the resistance of the soil material to erosion.

The derivation given here is primarily from Horton's analysis (1945). Horton emphasized the geometry of the drainage net and its origin, and gave less emphasis to the form of the hillslope itself. In addition, because only a single process was considered it is important to bear in mind that the derived hillslope forms are those which would be molded by running water alone. Horton's formulation provides an excellent framework for understanding the behavior of rainfall and runoff and their relation to incipient erosion by overland flow. As the illustrative examples show, however, hillslopes subject to overland flow exclusively are probably not characteristic of extensive landforms of the earth.

Runoff

Rain falling on an irregular surface disposes itself in a number of different ways. If the soil is not saturated, moisture will infiltrate into the ground at a rate controlled by the soil texture, vegetal cover, and degree of saturation. Initially the rate of infiltration is rapid, but after a period of time the rate approaches an asymptotic limit, defined as the "infiltration capacity" of a particular soil (Fig. 8-10). The infiltration capacity is the maximum sustained rate (measured in units of length per unit time) at which a particular soil will transmit water. If the infiltration capacity of the soil is exceeded and the rain continues, water will collect in depressions on the surface of the ground (Fig. 8-11). These depressions may be minute or large in size, but every natural land surface is irregular. These irregular depressions provide what is known as depression storage. When they overflow, water begins to accumulate in a layer on the surface of the ground. A layer of some finite thickness is required before flow will begin.

Several observers have noted in desert and tropics alike that after several minutes of rain the land surface will suddenly appear to shine or glisten like a slick as overland flow begins. The layer or film

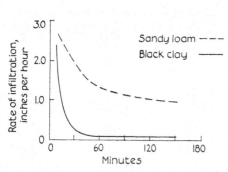

Figure 8-10.

Infiltration rates on sandy loam and on clay. Infiltration capacity of the clay is about 0.04 inches per hour, that of the sandy loam 0.9 inches per hour. The loam contains only 6% clay sizes (finer than 0.002 mm); the clay contains 63%.

at any moment is known as surface detention and represents the water in transit. The detention increases with increasing distance away from the crest of the hill. If the flow is turbulent the profile of variation of depth with distance is given by the equation

$$d \sim \left(r \frac{x}{ln} \right)^{3/5},$$

where r is the supply rate, roughly equal to the rate of rainfall minus infiltration rate in inches per hour, l is the total length of the slope over which overland flow occurs, n is a roughness coefficient, and x is the distance down the slope from the watershed divide.

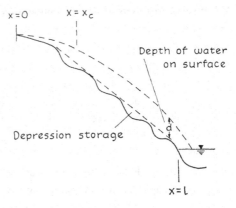

Figure 8-11.

Exaggerated figure to show mode of accumulation of water on hillslopes and development of overland flow. d is depth of flow, and x is distance from the divide. x_c is defined as the width of the hypothetical belt of no erosion. [After Horton, 1945.]

During the principal period of runoff in a storm an average of about two-thirds of the drainage basin is covered with surface detention. Averaged over the entire area, then, infiltration takes place at a rate equal to about one-half the infiltration capacity (Horton, 1936, p. 14).

The velocity of flow is given by the Manning equation:

$$v = \frac{1.5}{n} d^{2/3} s^{1/2}.$$

The flow produced under overland flow conditions provides the force necessary for removal and transport of particles on the soil surface. It can be seen that the form of the profile and the depth, and thus the available force of flowing water, is a function of six variables: (1) the intensity of the rain; (2) the infiltration capacity; (3) length of overland flow; (4) slope; (5) roughness of the surface; and (6) the degree of turbulence or the type of flow. It is here assumed that the velocity is proportional to the square root of the slope. This is the same assumption that is made when dealing with the flow of water in open channels. Where the depth of flow is exceedingly small and the flow is nearly laminar, as it might well be on hillslopes, the exponent of slope perhaps may be larger.

Quantitative study of the hydraulics of overland flow are surprisingly meager. Laboratory experiments using sprinkled surfaces of pavement and turf have been made by Izzard (1946), who found that length of overland flow was as much as 72 feet. Most data on the subject, however, are from sprinkled plats of natural land surface, limited in size to the 6 × 24 feet of the apparatus designed by the U. S. Soil Conservation Service. From a geomorphic point of view, knowledge of erosion processes in overland flow would be enhanced by extending this type of experiment to greater lengths of flow.

The microrelief on any land surface is highly irregular, and flow over the surface is usually not steady and uniform but consists instead of intermittent slugs of flow, or surges. The overland flow then moves in trains of waves across the surface. These may be effective in increasing the erosive ability of the flow. Irregularities on the surface also concentrate the flow in rills or in anastomosing paths, which on a microscale join together the portions of the slope in depression storage.

As Horton (1945, p. 309) pointed out, it is important to recognize that the depth of overland flow is likely to be extremely small, measured in fractions of an inch and not in feet. This can be seen from Horton's example: 1 inch per hour equals approximately 1 cubic foot per second per acre; 1 acre equals 43,560 square feet. The intensity of runoff, q_1, in cubic feet per second from a unit strip one foot wide and a slope length l will be

$$q_1 = q_r \left(l \frac{1}{43,560} \right) = 0.000023 \, lq_r,$$

where q_r is the runoff in inches per hour. Since discharge in the unit strip is also depth times velocity, then

$$q_1 = \frac{dv}{12},$$

where depth is measured in inches and velocity in feet per second. Then

$$d = \frac{0.000277 \, lq_r}{v}.$$

On a slope 100 feet long a velocity of a quarter of a foot per second at a depth of 0.11 inch will produce one inch of runoff per hour. Both the velocity and depth are exceedingly, even imperceptibly, small, yet the flow is large and its capacity to work upon the landscape is of major importance.

By substituting the Manning expression for velocity in the last equation, we can obtain an expression for depth of overland flow in terms of distance x from the watershed divide, intensity of runoff q_i, angle of slope α, and surface roughness n:

$$d_x = \left(\frac{q_i n x}{1020}\right)^{3.5} \frac{1}{\tan^{0.3} \alpha}.$$

Depth is directly proportional to length of slope, intensity of runoff, and surface roughness, and inversely proportional to the angle of slope.

It is somewhat surprising that the magnitude of the roughness factor, n, is similar to that applicable to river channels. Values for overland flow in sprinkled-plat experiments vary from 0.10 to less than 0.05 (Beutner, Gaebe, and Horton, 1940, p. 26). It appears that the true boundary resistance due to surface rugosity is similar to that for open-channel flow of streams, but highly increased resistance may occur because of surface obstructions of vegetation, plant debris, stones, or gravel.

Erosion

Erosion will take place on the surface when the force provided by the flow exceeds the resistance of the soil. Erodibility, as we have seen, is a function of a number of factors—intensity of the rain, the infiltration capacity (permeability) of the surface, the chemical and physical properties that control the disintegration of the rocks and determine the cohesiveness of the soil, and the vegetation, which directly affects both the stability and the infiltration capacity of the soil.

The available force or shear stress provided by the water is a function of raindrop impact, slope, and depth; and the depth in turn is controlled by the relative rates of rainfall and infiltration, by the velocity of flow, and by the length of slope. Raindrop impact may have two effects—one direct, the other indirect. First, the force of impact serves to dislodge particles, much as bombs eject debris upon impact. On a slope particles dislodged at an angle will be displaced downslope. Second, on some surfaces the impact of the drops helps to seal the surface, thus reducing infiltration and promoting runoff. In this case erodibility may also be reduced if the surface is sufficiently compacted to overcome the effect of increased runoff. As a rule, however, the effect of the runoff exceeds the compacting effect.

One should expect the infiltration capacity of sands and gravels to be exceedingly high. The total amount of overland runoff should therefore be low, and as a result erosion by surface runoff should be low. An extreme example is the absence of drainage channels in the Sand Hills of Nebraska.

On the other hand, heavily grassed surfaces such as one might find in pastures or on golf courses should be expected to have only modest erosion by surface runoff, for additional reasons. Grass itself reduces the impact of the falling raindrops. Infiltration capacity of the grass surface is high. A mat of vegetation is itself physically resistant to erosion. The conditions on bare areas, in contrast, are precisely the reverse of those on sand and grassed areas, and hence bare areas under comparable rainfall should be readily subject to erosion by running water. This is, of course, a common observation. It forms the basis of the conservation practice in agriculture designed to reduce erosion by substitution of pastures for cultivated and bare fields.

Horton (1945, p. 319) states that the eroding force per unit area of soil surface is given by the equation for shear stress (Chapter 6):

$$F_e = w \frac{d}{12} \sin \alpha,$$

where F_e is the eroding force in pounds per square foot, w is the weight of a cubic foot of water, d is the depth of the flow in inches, and α is the slope. For steep slopes, as on hillsides, tan α, s, is not equal to sin α, however. By substituting in the last equation the value of the depth as a function of runoff, q_r, in inches per hour, given earlier, the eroding force at a distance x below the crest is given by the equation

$$F_e = \frac{w}{12} \left(\frac{q_r n x}{1020} \right)^{3\,5} \frac{\sin \alpha}{\tan^{0.3} \alpha}.$$

Because the depth of flow is zero at the crest and increases downslope, under these controls alone there should be no erosion until the depth and slope have attained a value sufficient to provide the necessary erosive force. Horton (1945, p. 320) has called this zone a "belt of no erosion." The width or distance from the crest of this belt of no erosion is found by equating a threshold value of the resistance or shear stress with the eroding force F_e in the last equation. If we call this resistance R_e and solve for x equal to the width of the belt of no erosion, assuming that w equals 62.4 pounds per cubic foot, we obtain

$$x_e = \frac{65}{q_r n} \left(R_e \frac{\tan^{0.3} \alpha}{\sin \alpha} \right)^{5\,3}.$$

The width of the hypothetical belt of no erosion is then inversely proportional to the intensity of runoff and the roughness, and directly proportional

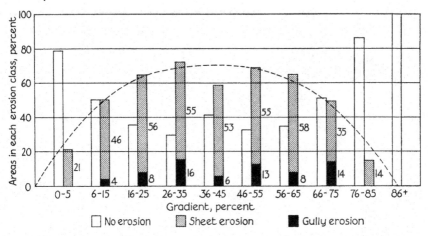

Figure 8-12.   *Field data (bar graphs) showing percentage of area eroded. Boise River basin, Idaho, compared with Horton slope function for surface erosion (dashed line) given by the equation. [After Horton, 1945.]*

to the $\frac{5}{3}$ power of the resistance. For slopes up to 20° the width of the belt of no erosion decreases, or erosion increases with increase in slope.

This formulation may be confirmed by field data on the percentages of areas having different slope angles which were subject to erosion in the Boise River basin, Idaho. The form of the Horton slope function for surface erosion accords roughly with the field data shown in Fig. 8-12. It will be noted that erosion on a given slope increased to a maximum on a 40° slope and thereafter decreased to zero as the slope angle approached 90°.

Erosion by surface runoff might be expected to be greatest on long steep slopes with low infiltration capacity. Empirical data developed by the U. S. Department of Agriculture (Musgrave, 1947) appear to verify this. The rate of erosion measured in tons per acre is proportional to rainfall intensity and length of slope according to the empirical equation

$$E \approx (P_{30}{}^{1.75}s^{1.35}l^{0.35}),$$

where P_{30} is the maximum 30-minute rainfall of 2-year frequency, s is the slope in percent, and l is the length of slope in feet. All of these results show the dependence of erosion upon the same variables, but as yet the specific exponents and variables are not evaluated with precision.

From the standpoint of hillslope form, these equations suggest that the crests of divides—where running water is the responsible erosive agent—

Figure 8-13.

Above: *"Belt of no erosion" and rimmed by rills developed on fresh ash at Pari-cutin, Mexico. Belt of no erosion approximately one-foot wide. [Photograph by K. Segerstrom.]* Right: *Minute dissection by rainwash forming drainage chan-nels on ash at Paricutin [Photograph by K. Segerstrom.]*

will be narrow where either erosive force is large or resistance to erosion is low. To the extent that such crest slopes are formed by running water they would presumably be straight, thus meeting in a knife-edge at the crestline. Raindrop impact and splash would tend to flatten slightly such a knife-edge, but the crest slopes themselves would be straight. At the point where the erosive force of the flowing water exceeds the resistance of the soil, rills should begin to develop on the surface. These rills would then coalesce to form larger channels.

A slope dominated by surface runoff, however, should display a relatively smooth unrilled crest slope associated with a rilled midslope. A particularly good example of such a slope—developed on fresh ash from the eruption of the volcano Paricutín in Mexico in 1943—is shown in Fig. 8-13. This is actually a kind of microtopography and the "belt of no erosion" from

crestline to the head of the rills is about 1 foot, indicating highly erodible materials and high runoff intensity, conditions which do prevail in the photographed area. Observations indicate, however, that the crests or divide areas of the ash also experience mudflows or creep and hence are not strictly loci of no erosion. Their convex or rounded form then is presumably related to creep rather than to rainwash, although the denudation of the crests is exceedingly slow compared to erosion on the slopes.

According to the equation for x_c above and assuming a runoff of 2 inches per hour on a slope of 10 to 20°, the distance from crest to head of rill would be about 1 foot in material having a resistance of 0.1 pound per square foot. This agrees with experimental values for the critical tractive force of bare silty materials. As a comparable example of high resistance, bermuda grass will have a value of R_c at about 1.0 pound per square foot.

The equation for x_c also illustrates the significance of a concept of the effectiveness of events of different frequency. In the Middle Atlantic region of the United States a 1-hour rainfall of 2 inches per hour recurs on the average about once every ten years, whereas a 1-hour rainfall of 1 inch per hour recurs on the average of once every 1.5 years. A landscape controlled by the latter rains could have crest slopes twice as long as the former. But if rills, once they were formed, never healed, then the distance from the watershed line to rillhead would be determined by the maximum runoff rate experienced at each specific point on the landscape.

Hack and Goodlett (1960) observed the results of such local extreme events in the Appalachian mountains of Virginia, where fresh erosion scars were formed by debris avalanches and first-order valleys were severely trenched by a single storm having an estimated return period of several hundred years. In the valleys the distance from the head of a first-order valley to the beginning of a channel is on the order of 150 to 200 feet. This short distance could be obtained on slopes of 15 to 20° for high runoff intensities of 5 inches per hour, even on extremely rough surfaces. Unfortunately, we cannot work backward and evaluate with any precision the relative effectiveness of individual events, because (1) here as elsewhere a single process such as the debris avalanche is not the only process active in forming the slope, and hence a given geometry cannot be readily attributed to a single agent; (2) the hydraulics of runoff on such a mountain surface is itself unknown, and thus reconstruction of the flow and its profile still remains conjectural.

Nevertheless, by focusing on the equation for x_c it can be seen how such factors as overgrazing, which increases runoff or lessens resistance, will promote erosion, and how variations in slope may promote erosion on steep segments of slopes and deposition on flatter portions.

Another example from the badlands of South Dakota illustrates the principle of slope development by running water but also indicates that—even on a slope controlled primarily by this process—erosion may take place at the watershed line, creating a knife-edge divide. These badlands occur principally on two formations, Chadron and Brule shales. On the Chadron, creep appears to be dominant, but on the underlying Brule formation Schumm's (1956) observations indicate a low infiltration capacity and hence a rapid rate of runoff. Runoff on the Brule shale causes rapid erosion of the surface. Landforms on that formation are steep and relatively sharp (Fig. 8-7), providing perhaps the best illustration of the divide form associated with running water in accord with the derivation given above. The belt of no erosion, however, is not apparent. Measurements and observations by Schumm on a fill at Perth Amboy, New Jersey, show that the amount of erosion, on the order of 1 inch per year, is directly proportional to the slope at each position on the hillslope. The surface flow consisted of small surges which were built up behind small obstacles and isolated plants, and the surges consisted of water temporarily stored and then released over or around such obstacles.

Although rainwash accounts for relatively small amounts of movement in areas dominated by frost and snow, it may be influential in fixing the form of the landscape. Rapp (1960) reports that in Spitsbergen rain and meltwater are important in the removal of fines on slopes ranging in angle from 7 to 15°. The removal of fines is important even though the total depth of removal by the mechanism may be less than a millimeter in 150 years because the elimination of fines reduces the significance of solifluction and mudflows which are unable to occur on coarse, dry debris slopes devoid of fines.

Such single isolation of activity is obviously rare in nature, and although they illustrate—much as a laboratory study might—the action of specific processes on specific rock types, the results cannot be applied with equal simplicity to most field examples. In the following section we consider some more complex combinations of processes and hillslope forms, in different environments and in different lithologies.

Lithology and Form

Some general characteristics of hillslope forms in many regions through-out the world can be described. Table 8-4 is taken from the work of

Table 8-4.   Characteristic hillslope angles in Natal, the United States, and the Alps.

Author	Location	Hillslope Angles (in degrees)									
Fair	Natal[1]	45	42	36		33	27	23	18	12	4
Strahler	United States[2]	45	42	38	35	33	26		15		
Piwowar	Alps[2]		43	36		32	28				

[1] Mainly sandstone capped with dolerite.
[2] Diverse lithologies.

Savigear and shows a characteristic set of angles of facets of slopes in different regions of the world. These measurements represent diverse lithologies, and it is relatively clear that despite the climatic variations represented by the data there are no marked differences in the characteristic angles. On the other hand, the differential effect of lithology is clearly shown by the compilation of data given in Table 8-5. In the study by

Table 8-5.   Selected examples showing comparative morphometric

| Location | Order | Area of Basin (sq. mi.) | Average Length of Stream (miles) | | | No. of Streams |
			First Order	Second Order	Third Order	
Arizona, Chinle Badlands, Cameron Quadrangle	3	0.000075	0.0022	0.0054	0.00097	8
Arizona, Mt. Hughes Quadrangle, Tributary to Harshaw Creek NE 1/4 20 R16 E, T22S	3	.150	.085	.303	.189	18
Arizona, Tucson Quadrangle Sagard National Monument (SIV)	3	.0013	.0078	.0065	.031	34
Colorado, Black Hawk Quadrangle, Dury Hill Basin (III)	3	.76	.263	.415	.66	14
New Mexico, Santa Fe Quadrangle, Camino Basin (III)	3	.0215	.0425	.0425	.0350	15
Indiana, Leavenworth Area[a]	3	.37	.126	.248	.52	16
Indiana, Unionville Area[a]	3	.18	.065	.092	.596	26
Indiana, Jasper Area[a]	3	.105	.077	.096	.241	16
Virginia, Pennington Area[a]	2	.0363	.100	.435		
Virginia, Copper Ridge Area[a]	2	.0760	.139	.705		
Virginia, Blountsville Area[a]	2	.0469	.106	.496		

[a] Values represent mean of ten localities in each area.

Coates the valley sideslopes on sandstone and coarse-grained limestone are 20°, those on siltstone are 38°, and those on the shale are 15°. Similarly, Melton's data on New Mexico and Colorado show considerable variation of angle of slope and lithology. These differences, however, are not exclusively related to lithology.

In southwestern Virginia, slopes on dolomite are significantly lower than they are on shale or sandstone. These results certainly come as no surprise, although the lack of truly comparable observations on varied lithologies in different climates still prevents a completely clear differentiation due to rock type alone. Under some conditions other factors have much more influence on the angle of slope. Strahler found in the Verdugo Hills of California, for example, that angles of hillslopes in similar rocks differed significantly, whereas those in different rocks were the same. It appears in this case that erosion by stream channels at the base of the slopes is a more significant factor in fixing the angle of slope than are the processes occurring on the slopes themselves.

Using the graphical relationship described earlier, Hack and Goodlett (1960) compared profiles of hillslopes on different types of rocks in the properties of small-order drainage basins.

Bifurcation Ratios		Length Ratios		Drainage Density: Miles	Drainage Density of First Order: Miles	Maximum Valley Side Slopes	Lithology	Source
Order 1 / Order 2	Order 2 / Order 3	Order 2 / Order 1	Order 3 / Order 2	Sq. Mi.	Sq. Mi.	(degrees)		
2.5	2.0	2.47	0.18	303	145.7	21.2	Flat-lying shale	Melton (1958)
7.5	2.0	3.58	0.62	13.8	8.5	37.0	Igneous	Melton (1958)
3.7	7.0	.83	4.72	210.1	152.8	12.8	Schist and granite	Melton (1958)
5.5	2.0	.16	1.59	5.8	3.8		Gneiss and schist	Melton (1958)
6.0	2.0	.82	4.28	34.0	23.7	11.4	Indurated sand and gravel	Melton (1958)
4.4	2.8	2.00	2.45	8.0		20.	Sandstone-limestone	Coates (1958)
4.1	4.9	1.46	7.53	13.15		38.	Siltstone	Coates (1958)
4.2	3.1	1.22	3.18	14.		15.	Sandstone and some shale	Coates (1958)
3.8				10.8	12.0	33.	Shale and fine sandstone	Miller
3.7				8.4	9.0	23	Dolomite	Miller
3.8				9.7	9.7	32	Sandstone	Miller

eastern United States. Their data, generalized in Fig. 8-2, **B,** indicate that in general slopes on sandstone mountain ridges are both steeper and straighter than slopes on either shale or sands, silts, and gravels of the Coastal Plain of Maryland. Convex slopes on shale and limestone in this region are roughly similar to those in the Coastal Plain, although some of the latter are less steep and have higher curvatures. Both the steepness and the form are related to the greater resistance of the sandstone. The steepness itself is a function of relief and hence a function of both the base level provided by streams at the foot of the side slopes and the slope processes active at the crest.

Curvature is clearly not solely a function of processes active on the slope itself. Savigear's (1956) detailed measurements of changes in hillslopes on the coast of Wales, discussed in Chapter 12, show the effects of wave action on hillslope form, and the steepening caused by undercutting of slopes by stream channels is a commonly observed effect of base-level control.

Reasoning from the resultant forms, one may argue that infiltration is greater on the Coastal Plain sands and gravels, and hence the flow at a given distance from the crest is less, providing less erosive force and therefore no steepening of the slope. In a like manner, because the sizes and quantities of materials to be moved on the semiconsolidated lithologies are less than on the sandstones, less gradient is required at a given distance below the crest. The increasing steepness with increasing distance represents an increasing gradient required to transport an increasing quantity of material downslope without benefit of an equivalent or adequate increase in flow.

Observations in one area in Belgium (De Béthune and Mammerickx, 1960) indicate that 75% of the slopes in argillaceous material are less than 4°, while a similar percentage in the valley sections of the sandy silicious Brussels formation are 8°. Maximum slopes in the latter formation are 26°, whereas those in the argillite are closer to 7°.

It is interesting to note that some measurements of slopes in Belgium indicate that the slope angles on shales and sands are markedly less than the slopes on limestone and sandstone and that in this case the latter two are quite similar. Based on reasoning relating both infiltration and erodibility, one should have expected that the less erodible sandstone would have produced the steeper slopes. Macar and Fourneau (1960) do suggest that, in fact, the increased infiltration capacity of the limestone compensates in a crude way for its lesser resistance, resulting in a similarity of slope angle.

Melton (1958) adopted a somewhat similar explanation in discussing a direct correlation between the angle of valley side slopes and infiltration capacity. He suggested that where the infiltration capacity is high, (a) surface runoff would be low, (b) erosion rates would similarly be low, and (c) the adjacent channel receiving clear runoff from beneath the surface would erode its channel, thus steepening the side slopes. In contrast, where rapid surface runoff promotes erosion, channels will fill with debris, which will tend to lessen the angle of the valley side slopes. A subsurface section through the valley would be required to substantiate this hypothesis but the rationale is reasonable and appears to agree with the Belgian observations.

Measurements of the relations between the dip of underlying bedding and the angle of the hillslope indicate that, in general, slopes dipping in the reverse direction from the bedding (scarp slopes) are steeper than those paralleling the bedding. Observations in Belgium indicate that this difference is greatest for slopes in limestone. In neither case are these the simple dip slopes of a cuesta. Where the bedding dips from 30° to 60°, differences in the mean angles of the hillslopes are on the order of one to two degrees.

The earlier analysis of the interrelation of sediment yield and vegetation suggested that the inhibiting effect of vegetation should influence the form and declivity of slopes. The evidence with which to evaluate this single factor is inconclusive, although data from Belgium indicate that on comparable rock types in a single region there is no significant difference between the declivity of slopes under forest cover or in agricultural fields. As a rule, elimination of vegetation leads to acceleration of erosion and, in accord with the principles developed by Horton, increased erosion results in an increased development of the network of drainage channels. Incision of the drainage channels and the increased density, as at Paricutín (Fig. 8-13), dissect the land surface. By lowering the elevation of the channels the same incision causes steeper slopes to form when the divides at the crest of individual slopes are not reduced as rapidly as the channels at their foot slopes.

Climate and Form

It is common to observe, even in diverse climatic regions, an asymmetry of profiles of valley slopes on opposite sides of a given valley. Such observations have been noted in virtually all climatic regions, from the periglacial zone to the semiarid regions. As a rule, the asymmetry can be related to the differential climate on opposite sides of the valley. These asymmetric profiles provide perhaps the best examples of the effect of different climates on similar rock types. Where snow accumulates it will melt more often and more rapidly on south-facing slopes. Similarly, diurnal fluctuations of temperature will produce alternate freezing and thawing on the south-facing slopes. Such differences, plus differences in insolation on north- and south-facing slopes, may be comparable to differences of several degrees of latitude so far as the local climate is concerned. The climatic variation is associated with significant differences in vegetative cover on opposite sides of a valley. Thus, for example, drier southwest-facing slopes in the Shenandoah Valley of Virginia are characterized by yellow pine, pitch pine, and table-mountain pine; on the moister slopes several species of oak are present and both pitch and table-mountain pine are absent. The Shenandoah example is particularly pertinent because in some valleys the dip of the strata produces differential moisture conditions, with asymmetries similar to those controlled by climate. This gives added weight to the explanation of slope control by moisture regimen.

Characteristics of asymmetric valleys, summarized by Hack and Goodlett (1960), are given in Table 8-6. From the standpoint of slope form, note

Table 8-6.   Summary of characteristics of opposite sides of asymmetric valleys in the Little River Basin. [From Hack and Goodlett, 1960.]

Characteristics	Northwest- or Southwest-facing Slopes[a]	Northeast- or Southeast-facing Slopes[a]
Declivity	Gentle	Steep
Moistness	Dry	Wet
Surface mantle of stones	Coarse	Fine
Predominant vegetation	Yellow pine forest unit	Oak forest unit
Density of cover	Dense, many shrubs	Open, few shrubs
Drainage density	High	Low
Drainage network	Well developed	Less well developed
Postulated most important process	Slope wash and channel erosion	Creep

[a] Differences between northwest and southeast characteristics are related to the prevailing southeasterly dip of the rock strata. Differences between northeast and southwest characteristics are related to exposure.

that the drier slopes are longer and the declivity less, whereas the reverse is true on the moister slope. Vegetation attests to the differences in moisture, and such differences indicate that on the wetter slope, where moisture is retained, movement of soil is more continuous under an open cover. The coarser debris and denser vegetative cover on the drier slope suggest that movement on it is associated with intermittent surface runoff. These differences suggest that the more continuous process of creep may dominate in the moister area, while surface runoff controls the drier slopes. These differential processes suggested for slopes in the Shenandoah Valley have their counterpart in the differential behavior and form of the Chadron and Brule formations, due there to differences in infiltration.

In the semiarid regions there appear to be two mechanisms that demonstrably lead to asymmetry of slopes on different sides of the same valley. Reasoning from field observations and the association of decreasing differences in slope angle (lessening asymmetry) with increasing channel gradient, Melton (1960) has suggested that the entrance of tributaries to a main stream forces the channel to undercut the hillslope opposite the channel entrance and that the resultant steeper slopes only occur where the main channel is unable to remove all of the debris contributed by the tributary. Where there are steep channels and all of the tributary debris is removed, the asymmetry of hillslopes does not occur. To the extent that both slopes and tributaries with south-facing orientation are intensified in their activity by differences in climate, this form of asymmetry may be said to be climatically controlled. Melton's data do not support this conclusion for his examples.

In contrast to this type of channel—controlled asymmetry—observations in valleys in New Mexico, where Mesa Verde sandstone overlies weaker shale, show marked asymmetry in the valley side profiles, even in the absence of tributaries. Here the moister north-facing slopes on which snow accumulates are gentler. The differential "climate" of opposite sides of the valley is also reflected in the vegetation. Observations indicate that when the shale is moist the upper sandstone blocks are easily removed as a result of sapping and weathering of the shale beneath. The increased mobility of the shale, whether the result of mass movement or running water, leads to a gradual decline in slope angle and the resultant marked asymmetry.

In areas subject to intense frost action a difference in exposure of hillslopes will lead to a difference in frost activity. Mass movement is enhanced by the presence of soil moisture and by the action of ice crystals, as well as by the presence of an underlying layer of frozen ground. Where differential exposure accelerates mass movement on one side of a valley, slope angles will be reduced and frost action at the crest will lower the divide more on one side of the valley than on the other. A particularly good example of asymmetrical valley side-slope development controlled by weather elements is in Lappland, where Rapp (1960) observed differential effects on dry east-facing and wet west-facing slopes. In this area in northern Sweden moisture-laden winds from the ocean to the west deposit snow on the mountainsides. The snow is driven by the wind up over the ridge crests and lodges in cols on the west-facing uplands. The melting snow promotes increased snow avalanching on these slopes, thus increasing both the overall rate of erosion and depressing the height to which talus or scree is built at the base of the slopes.

These examples of asymmetric valleys demonstrate some of the mechanisms through which climate exercises its control over slope form. This control may be primary, through control of processes on the slopes themselves, such as runoff or creep. It may be secondary, through erosion by stream channels which are either accelerated or reduced, depending on whether infiltration capacity is large or small and runoff is clear or sediment laden.

Let us look briefly at a few examples of slopes and processes active in different regions, specifically chosen to represent several different climates. We will disregard the theoretical or supposed significance of the climatic differences.

Examples of Slope Forms in Different Regions

Much has been written about landscape forms in various areas. It is surprising, in view of the extensive literature, how few examples have been described in quantitative detail—including slope angles, measured amount of relief, measured size of debris, and other pertinent quantities.

From the examples which have been described with some quantitative detail a few are chosen for review here. They represent a very small sample, but they indicate certain similarities in the landscape forms that do exist in the most diverse climatic zones.

Spitsbergen and Scandinavia

In Spitsbergen, Jahn (1960) has described a profile of what he terms a typical slope for that region, one characterized by snow and ice. The slope consists of four zones described by him as follows.

1. A zone of weathering rock walls and rock slopes with inclination greater than 40°, covered by a thin talus layer inclined at 25 to 40°.
2. A zone of dry gravitational talus cones and scree slopes at an angle of repose of 30 to 40°, and of humid slopes (from avalanches at the outlets of gullies) with inclinations of 15 to 25°.
3. A zone of solifluction terraces consisting of (a) high talus terraces and stone garlands of short length at 15 to 25°, (b) terraces of medium length and rock streams with inclinations of 10 to 15°, (c) long and low terraces and rock streams, inclined at 3 to 10°.
4. A zone of sedimentation by slope water and of disappearing solifluction features, with inclinations of 2 to 5°.

The proportions of the various parts of the slopes vary according to the stratigraphy or geologic structure, inherited forms, microclimate, and basal erosive activity of rivers or ice. (See Fig. 8-14.)

Figure 8-14.
Slope in Spitsbergen, showing rock wall, talus, and portion of convex upper slope.
[*Photograph by H. Amien.*]

In what Rapp (1960) calls the zone of block fields and tundra, he asserts that there are three principal processes active on the upper slope: rockfalls, the debris from which is deposited on the talus slopes; mudflows, which occur on the wet talus and in turn build alluvial cones; and dirty-snow avalanches, which produce tongues at the base of the slope. Comparative photographs taken in 1873 and in 1959 show very little movement on the dry talus in Spitsbergen, and retreat of the rock wall itself appears to be greater than gullying of the talus. On the moderate slopes the principal processes appear to be slides and mudflows. On slopes of 7 to 15° there are small solifluction tongues and terraces and rates of movement, in the 20-year period 1938–1958, varying from highs of 5 to 12 cm per year to no movement. On gentle slopes of less than 2°, rainwash removes fines to the valley bottoms. The bottoms are coursed by innumerable braided channels, which are inundated during periods of floods with little erosive effect.

In the forest zone the steep slopes in Scandinavia are primarily relic glacial forms. Again the process is one of rockfalls to talus surfaces below. Mudflows in spring occur on the more moderate slopes with declivities of about 30°. Earth-flows and avalanches have been observed even in coniferous forest zones. On gentle slopes there are still bare areas exposed, as well as undisturbed landforms composed of till and glaciofluvial materials upon which there appears to have been no movement since deglaciation. On cultivated land there is characteristic sheet erosion.

Central Appalachian Mountain Region

There is a surprising lack of measurement and observation on hillslopes in the temperate regions, even though the majority of workers in the field probably live in this region and for many years the temperate latitudes have been considered by many to be the "normal" geomorphic environment. The following illustrations of slope forms and properties are drawn from the work of Hack and Goodlett (1960).

Figure 8-15, **A,** is a topographic map of a first-order tributary valley on Crawford Mountain in western Virginia. In the explanation on the figure, portions of the side slopes have been designated according to their topographic form. The inferred relation of this form to runoff is also given in the explanation. A logarithmic plot of the profiles along the lines A, B, and C shows that the crest slopes of each of the profiles is convex upward (Fig. 8-15, **B**). Only the profile along the axis becomes concave, at a distance of about 400 feet. In this particular valley the side slopes have no basal or foot slope but meet the channel at a sharp angle. This is not so in all first-order valleys in the area, however. Some have small valley floors which join the side slopes in smooth concave curves. The north-facing slope is appreciably gentler although of the same form as the head of the valley and south-facing slopes. This was noted earlier in the discussion of the asymmetry of these valleys.

Over much of the surface the debris mantle is thin, on the order of 3 feet, but in places bedrock crops out and in others the mantle may be 10 to 20 feet thick. Development of soil horizons is minimal, which Hack and Goodlett attribute to the high rate of activity on the slopes. Particles on the convex bulges (noses) and side slopes are small (generally less than 22 mm) and poorly sorted; those in the hollows or channelways are coarser

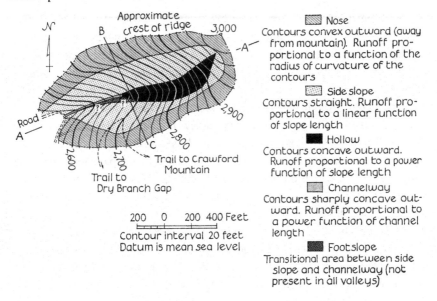

Approximate crest of ridge

Nose
Contours convex outward (away from mountain). Runoff proportional to a function of the radius of curvature of the contours

Side slope
Contours straight. Runoff proportional to a linear function of slope length

Hollow
Contours concave outward. Runoff proportional to a power function of slope length

Channelway
Contours sharply concave outward. Runoff proportional to a power function of channel length

Footslope
Transitional area between side slope and channelway (not present in all valleys)

200 0 200 400 Feet
Contour interval 20 feet
Datum is mean sea level

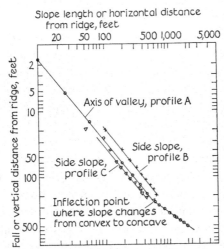

Slope length or horizontal distance from ridge, feet

Axis of valley, profile A

Side slope, profile B

Side slope, profile C

Inflection point where slope changes from convex to concave

Fall or vertical distance from ridge, feet

Figure 8-15. *Topographic map and profiles of first order valley in the central Appalachian Mountains.* [*From Hack and Goodlett, 1960.*]

(on the order of 100 mm) and better sorted. Particle size appears to be a function of the supply of water, for both increase with drainage area. It is interesting that the forest types correspond closely to the particle sizes of the surficial mantle within a given first-order basin as well as on the asymmetric slopes. Finer sizes are associated with the drier oak-type vegetation on side slopes and pine-type vegetation on noses, while coarser particles are associated with the northern hardwood forest type found in the hollow and channelway. The factor controlling vegetation, however, appears to be differences in soil moisture, but the control of particle sizes appears to be through winnowing of fines by surface runoff.

As the discussion of valley asymmetry indicated, the steeper wetter slope (C in Fig. 8-15, **B**) appears to be less stable and subject to creep. That the particle size is somewhat smaller on this steeper slope also suggests that creep, which moves large and small particles alike, may be more significant here than on the longer gentler slopes from which rainwash has winnowed the fines, leaving larger particles behind. Both slopes are adjusted to the kinds of processes occurring on them, inasmuch as the longer slope provides a higher discharge or runoff to compensate for the flatter gradient, and on the steeper gradient a shorter slope provides less discharge. Although the convex form of the crest slope of the longer slope also suggests erosion by creep, there is no evidence to support or controvert this possibility.

In addition to creep and rainwash, which are considered to be nearly continuous processes on the slopes, the rare rainstorms of unusual intensity also contribute to erosion of the hillslopes by the production of large debris avalanches; these strip the forest cover from the surface, leaving shallow trough-shaped scars on the surface, flanked by levee-like bordering ridges.

The close tie between channel and valley processes and hillslope characteristics is particularly apparent in the first-order valleys in the Appalachian highlands. Slope form and process vary with their position relative to the drainage ways. As major rare floods scour and widen the valley bottoms, side slopes are undercut and steepened. Over time creep and rill wash on the slope maintain a consistent form, representing an adjustment between slope and channel processes. It is possible that the devastating effects of the floods are partially erased by more frequent lesser events, as Bryan suggested for the veneered slopes in the southwestern United States.

Hawaii

Porous basaltic lava flows in the Hawaiian Islands, combined with high precipitation, give rise to some particularly spectacular and instructive landforms. Virtually the entire island group is composed of lava flows, which weather rapidly by chemical disintegration in the presence of water, and rainfall in many areas exceeds 300 inches per year on the windward flanks of the high mountains. However, because the volcanic rocks are extremely porous the water table is low, and only the deeper valleys maintain perennial streams.

Major valleys grow at the expense of others as they tap ever larger supplies of groundwater. Because weathering is most intense at or just below the water table, incision of the stream channel and sapping of the

Figure 8-16.

U-shaped valley in Hawaii showing steep walls and flat floors typical of regions in permeable basalt in which dry walls are subject to little or no removal while valley bottom is widened and lowered at the water table by rapid chemical disintegration.

valley walls by weathering takes place near the level of the valley floor, either perennially or during intermittent wet periods. Physical breakdown and removal of debris derived from the fine-grained basalts on the dry upper slopes is slow. Concomitantly intensive chemical disintegration at the water table produces and maintains steep-walled valleys (Wentworth, 1928). Due either to sapping by weathering of the basal slopes or by planation of streams in the valleys, the valley floors tend to be flat. Where small accumulations of debris form at the base of the straight slopes, valleys tend to be U-shaped in cross section, similar in form to the typical U-shaped glaciated valley (see Fig. 8-16). Only at the very crest of the slope are small areas rounded off and made convex.

In regions where the differential rates of weathering and erosion between valley bottom and valley crest are less striking, more debris may accumulate in the valley bottoms in the form of talus and colluvium. Such detritus varies from large blocks to silt and mantles the basal portions of the bedrock surfaces. If removal by slides or streams is insufficient, cliff faces will be obliterated by encroaching debris.

As a rule a composite slope is formed (Fig. 8-17) with a crest slope cut in rock at an angle of 35 to $60°$, a gentler slope below cut on the talus and colluvium, and at the base a more gradual slope of finer debris, which grades onto the surface of the valley below. The cover of talus and colluvium (named "taluvium" by Wentworth, 1943) is often only a relatively thin veneer on the bedrock beneath it. Over a period of time debris accumulates on the upper segments of the bedrock slope as a result of weathering and creep from above. The mantle of fines helps to retard infiltration and vegetation becomes established on the surficial mantle, the

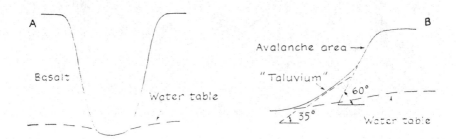

Figure 8-17.   *Forms of slopes in basalt in Hawaii.* **A.** *U-shaped valley intersecting typically low water table with dry upper slopes preserved at steep angles.* **B.** *Area of lower relief subject to avalanching and accumulation of talus.*

roots of the vegetation helping to bind together the debris of weathered rock and soil.

On the steeper slopes, however, the surficial mantle is usually no more than 1 to 2 feet thick. Heavy and prolonged rains which regularly recur in the region soak the surficial mantle, increasing its weight while at the same time lubricating its contact with and frictional hold upon the rock below. Under these conditions, at angles of 45 to 48° the mantle periodically gives way as a soil or debris avalanche, leaving a scarred surface of exposed rock. The surface of the scar remains roughly parallel to the former surface of the surficial mantle. As the avalanches occur beneath the crest of the slope the uppermost portion of the slope maintains a steeper angle, up to approximately 60°, while on the slopes below rainwash and gullies transmit the debris to the valley floors, where it is removed by streams. The process of removal of fines on the taluvium-covered slopes is aided by percolating waters, which carry clays and silts into the rubble mantle, from which they reappear on the surface lower down the slope. Between storms the silts and clays dry out on the surface, only to be removed again in the next downpour (White, 1949). Wentworth estimated that avalanching removed approximately 1 foot of material in 400 years, averaged over an area of 15 square miles. It is thus an effective agent of erosion as well as a contributor to renewal of exposure to weathering of bare rock surfaces.

It is perhaps worth noting that many of the valley features in the basaltic lavas are similar to those of karst regions, which they closely resemble in their susceptibility to chemical weathering and in the low level of the water table.

Southern Sudan

In the southern Sudan, a region experiencing a rainfall of about 26 inches in thunderstorms primarily between May and October, a granite landform has been described by Ruxton (1958), which resembles those described by Bryan and by King but has some especially significant differences. A cross section and plan of this form is given in Fig. 8-18. This hillslope is separable into several parts: (1) a rocky cliff with boulders about 9 to 20 feet in diameter; (2) a boulder-controlled slope at an inclination of 32°, with boulders on the order of 6 feet in diameter derived from the cliff above; (3) the major segment of the slope at 24°, mantled by disintegrated granite of compound fragments and much feldspar to a

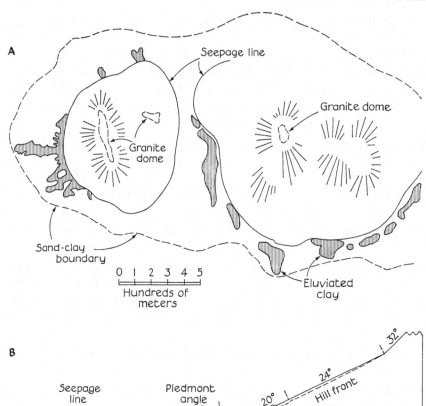

Figure 8-18.   *Map and profile of granite domes and associated pediment,
at Jebel Balos in the Sudan. [After Ruxton, 1958.] **A.** Planimetric sketch. **B.** The slope profile and excavations on the
southwest side of Jebel Balos, drawn to natural scale.*

depth of 2.5 feet, with scattered boulders on the surface; (4) a bedrock
outcrop on a slope of 20°; (5) a lower slope of 13°, mantled by weathered
material in transit with median diameters of 40 to 50 cm and containing
about 15% feldspar. Beneath this moving mantle the section contains more
than 65% quartz, less than 15% feldspar, and less than 25% clay; (6)
below the break in slope, which occurs within a distance of 2 m, there
is an upper pediment segment declining from 5 to 3°, containing particles
markedly smaller than those above, about 20 centimeters in diameter;

(7) the distal pediment surface has a gradient of about 1° and meets a clay plain abruptly at its margin. In a section 5 feet below the surface the segment above the break in slope (see Fig. 8-18) contains about 4 to 8% solid rock, but in the segment below the break there is an abrupt drop to less than 1% solid rock.

Ruxton presents good evidence to indicate that the break in slope is maintained by accelerated weathering, which occurs in the zone just downslope from the break in gradient. Granitic boulders appear to break up suddenly—that is, in a very short distance. Winnowing of the fines produced by this breakup will initially produce a slight break in slope. If deposition occurs on the flatter downslope segment, subsurface flow will persist there longer. Weathering is most intense at this site in the region of seepage. During a rain, infiltration occurs on the slopes above, and lateral movement of groundwater transports particles through the ground toward the seepage zone shown in Fig. 8-18. These are the "eluviated" materials of the clay plain. Differential subsurface weathering tends to persist where the thickness of the mantle is greatest. Coupled with removal this weathering tends to maintain the abrupt break in slope. Material rich in feldspar is moved over the surface by both rainwash and creep from the upper slopes and is carried out over the pediment. Its origin on the slopes above and not from the pediment beneath is suggested by the fact that the weathering zone of the pediment itself is impoverished in feldspar. Weathering of the pediment surface causes a reduction in volume and in elevation due to compaction. Further lowering appears to be by rills and gullies dissecting the surface.

Southwestern United States

In his classic paper on erosion and sedimentation in the Papago Country, Bryan (1923) described three classes of slopes found in this arid region of Arizona where rainfall ranges from about 3 to 5 inches annually: (1) cliffy slopes in massive rocks at inclinations from 45 to 90°; (2) debris-mantled or boulder-controlled slopes in well-jointed rocks at inclinations of 20 to 45°; and (3) rainwashed slopes in minutely jointed rocks at angles of 15 to 20°. Inclination of each slope was found to be a function of the jointing and mode of breakdown of the bedrock.

Cliffy slopes are found on massive rocks with widely spaced joints. Because the joints control the boulder sizes which are produced, the wider the joint spacing, the larger the boulders and the steeper are the slopes.

On many such slopes the blocks are so weathered that when rockfalls occur the boulders disintegrate on impact and little or no talus is formed.

Boulder-controlled slopes predominantly have slope angles of 30 to 35°. In bedded lavas, blocks 2 to 6 feet in diameter are relatively easily dislodged but weather slowly, thus forming extensive talus, which may mantle the entire slope. The resulting slope then is at an angle equal to the angle of repose of the weathered debris.

Figure 8-19.

Slope form and pediment in the Mojave desert area, California, showing boulder-controlled slope at steep angle. Bedrock projects above pediment surface and shows that bedrock underlies the surface alluvium at shallow depth. It is not possible to tell from only surface data whether this pediment is being dissected and this has exposed the bedrock knobs.

In places gullies dissect the talus to bedrock. The lack of severely weathered boulders in the gully scars and the extensive coverage of weathered boulders on ungullied slopes indicate that the scars "heal" gradually when uncut by channels.

Boulder slopes on granite range from 20 to 45°, with the angle of slope being roughly proportional to size of the boulders. Unlike the talus on lava the granite mantle is simply a layer of boulders. These as a rule are exceedingly weathered, as is the immediate rock surface beneath them. When the boulders are dislodged they disintegrate rapidly, rarely reaching the base of the slope. The surface of the slopes is often irregular as a result of weathering and subsequent removal in joints and along bedding plains, leaving temporary niches, rock shelters, and other microfeatures.

The lowest slopes in the Papago region are rainwash slopes on gneisses, phyllites, tuffs, and shales, which yield fine-grained particles upon weathering. As these particles are readily eroded and transported, residual landforms on these lithologies form modest hills compared to the steep mountains associated with cliffs and boulder-controlled slopes.

In such desert topography hill or mountain slopes make an abrupt contact with the bedrock erosion surface of the surrounding pediment (Fig. 8-19). Bryan (1923, p. 54) attributed the difference in angle between mountain front and pediment surface to a difference in process on each surface. He concluded that "fine rock debris is moved down the mountain slope by rainwash [in this book termed overland flow] and carried away from the foot of the slope by rivulets and streams that form through the concentration of the rainwash." The lesser slope of the pediment is thus attributed to the greater capacity for transport of the concentrated flow. In general the slopes in the arid region of the southwestern United States are controlled primarily by the weathering characteristics of the rock and by the type or relative concentration of flow.

South Africa

L. C. King (1953) and others have described the form and processes active on slopes and pediments. King, following Wood (1942), suggests that every slope is divided into four essential elements, or what we have called segments. This fourfold division is shown in Fig. 8-20. The rounded convex crest slope is primarily determined by creep. The free face is the outcrop zone where bedrock is exposed. Below this is the debris slope, controlled by the size of fragments which come to rest at their angle of repose.

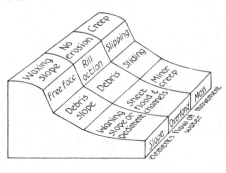

Figure 8-20.

Slope elements, processes, and forms of hillslopes in semiarid regions; and in all regions subject to erosion by running water. [After King, 1962.]

King (1953, p. 728) notes that boulders 3 to 4 feet in diameter may have a declivity of 35°.

The waning slope or pediment exhibits the concave profile usually attributed to erosion by running water. Beneath the surface wash of the waning slope may be a mantle of weathered material grading into hard bedrock at depth, or solid rock sharply truncated by erosion may lie beneath a thin mantle of debris in transport. The latter is the typical pediment, a planed surface on bedrock overlain by a veneer of material awaiting its turn for running water to set it in motion. All gradations of these two conditions may occur in nature.

As runoff increases during a rainstorm, overland flow becomes dominant and flow occurs over the free face and onto the debris slope below in tiny closely spaced rills. If the rills coalesce, guided by irregularities in the rock, they may create gullies on the surface of the debris slope. Water flowing in the rills over the hillslope is checked on the flatter slope at the base of the hill (the pediment surface) where debris may be temporarily deposited. As flow continues to build up on the pediment surface, according to King the sheet of water becomes more turbulent and channeling of the surface may take place, leading to the development of gullies on the pediment surface.

As in the Papago country, the steepness of the rock slopes is a function of the particle size of the weathered material, in turn controlled by lithology and jointing. King states that the break in slope of erosional remnants and pediment surface is most abrupt where the amount of detritus being shed from the slopes is small. Although he does not indicate precisely how particle size is related to slope angle, he does state that hillslopes on jointed granite stand steeply.

Examples of the slope segments of King are described by him (1953, p. 744) in a typical landscape of the east and south sides of Manda (Concession), Southern Rhodesia. Pediments more than a mile in width sweep down to a narrow strip of swampy flood plain 200 to 400 yards wide. Crowning the pediments near their highest points are rocky koppies (tors or inselbergs), the side slopes of which incline at 30° although the lower 300 to 500 feet may approach a slope of 45°. The slopes of the pediment surface decrease streamward, beginning at 1°40′ near the koppies, 1°10′ in middle parts, and declining to 0°30′ where the pediment meets the swampy area of the stream. Where the pediments enter into the mouths of narrow side valleys, rock fans occur that slope 5° or 6°, but the pediments themselves have slopes not exceeding 3° even when they are narrow.

Some Generalizations on the Forms of Hillslopes

These very brief case histories of slopes in several different areas of the world and the isolated examples of specific processes on hillslopes all indicate that slope forms, except in completely dry areas or ones covered by glacial ice, are not uniquely related to climate. All kinds of hill forms are found in all kinds of climates. The inselbergs of the tropics are also found in the semiarid desert. The cliff face, debris face, talus-veneered rock surface, and concave foot slope are found in the arctic regions of snow and ice, the tropical forests of Hawaii, and the semiarid regions of the Sudan and southern Africa. These are, in fact, the same four segment slopes described by King, with the crest or waxing slope exceedingly small or missing.

The cliff or free face is credited by most observers to rockfalls, with or without various mechanisms of basal sapping, and not to rill wash. King, on the other hand, holds that the processes of scarp retreat are gully-head incision, sheetwash, downhill movement under gravity, and weathering, of which the most important is the attack of gully heads. The similarities of forms in diverse climatic regions and the differences of forms in similar climatic environments, however, emphasize the need not of classification but of understanding the interrelation of climate, lithology, and process. Keeping in mind the diversity, some generalizations can be drawn about hillslope forms and their relation to process and control. Although some are quite elementary, they direct attention to the principles and mechanisms responsible for slope form rather than to real or apparent regional correlations.

The following statements are intended as a summary of concepts and implications derived from the preceding discussion of processes and factors controlling hillslope form.

Hillslope forms are controlled by lithology (including erodibility and jointing), surface and subsurface flow of water, mass movements, and base level. Similar hillslope forms may have wholly different origins. Thus tectonic movements, climatic changes, and changes in base level may all produce complex hillslopes of similar or different forms.

Differential geologic characteristics such as mineralogic composition, structural attitudes, and jointing have a dominant influence on the forms of hillslopes in any region, partly through the number and orientation of zones of weakness and also by the type and rate of weathering.

Erosion by running water and mass movement has been observed on slopes in virtually all climatic regions; most slopes are formed under the combined rather than the unique action of mass movement and running water. The evidence suggests that rounded or convex hill-crests are formed by creep and that narrow sharp divides are associated with erosion by running water.

Climate affects hillslope forms through the action of such factors as precipitation and temperature on weathering and removal. Specific climates do not produce unique landforms; there are innumerable combinations of climatic factors and lithologies and structural or stratigraphic relations which will produce the same forms. The evidence does indicate, however, given the same structural relations, that mass movement can operate on slopes of only a few degrees, if moisture and freezing and thawing are prevalent. This combination will produce comparatively flat slopes, other factors being constant.

Low relief may limit hillslope form by (a) controlling the available length of overland flow, or (b) limiting the forces activated by the available potential energy to values less than those required to overcome the shear resistance of the rock materials.

The law of continuity applied to movement of debris on slopes requires that in every successive segment on a slope the inflow of debris must equal the outflow plus or minus the change in the amount of debris stored.

The maintenance of cliffs or rock walls requires as a minimum that the rate of removal of debris at the base of the slope equal the rate of fall from the cliff face. More often than not a cliff (the "free face" of King) is maintained by basal sapping, which may be due to heterogeneity of materials, to gullying, or to controls of base level. The difference in properties of surface materials and those below the surface has its effect principally in the differential ability to transmit water. This heterogeneity may lead to increased weathering and compaction; it may be the cause of slide or glide planes on which slippage or sliding occurs, and may cause lubrication of such planes by concentration of water.

Present hillslope forms may reflect antecedent history to the extent that modern controls have not obliterated the earlier combinations of process and substrate.

Most of these generalizations are self-evident. Their inclusion here is a plea for simplicity rather than complexity in the analysis of slope forms and processes. It appears to us that there is no most typical hillslope form. Landforms are affected by the modes and differential rates of deformation of earth materials, and this applies also to hillslopes. Finally, the dearth of descriptive measurement of both form and process on slopes causes any set of generalizations to seem too broad and hence too vague to be readily useful in the field.

REFERENCES

Beutner, E. L., Gaebe, R. R., and Horton, R. E., 1940, Sprinkled plat runoff and infiltration experiments on Arizona desert soils: *U. S. Dept. Agric. Soil Cons. Serv. Tech. Paper* 38, 30 pp.

Bryan, K., 1923, Erosion and sedimentation in the Papago country: *U. S. Geol. Survey Bull. 730,* pp. 19–90.

Coates, D. R., 1958, Quantitative geomorphology of small drainage basins of southern Indiana: Technical Report No. 10, Project No. 389–042, Office of Naval Research, 67 pp.

De Béthune, P., and Mammerickx, 1960, Études clinométriques du laboratoire géomorphologique de l'Université de Louvain (Belgium): *Zeit. für Geomorphologie,* Suppl. Bd. I, pp. 93–103.

Everett, K. R., 1962, Quantitative measurement of soil movement (abstr.): *Geol. Soc. Am. Special Paper No. 73,* p. 147.

Gilbert, G. K., 1909, Convexity of hilltops: *Journ. Geol.,* v. 17, pp. 344–350.

Hack, J. T., 1957, Studies of longitudinal stream profiles in Virginia and Maryland: *U. S. Geol. Survey Prof. Paper* 294-B.

Hack, J. T., and Goodlett, J. C., 1960, Geomorphology and forest ecology of a mountain region in the central Appalachians: *U. S. Geol. Survey Prof. Paper* 347.

Highway Research Board, 1958, Landslides and engineering practice: Highway Research Board Special Report 29, Public. 544.

Horton, R. E., 1936, Hydrologic interrelations of water and soils: Joint Program American Soil Survey Assoc. and Soils Section, *Am. Soc. Agron.,* Washington, D. C., 60 pp. mimeog.

Horton, R. E., 1945, Erosional development of streams and their drainage basins; hydrophysical approach to quantitative morphology: *Geol. Soc. Am. Bull.,* v. 56, pp. 275–370.

Izzard, C. F., 1946, Hydraulics of runoff from developed surfaces: 26th Ann.

Mtg. Highway Research Board, Proc., 17 pp. (repr.).

Jahn, A., 1960, Some remarks on evolution of slopes on Spitsbergen: *Zeit. für Geomorphologie*, Supp. Band 1, pp. 49–58.

King, L. C., 1953, Canons of landscape evolution: *Geol. Soc. Am. Bull.*, v. 64, pp. 721–752.

Leopold, L. B., and Langbein, W. B., 1962, The concept of entropy in landscape evolution: *U. S. Geol. Surv. Prof. Paper* 500-A.

Macar, P., and Fourneau, R., 1960, Relations entre versants et nature du substratum en Belgique: *Zeit. für Geomorphologie*, Supp. Bd. 1, pp. 124–132.

Melton, M. A., 1958, List of sample parameters of quantitative properties of landforms: their use in determining size of geomorphic experiments: Off. of Naval Research Tech. Rept. No. 16 (Project N.R. 389–042). 17 pp.

Melton, M. A., 1960, Intravalley variation in slope angles related to microclimate and erosional environment: *Geol. Soc. Am. Bull.*, v. 71, pp. 133–144.

Miller, J. P., Montgomery, A., and Sutherland, P., Geology of the Sangre de Cristo Range, New Mexico: *New Mex. State Bur. Mines and Min. Resources, Memoir* (in preparation).

Musgrave, G. W., 1947, Quantitative evaluation of factors in water erosion —first approximation: *Journ. Soil and Water Conserv.*, v. 2, no. 3, pp. 133–138.

Penck, W., 1953, Morphological analysis of land forms: ed. H. Czech and K. C. Boswell, Macmillan, London, 429 pp.

Rapp, A., 1960, Recent developments of mountain slope in Kärkevagge and surroundings, northern Scandinavia: *Geog. Annaler*, v. 42, no. 2–3.

Ruxton, B. P., 1958, Weathering and subsurface erosion in granite at the Piedmont angle, Balos, Sudan: *Geol. Mag.*, v. 95, pp. 353–377.

Savigear, R. A. G., 1956, Technique and terminology in the investigation of slope forms: *Union Géog. Internatl.*, Premier rapport de la Comm. pour l'étude des versants, Rio de Janeiro, 1956, pp. 66–75.

Schumm, S. A., 1956, Evolution of drainage systems and slopes in badlands at Perth Amboy, New Jersey: *Geol. Soc. Am. Bull.*, v. 67, pp. 597–646.

Schumm, S. A., 1956, The role of creep and rainwash on the retreat of badland slopes: *Am. Journ. Sci.*, v. 254, pp. 693–706.

Segerström, K., 1950, Erosion studies at Paricutin, State of Michoacan, Mexico: *U. S. Geol. Survey Bull.* 965-A, 164 pp.

Sharp, R. P., 1942, Soil structures in the St. Elias Range, Yukon Territory: *Journ. Geomorph.*, v. 5, pp. 274–301.

Wentworth, C. K., 1928, Principles of stream erosion in Hawaii: *Journ. Geol.*, v. 36, no. 5, pp. 385–410.

Wentworth, C. K., 1943, Soil avalanches on Oahu, Hawaii: *Geol. Soc. Am. Bull.*, v. 54, pp. 53–64.

White, S. E., 1949, Processes of erosion on the steep slopes of Oahu, Hawaii: *Am. Journ. Sci.*, v. 247, pp. 168–186.

Williams, P. J., 1959, An investigation into processes occurring in solifluction: *Am. Journ. Sci.*, v. 257, pp. 481–490.

Wood, A., 1942, The development of hillside slopes: *Geol. Assoc. Proc.*, v. 53, pp. 128–138.

Young, A., 1960, Soil movement by denudational processes on slopes: *Nature*, v. 188, no. 4745, pp. 120–122.

Part III

THE EFFECTS
OF TIME

Chapter 9

Geochronology

Introduction

The record of climatic conditions is very short, considering all the needs for estimates of the frequency of various events. Engineering structures—the usefulness and longevity of which are importantly affected by geomorphic processes—may require estimates of rates of sediment transport and of the average frequency and magnitude of events such as floods. Engineering design is often based on the hydrologic or climatic record, which for many factors is considerably less than a century in length. Thus any geochronologic techniques that allow extension of these records back in time, even if the extension is approximate and relative, can be of immense practical value.

These backward extensions of hydrologic, climatic, and geomorphic records potentially allow improvement in the accuracy of assessing the frequency of various events and rates of various processes and in the chronologic sequence of events.

The succeeding chapter is concerned with the effects of time. Included are brief descriptions and discussions of the several methods now available for measuring or estimating the date or duration of past events.

Historical Records

Actual records of weather data, stream flow, sediment movement, and levels of lakes and seas are highly variable in both quality and duration. Data on flood stages for the Nile near Cairo extend back to 622 A.D. Hurst (1952, p. 260) discusses the remarkably consistent record for Nile stages, covering a period of about a thousand years, the longest hydrologic record in existence. The level of the Caspian Sea has been recorded on the gage at Baku since 1839, but data from miscellaneous sources extend a partial record back to about 1550. In the United States systematic weather records for a few stations cover slightly more than a century. The longest U. S. Geological Survey record of river flow in the United States—the Rio Grande at Embudo, New Mexico—began in 1889. Sediment load measurement, except at a few stations, began during the last three decades. Variations in the level of Great Salt Lake have been recorded since 1850. Tide gage data for the coasts of North America mostly date from about 1900. Annual or other frequent measurement of glacier fluctuations mostly postdate establishment of the National Park system.

Information from miscellaneous historical sources may in some cases extend actual measured records, especially for those cases which involve variation of level (peak floods, tides, lake surfaces). More commonly, however, such information is merely supplementary and qualitative. Quantitative observations are needed on the nature and density of flora and fauna, initiation of gullying, appreciable advances or retreats of glaciers, and changes in the dimensions of depositional features.

One distinct drawback of historical accounts bearing on specific geomorphic processes is that spectacular events of large magnitude are the only ones mentioned. Impressive as the catastrophes are, they are not necessarily the only or even the most significant events from the geomorphic viewpoint.

It is easy enough to speak of using historical records for dating geomorphic events, but an investigator facing an actual problem may find it difficult to begin. This is illustrated by our inquiry in Chapter 4 into the problem of rates of weathering. The conclusions in various studies were nearly all presented in different form, depending on the types of evidence available. Few of the examples could be reduced to units of a certain number of inches of weathering per year or per century. The definition of what constitutes a measure of weathering varies with the circumstance. Thus one who would tabulate such diverse observations is frustrated by lack of comparability and nonuniform criteria. For this reason a final tabulation was not included.

In the use of historical records a few examples can be cited in which the evidence is somewhat clearer and in which the usefulness of the method is undoubted. The problem is the separation of the effects of climate from those of man's activities on causing or promoting the recent epicycle of valley-trenching or arroyo-cutting of alluvial valleys in southwestern United States in the 19th century. These influences are difficult to separate. The effect of slight variations in climatic factors takes the form of change in rainfall-runoff relations which in turn upsets the quasi-equilibrium between forces of erosion and deposition in these alluvial valleys. The effects of man, exerted through overgrazing by animals, fire, and trampling, also operate through a change in rainfall-runoff relations. That is, a given rainfall results in a different runoff from an overgrazed or trampled area than from one in which the vegetation and soil are unimpaired. In this manner the effects of man appear in the same form as the effects of climatic variation. Both, in a similar way, alter the balance mechanism of the soil-vegetation complex.

Man's land use or misuse in the Southwest culminated in the period 1880–1890, at which time the number of livestock using western ranges was at an all-time high. The rapid erosion of alluvial valleys was particularly marked at that same time. The coincidence of these events led many observers to conclude that the erosion epicycle was the result of overgrazing, essentially uncomplicated by climatic variation (for example, Cooperrider and Hendricks, 1937; Thornthwaite et al., 1942).

This interpretation was strongly held, even though it had been conclusively shown by Bryan and others that in the pre-Christian era the same valleys had suffered similar gully erosion, which surely could not have been caused by grazing of livestock. Therefore, it became important to ascertain whether the erosion epicycle of the 19th century had actually begun, at least in some places, prior to the great livestock concentration of the 1880's.

This question led to Bryan's study of the historical records of military expeditions through the area, which were particularly concentrated in the decade 1846–1856, and to the publication in 1925 of his paper in *Science* —a brilliant model of the application of historical records to a specific geomorphic problem. As an example of particularly enlightening evidence, there is the quotation from the published journal of Lieut. J. H. Simpson, who was the topographer on a cavalry contingent which traveled from Santa Fe to Canyon de Chelly to negotiate a treaty with the Navajos. Traveling over the alluvial valley of the Rio Puerco, which suffered the main arroyo-trenching in the 1880's, he noted on August 24, 1846:

"The Rio Puerco, which, from its great length upon the maps, we had conjectured to be a flowing stream of some importance, we found to contain water only here and there, in pools—the fluid being of a greenish, sickening color, and brackish to the taste. The width of its bottom, which is a commixture of clay and gravel, is about one hundred feet. Its banks, between twenty and thirty feet high, are vertical, and had to be graded down to allow the artillery and pack animals to cross them. The six-pounder had to be unlimbered and dragged up on the west side by men at the prolongas" (1850, p. 64).

This was by no means the only such record Bryan found. The accumulated evidence showed without doubt that gully erosion had begun in some alluvial valleys by 1850 or at least that the presence of some gullies at any one time has always been a characteristic of the area.

There is much, of course, which even pertinent historical evidence does not explain in the problem of erosion, but the historical search made an indubitable contribution.

On the same problem, Leopold (1951b) extended Bryan's analysis of historical records to additional sources. Also utilized were a series of early photographs, the scenes of which were rephotographed from the same location 40 to 50 years after the originals. An ecologist analyzed the early pictures for identification of vegetal species, and these plant lists were compared with field collections made when the second photos were taken. In some photograph pairs, it was possible to count the numbers of trees of certain species. The problem of interpretation in the use of before and after photograph pairs lies principally in ascertaining what a noted change in vegetation means in terms of climate, use, or circumstance. The comparative photographs were interesting but yielded negligible quantitative data of any usefulness.

The analysis of historical records usually requires techniques for obtaining the maximum useful inference from limited data, for the nature of old records usually means paucity of volume. An example is the analysis of the longest record of rainfall in the United States for possible inferences regarding the problem of erosion. One record of 100 years and several somewhat shorter exist for New Mexico, where early records were taken by medical officers attached to the military expeditions. Leopold (1951) analyzed the 100-year record at Santa Fe in terms of the daily rainfall, rather than the usual measures of monthly and annual averages. It was shown that statistically significant secular changes occurred in the frequency of large values of daily rainfall. The period characterized by unusually large daily fall was shown to coincide with the decades in which western gully erosion was proceeding at greatest rapidity.

The inference was drawn that though significant secular changes in monthly and annual totals did not occur, unusually heavy summer rainfalls characterized the decades of greatest erosional activity. This finding is in qualitative agreement with the generally held opinion that gully erosion in semiarid valleys is related to intense and local summer thunderstorms and not to the low-intensity, widespread precipitation of winter.

Since that analysis was published, a subsequent observation has provided a line of inference which supports the conclusion reached on the basis of the historical rainfall record. Firsthand observation of loess-covered valleys and hillslopes in the north-facing foothills of Uzbekistan, Central Asia, in a climate similar to that of semiarid western United States, shows that there was no epicycle of gully erosion in the last century. Smooth, unchanneled swales characterize the foothills area near Samarkand. A comparable area in New Mexico would be cut by a deep gully system of historical origin.

The answer appears to lie in the nearly complete absence of summer thunderstorm rainfall in the steppe area of Central Asia. Thus it may be surmised that a climatic variation experienced in the southwestern United States, and which affected the frequency and intensity of thunderstorm rains, could not operate on that mechanism in the steppe of Uzbekistan.

Dendrochronology

Dendrochronology, the study of tree rings to determine time intervals or past variations in environment, is capable of yielding precise dates, and hence the method may be considered as an absolute one in the geochronological sense.

In the first quarter of the present century Douglass recognized the possibility of obtaining climatic histories from tree-ring sequences, and it was he who established a scientific basis for tree-ring investigations. The most spectacular result of Douglass' research was the use of tree-ring dates in providing a time scale for prehistoric Indian cultures of the southwest. Applications of dendrochronology are not of course limited to archeological problems, but include a diverse group of fields, including geology, botany, hydrology, and climatology.

Special techniques are required for the collection and preparation of samples for tree-ring dating. The usual procedure is to measure the entire ring sequence on one or more radii in the log. The relative thickness of the ring sequences provides the basis for correlation with other specimens. By counting the rings and measuring their thicknesses in many specimens, standardized plots have been developed, which are incorporated into composite plots; these in turn become a kind of master diagram for a whole area. To develop a chronology by this system requires starting with modern specimens of known date and continuously cross-dating successively older specimens. This depends on coincidence of a certain number of peaks and troughs in the various graphs.

The establishment of a satisfactory tree-ring chronology depends on certain conditions.

1. The trees must produce clearly defined annual rings during a definite growing season. Some trees—monocots, for example—lack these characteristics. Eucalyptus trees growing on East Maui, Hawaii, in an annual rainfall of about 350 inches, have no visible growth rings.
2. Tree growth must be sensitive to some controlling factor, usually climatic, the operation of which causes rings to vary in thickness. Cross-dating consists of the identification of the same ring pattern in several trees, so that a dated ring is considered equivalent in date to a particular ring in another tree. The overlapping of a particular period of time in two specimens allows the dating of individual rings to be extended successively back in time in a "tree-ring calendar."

In the southwestern United States, tree-ring dates presently cover the period from 59 B.C. to the present. Some sequoias provide a record of 3,200 years, but because of its habitat the tree does not provide a useful or sensitive record. Within the last decade, bristle-cone pines at elevations of about 10,000 feet in the White Mountains, California, have been found to be 4,000 years or more in age (Schulman, 1956). Because of their locations on steep mountain slopes these trees may provide a sensitive climatic record as well as a remarkably long chronology.

Tree-ring dates have been used in turn to date various kinds of archeological remains. In particular, the pottery made by Pueblo Indians in the Rio Grande valley has been dated quite accurately. The data given in the following table is a typical example of this kind. These dates (all A.D.) are considered reliable to 50 years or less.

Kwahe Black-on-white	950–1200
Santa Fe Black-on-white	1200–1300
Wiyo Black-on-white	1275–1400
Abiquiu Black-on-white (Biscuit A)	1350–1450
Bandelier Black-on-white (Biscuit B)	1425–1625
Sankawi Black-on-cream	1500–1625
Tewa Polychrome	1625–1700

The alluvial chronology of the southwest, mentioned in more detail later (pp. 407–408), is an example of geomorphic history derived in part from pottery and tree-ring dates. The sequence indicates that during the last

2,000 years there have been two phases of deposition and, likewise, two phases of erosion along many valleys in the west and southwest.

The study of tree rings is capable of furnishing absolute dates by the tree-ring calendar. It also has the possibility of furnishing information on the incidence, duration, and intensity of periods of relative aridity and humidity. The establishment of a calendar has been very successful, but the determination of wet-dry periods is fraught with problems. However, so important is the potential utility of dendrochronology as an indicator of past climate that it is desirable to discuss here, at least briefly, the nature of these problems—as a possible spur to research and to direct attention to the points on which additional understanding is necessary.

In the tree-ring calendar for the southwest a tendency for narrow tree rings in the period 1200–1400 A.D. was found in a sufficient number of group averages that it became generally accepted that these two centuries were relatively arid. The period was consequently referred to as the Great Drought.

The simple assumption that narrow rings are, in semiarid areas, related to comparative aridity deserves and apparently is getting some intensive review (Martin et al., 1961). The assumption is predicated on the idea that a tree with noncomplacent (variable width) rings is physiologically at the edge of its natural range, and therefore growth rate is altered by relative availability of moisture. This seems reasonable enough, but water availability is not necessarily measured by precipitation during a year or several years. To the tree, water availability is a function of soil moisture, whose level is affected not only by annual precipitation but also by intensity of storms, length of dry periods between storms, evapotranspiration rate—and thus cloudiness, windiness, moisture content of the air, and temperature. Thus a year with total precipitation during the year equal to the average may include a long dry period in which soil moisture fell below optimum, as a result of which the tree added an unusually narrow ring or even two rings separated by a period of no growth. This possibility implies that it is necessary to know what time patterns of precipitation during the growing season are related to wide as against narrow rings, a need long expressed by Glock (for example, 1955) and studied recently by Phipps (1961).

These factors are not the only ones affecting ring width. Most trees grow faster and put on wider rings in the early stages of life (the first 25 years or so) than in maturity. This differential rate of growth occurs at each level above the ground, and thus apparently the zone of rapid growth and wide rings progressively moves upward from ground level as the crown reaches higher levels. Whereas the data on measured ring width in a given cross section of the bole are usually corrected for this effect of tree age, the correction itself introduces into the data an extraneous effect. Inasmuch as tree-to-tree and site-to-site variability is high, any sources of additional variance must be analyzed.

There are other sources of variability, whose importance is not known. Width of a given ring differs among radii at the same cross section in the log, a matter which has been explored in some detail. But in the same radius at various levels above the ground the magnitude of ring variability is less well known. If this is important it would be well to know it because many logs usable for ring analysis will have come from an unknown height above the original ground surfaces.

Early work on tree rings as indicators of climatic variation involved much less sophisticated statistical techniques than are now available. Recently power-spectrum techniques have been applied to ring-width data, a method in which all rings differing in age by 1, 2, 3, . . . years are compared. The degree of correlation between rings different distances apart in time is a measure of the tendency for cyclical repetition. Preliminary analyses tend to discount the earlier view that the sunspot cycles figured prominently in tree-ring data.

Pollen analysis provides an independent method of studying the climate during the supposed drought of 1200–1400 A.D. in the southwest. This work suggests that some changes occurred which affected the plant species making up the vegetal association in that period. They may well have been climatic factors and probably were, at least in part. But the idea that the period was drier than preceding and following ones requires further examination.

Thus dendrochronology, though an important tool in dating, is as yet less reliable as an indicator of climate. It opens a broad field of needed research for teams on which would be, as required, ecologists, plant physiologists, geomorphologists, and archeologists. The research problems point to the interdisciplinary nature of studies of the natural and man-influenced environment.

Archeological Methods

At the present stage of development, geological information is as frequently used to date archeological remains as is the reverse. In time, an increasing number of types of artifacts will become useful horizon markers for dating geologic events, especially if dates established by other techniques such as radiocarbon and dendrochronology can be integrated with archeological findings.

The pottery chronology in the Rio Grande valley enabled Miller and Wendorf (1958) not only to date the periods of deposition and erosion in a tributary valley, but also to estimate certain environmental conditions which prevailed at the time. For the specific area considered, the rate of sediment production during the last 2,000 years has varied relatively little, the average rate of erosion being of comparable magnitude to the present one. Leopold and Miller (1954) reached essentially the same conclusion for some drainage basins in Wyoming.

Recent unpublished investigations of talus deposits along the sides of the Vezère Valley near Les Ezyies, France, which bury the remains of Paleolithic man, lead to the conclusion that the rate of talus accumulation has varied relatively little during the last 20,000 years, despite apparent climatic changes indicated by the character of the fauna.

Because an ever larger fraction of the absolute ages now being determined are for materials derived from archeological sites, the opportunities afforded for studying rates of geomorphic processes are increasing and should be exploited to the fullest extent.

Varves

Varves, the laminations in lacustrine deposits, were considered to be annual deposits for more than a century before there began detailed work directed at using them for chronological purposes. There is no single definite criterion which establishes the annual character of varves. Knowledge of varve origin and particularly of the period of rhythm involved remains somewhat tenuous. Some rhythmic banding may represent deposition other than annual, and this point is important for chronological purposes. Some varves of pre-Pleistocene age—for example, the Green River beds—apparently owe their laminations to a seasonal pulse in the supply of organic matter that was essentially independent of the rate of deposition of other materials in each layer.

DeGeer's work in Sweden through a north-south distance of nearly 1,000 kilometers established a chronology (with one gap and one extrapolation) extending from 1900 A.D. over a period of nearly 12,000 years. There is, however, some disagreement on the part of other Scandinavian geologists with the DeGeer chronology. In particular, the possibility of redeposition of sediment after being stirred up by storms in shallow lakes may have given rise to smaller laminations; these could cause the formation of false varves. However, the DeGeer chronology is corroborated at one point near the middle by radiocarbon dates that correlate closely with it.

Use of varves for chronologic purposes in North America is severely limited by their restricted occurrence. Only in eastern North America, particularly in the Connecticut Valley, do varved deposits have appreciable thickness and horizontal extent. These have been studied by Antevs, who made measurements along a north-south distance totaling more than 600 miles, but including certain gaps.

Changes in conditions of moisture provide a basis for correlation only within a limited geographic area, and not for trans-Atlantic correlations. Correlation between widely separated areas is possible only on the basis of concurrent experience in temperature regimens and on the similarity of deglacial and temperature histories (Antevs, 1955). On such bases Antevs has tied the Scandinavian chronologic work in with his own findings for North America, excellent summaries of which are given in his 1953 and 1955 papers.

Unfortunately, few recent investigators in this country have shown the patience, energy, and resourcefulness that characterize the work of Antevs. Though it is true that varved beds are not commonly exhibited in outcrops, many lake beds have been studied for pollen stratigraphy without any method being developed for the preservation and study of textural laminations. The usefulness of the pollen data themselves is likely to be enhanced if other sedimentologic characteristics of the same deposits were studied.

Pollen Analysis

Except for the advances in dating by use of radioisotopes, no other field of geochronology has moved ahead with such vigor and has attracted such obvious talent to it as has the field of palynology. Knowledge of the species frequency at any given time and place yields important inferences about climate and climatic factors and in some instances about rainfall-runoff relations, particularly as they affected sedimentation conditions.

The technique involves the collection of samples at various levels in a deposit, removal of the mineral material that constitutes the bulk of each sample, and study of the organic remnant for its included pollen. Individual pollen grains are identified and counted so that a frequency distribution can be drawn as a function of depth in the deposit.

When pollen analysis was first being developed, the samples analyzed were in nearly all instances peat or other highly organic material usually found stratified in pond, lake, or marsh deposits. Because lacustrine environments were so intimately associated with glaciation, palynology found its first usefulness in studies of the Pleistocene and immediately post-Pleistocene periods. Presently increasing attention is being paid to Tertiary and even pre-Tertiary materials, which also have yielded pollen, even from deposits that at first sight do not appear to contain appreciable amounts of organic material.

The pollen grain is not a fossil in the usual sense, for it is composed of the same actual material which lived in the past, being neither a cast nor a replacement. Such an infinite variety in microconfiguration, structure, and elaboration exists among grains of different species, one wonders how the minute details were preserved through long periods of time, despite transportation, pressure, and circulating ground water. The pollen wall is a highly stable material, insoluble in most of the acids that dissolve minerals, including quartz. And the grain is so small that it probably can lodge between harder sediment particles and thus often avoid direct deformation. A most interesting sidelight on the durability of pollen was discovered by Pauline J. Dunton of the U. S. Geological Survey. She purposely burned a sample of modern pine pollen to what appeared to be a black crisp and then soaked it in aerosol. When she looked at the pollen under the microscope it appeared that the majority of grains had absorbed liquid and expanded to normal size and shapes; even details of structure were preserved. This suggests that pollen obtained from organic shales and other materials which had suffered high pressure might actually have been crushed and heated and remained deformed for millions of years, but partially at least had regained original shape and structure during laboratory preparation—a remarkable biologic material, to say the least.

It is now known that pollen may be extracted from a variety of clays and even silts in alluvial deposits, colluvium, and soil. Tertiary pollen was found mixed with post-Pleistocene pollen in alluvial valley fill in Wyoming, the alluvium having been derived from local bedrock of Tertiary shale which had weathered and eroded and was redeposited in Holocene time. Modern flood-plain clay usually contains pollen contemporaneous with the deposition, though older reworked pollen grains may also be present.

Analytical pollen techniques and classification schemes for identification constitute a large literature (especially Erdtman, 1954) and will not be discussed here. However, some remarks on interpretation of pollen diagrams are in order in this general discussion of geochronology.

Some plants produce large amounts of pollen, others do not. And some grains are of such size, shape, and weight that they are transported by wind over long distances, whereas others are not. Such variations must be kept in mind when studying any pollen-frequency diagram.

Care must be exercised in assuming that a pollen suite from a given stratigraphic sequence represents local vegetal association of the vicinity of the deposit. In alluvial deposits it is possible that pollen can have been transported from an entirely different environment by running water. Pollen of plants characteristic of high mountains might be expected in an alluvium far distant if the river which deposited the alluvium heads in mountains. Muller (1959) has recorded pollen of Andes plants in recent deltaic sediments at the mouth of the Orinoco River, indicating long-distance transport by water.

It is usually assumed that pollen is primarily airborne. A pollen profile that shows a frequency distribution of pollen from different plant genera at each stratigraphic level is sometimes tacitly accepted as representing the relative frequency of genera in the actual source vegetation. More often it is recognized that the high pollen producing plants are overrepresented and the low pollen producers underrepresented or absent from the pollen assemblage, but no more precise interpretation than this deductive approach has been available. In the last decade renewed attention has been paid to the factors which determine relative frequency of pollen from different species deposited in the immediate vicinity—that is, modern pollen rain— an interest displayed by Scandinavian palynologists many years ago.

The most illuminating investigations on rain of modern pollen have been carried out by Margaret B. Davis and her colleagues. These studies emphasize that pollen deposited in a particular forest may not have a species-frequency distribution of pollen grains comparable to the species-frequency distribution of plants in the contributing local plant association (Davis and Goodlett, 1960). This work points up the need for much more intensive studies of pollen production, distance of transportation, and deposition in order to obtain a valid ecological interpretation of frequency distributions found in any locale of deposition. It may also lead to a major review of all New England quaternary pollen sequences. The vegetational interpretation previously made of them may need to be corrected for the type of anomalies now coming to light, or at least far more precise paleoecology may be inferred from them on the basis of studies of modern pollen rain.

The whole spectrum of generic representation in a pollen sequence need not always be used, and in such instances the difficulties brought out by studies of modern pollen rain are much less involved. An outstanding example is the construction of a map suggesting the approximate position and even some details of key isotherms in northeastern United States at the time of the last Wisconsin glaciation and at certain times during the postglacial period (E. B. Leopold, 1958). This geographic indication of certain temperature values in the glacial period was derived from the present relation of spruce to July temperature and the boundary of spruce habitat in Wisconsin time, based on dominance of spruce pollen in certain datable deposits.

Because pollen is a stratigraphic tool its interpretation should be made in the light of statistical relations which apply to all stratigraphic methods. Pollen data are not always so interpreted. It is a statistical axiom that a comparison of samples representing a much larger population involves the relative variance from sample to sample in a given population and the variance of samples from one population to another. A marker horizon— one that contains a given key fossil—is assumed to be similar from outcrop to outcrop and different from higher and lower beds in a given outcrop. That is, samples of the same population have less variance than samples from different populations. This implies that something is known about each of these sources of variance, even if the information is not expressed in the terms just stated.

Pollen analysis is time consuming. It requires much effort to recover, identify, and count pollen grains from each of several stratigraphic levels in a given core. It is not unusual, therefore, for an interpretation to be made on the basis of one core, with no study made to compare supposedly the same horizon in several cores. Indeed, the pollen diagram from a single core or hole is sometimes compared and even tentatively correlated with another single core some thousand miles away. Such a correlation, no matter how tentative, involves the tacit assumption that comparison of the same horizon in another core in the same area would show less variation than the variance from one level to another. That such an assumption is questionable is shown by the fact that pollen diagrams thought to include a span of 10 or more thousand years often have only one or perhaps two key levels, in which the pollen frequency is indubitably distinct from that of levels above and below.

To obtain the maximum value from a given amount of analytical work, it may often be better to analyze relatively few stratigraphic levels in several cores or outcrops than a large number of levels in a single core or outcrop.

Radioactivity

The underlying process of radioactivity is the natural transformation of an unstable nucleus to one which is stable. Development of chronologies involves the assumption that the proportions of certain isotopes in a sample are entirely the result of radioactive decay; gains or losses by any other process will lead to erroneous dates.

In general radioactive-decay systems of interest to geochronology, a parent nuclide (P) disintegrates with decay constant* λ to form daughter product (D). An initial number of atoms of a parent nuclide (P_0) will decrease, and the number (P_t) present at any later time t will be

$$P_t = P_0 e^{-\lambda t}.$$

The number of daughter atoms formed from P_0 in time t is

$$D_t = P_0 - P_t = P_0(1 - e^{-\lambda t}) = P_t(e^{\lambda t} - 1),$$

provided D is not radioactive. Hence, the age is

$$t = \frac{1}{\lambda} \ln \left(1 + \frac{D}{P} \right),$$

where D/P is the mole ratio of daughter to parent nuclide.

Natural radioactive nuclear species fall into three categories: (1) extremely long-lived nuclides that have survived since formation of the elements, (2) nuclides produced by decay of some of these primary nuclides, and (3) short-lived nuclides currently forming in nature as a result of cosmic-ray bombardment. The nuclides most important for age determination are listed below.

---

* λ is related to the half-life T by the relation $\lambda = (\ln 2)/T$.

ELEMENT	HALF-LIFE	RADIATION
Primary Natural Radionuclides		
U^{238}	4.5×10^9 years	α
U^{235}	7.1×10^8	α
Th^{232}	1.4×10^{10}	α
Rb^{87}	5×10^{10}	α
K^{40}	1.3×10^9	β (also e^- capture)
Secondary Natural Radionuclides		
Th^{230} (Ionium)	8.0×10^4	α
Ra^{226}	1620	α
Induced Natural Radionuclides		
C^{14}	5730 ± 40	β
H^3 (tritium)	12.4 ± 0.2	α

The first dates ascertained by radioactivity, published in 1907, were based on the complex disintegration of uranium to lead. During the next four decades variants of this method were developed and improved gradually, but age determination on a major scale and involving a variety of methods is mostly a post–World War II development, made possible by advances in mass spectrometry techniques for isotopic analysis of very small samples.

Amounts of U, Th, Pb, Rb, Sr, K, A, etc., can be determined with a precision of 2% or less. In general, errors due to analytical techniques can be resolved with much greater certainty than variations in age caused by unknown geological factors.

The complex decay of uranium and thorium isotopes offers alternative dating methods, which have been exploited successfully for some time. Potassium and rubidium dating are more recent developments, and both show great promise. In particular, the potassium method should fill some of the gap between the upper limit of the radiocarbon method and the lower limit of uranium and thorium methods. Even more significant is its potential applicability to authigenic minerals of sediments. Because all other methods are restricted to igneous minerals, this should lead to major advances in time calibration of the geologic column.

Radiocarbon is produced in the upper atmosphere by bombardment of nitrogen with cosmic ray neutrons,

$$N^{14} + n = C^{14} + H^1,$$

and reverts back to nitrogen by radioactive disintegration,

$$C^{14} \longrightarrow N^{14} + \beta,$$

with a half-life of 5730 ± 40 years.

As nitrogen is very abundant in nature the ratio of daughter to parent nuclide in the general age equation cannot be determined. Rather, it is assumed that the present ratio of C^{14} to C^{12} in the atmosphere has been constant for a long period. So long as an organism or mineral exchanges carbon with the atmosphere this ratio prevails, but upon death or burial it begins to decrease exponentially with time. Therefore analysis of C^{14} in wood, charcoal, bone, shell, and other organic materials can give an accurate measure of time, with the maximum possible age determined by the minimum amount of C^{14} that can be measured. A few samples up to 70,000 years old have been dated, but at most laboratories 30,000 years is the practical limit.

It is well to keep in mind the underlying assumptions of radiocarbon dating. One of the most fundamental assumptions is that cosmic radiation has been constant throughout the period of time covered by the radiocarbon method. This is essential because a steady state is assumed—namely, that the rate of decay of C^{14} is equal to the rate of its formation. Reasonable though it is, no conclusive proof of this assumption exists.

The method also assumes that after cessation of atmosphere exchange the C^{14} content is altered only by radioactive decay. Geologists have attacked some published dates on the grounds that percolating waters may have affected the materials used as samples. Experimentation of various kinds and for samples occurring under a wide variety of circumstances has thus far shown this to be a minor source of error.

So long as an organism is alive, the carbon in it is in equilibrium with the atmospheric reservoir, and only at the time of death of the organism does this equilibrium fail. One difficulty encountered with some samples is the incorporation of old carbon; for example, some cases of modern shells give dates of a few thousand years. This is a problem not ordinarily associated with other kinds of organic materials.

The magnitude of variations in the so-called initial activity or contemporary assay of radiocarbon is the source of considerable uncertainty. For example, different kinds of trees and particularly trees from different climatic zones contain different amounts of radiocarbon. Incorporation of nonradioactive or older carbon from organic constituents of the soil and the contemporary increase in atmospheric carbon (from fuels) serve to complicate the assays. Any error in choice of value of the contemporary assay obviously results in an error in age. The absolute value of this error in age is independent of the age of the sample and amounts to approximately 80 years for each percent error in contemporary assay.

Statistical error in counting is a matter of considerable importance. For example, the usual counting time of 2 days gives a standard deviation of 1 to $1\frac{1}{2}\%$ for contemporary carbon. This corresponds to an error in age of 80 to 120 years. The error doubles roughly for each half-life in age; thus the greatest utility of the radiocarbon method lies in the region of 2,000 to 20,000 years. For the younger and older samples the error is relatively large and a cause of considerable uncertainty. Techniques have improved and the range of application of the method has increased through the years. There is good reason to believe in the possibilities of even greater precision, and hence more accurate dates, although many problems of field samples and contamination will require further study.

Radioactivity of tritium provides opportunities for dating in the range of less than 100 years. Agricultural products, stored water, rain, snow, and ice are the principal datable materials. Opportunities for studies of processes involving rapid movements and exchanges have been greatly restricted by additions of artificial tritium from nuclear explosions, but in some cases the artificial tracer has proven useful.

Finally, it should be mentioned that other dating schemes dependent on radioactivity for calibration are under development. Recent investigations indicate that such effects as hydration of volcanic glasses and radiation damage (measured by thermoluminescence) of pottery and many other kinds of materials may be calibrated accurately enough for routine age determination.

An Example of Geochronologic Problems

The geochronologic methods discussed briefly here have been used in combination in few places as intensively as in the western United States. Some of the most comprehensive studies in this subject area have been done by Ernst Antevs.

As our own purpose here is both to cite problems and to provide a base from which it is hoped new researches will spring, it may be helpful to present in brief some salient points made by Antevs in connection with his syntheses. Figure 9-1 presents his reconstruction of the march of July temperature from about the time of the last major glaciation to the present, and, on an expanded scale, postulation of the occurrence of major drought periods for the southwest during the latter part of the Neothermal period.

As he has argued, droughts, more than periods of ample moisture, impress their effects on channels, soils, and vegetation in ways which are discernible in the recent geologic record. Droughts also had particular effects on early man as well.

Antevs reads from the evidence a somewhat more detailed set of relationships among climate, vegetation, conditions of erosion, deposition, and soil formation than have others. His ideas are summarized in Table 9-1. Kirk Bryan and his students have summarized the chronology for the southwest in a form typically represented in Table 9-2.

Comparison of the two tables shows that they serve somewhat different purposes. But the Bryan chronology, being the summary of several investi-

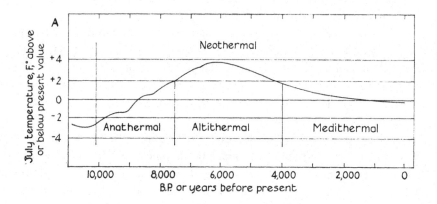

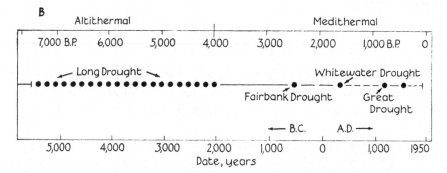

Figure 9-1.   *Climatic events in southwestern United States since the last glacial, 11,000 years ago. [After Antevs, 1955.]* **A.** *Tentative graph of July temperature showing relation to present value.* **B.** *Droughts (dots) and comparatively moist ages during the Altithermal and Medithermal.*

Table 9-1. Relationship of moisture, vegetation, and geologic processes in valleys of nonglacial streams. [After Antevs, 1955.]

Climate: Phases and Fluctuations	Plant Cover	Process in Upper Reaches of Streams	Process in Middle Reaches
Subhumid	Ample	Streams relatively clear; erosion of stream valleys, soil formation where no erosion	Erosion
Changing from subhumid to semiarid	Good	Filling of stream channels, followed by soil formation on fill	Deposition
Semiarid	Good	Neither appreciable erosion nor deposition; soil formation	Entrenchment; soil formation
Changing from semiarid to arid	Sparse to fair on uplands; fair to good in valleys	Deposition	Deposition; soil formation
Arid upon change from semiarid	Impaired or poor	Accelerated erosion, arroyo-cutting, locally leaving coarse residue; local deposition of fine charco clay	Deposition; main or principal valley alluviation
Changing from arid to semiarid	Sparse to fair on uplands; fair to good in valleys	Arroyo-filling	Deposition; soil formation

gations, implies that field evidence usually is insufficient to demonstrate the detailed differences in climate and vegetation which Antevs suggests were characteristic of the several phases and the transition periods between them. Antevs' interesting postulations require more study of field examples.

Antevs, in his 1954 paper on the glacial climate of New Mexico, makes an important point that seems to clear up some difficulties previously experienced. Certain pluvial lakes—for example, Lake Estancia in New Mexico—which were not fed by waters from melting glaciers, required both a lower temperature and a higher precipitation than that experienced at present. Water levels in these lakes appear to have followed closely contemporaneous changes in climatic conditions responsible for glaciation in nearby mountains and did not lag behind glacial advances. In contrast,

Table 9-2.   Generalized alluvial sequence in the western United States.
[After Bryan and students.]

Deposition 1	Contains extinct fauna. Probably correlates with a late-Wisconsin glacial substage. The few available dates suggest an age of 7000–8000 B.P.[2]
EROSION AND WIND ACTION (ALTITHERMAL?)	
Deposition 2	Twofold in several areas; modern fauna. Upper part at most places is no younger than 1100–1200 A.D.[1] Lower part of Deposition 2a at a few places dates 2200–2400 B.P.[2]
EROSION AND WIND ACTION	
Deposition 3	Began in most places 1200–1500 A.D.[1] and continued until 1880 or later
MODERN EROSION AND WIND ACTION	

[1] Pottery and dendrochronology.
[2] Radiocarbon (years before present).

pluvial lakes of the Great Basin, fed by waters originating in glaciated basins, tended to lag behind variations in glaciers at their headwaters.

Another important point postulated by Antevs also relates to variation in conditions within a given region. He points out that the moisture history of eastern and central Texas may not agree in all respects with that of New Mexico, only a few hundred miles away.

An example may illustrate in what respect there remain differences of interpretation. In Table 9-1 Antevs postulates that the main valley fill in the southwest occurred during a period when the climate was "arid upon change from semiarid." In such a transition he reasons that the headwaters of tributaries undergo accelerated erosion and arroyo-cutting, which presumably furnished the principal source of the alluvium simultaneously deposited in the main valleys.

In contrast, Leopold and Miller (1954, p. 83) noted that in many alluvial valleys of Wyoming the alluvial mantle extended in a smooth unbroken surface from main valley fills up into adjacent tributary valleys. We argued that aggradation in the main valleys was accompanied by deposition in tributary valleys and draws, the material being derived from mass movement and sheet erosion on the upland slopes.

In addition, geochronologic studies shed light on the environments of the past, they provide essential data on the rates of geomorphic processes, and, by extending in time our understanding of climatic fluctuations, they provide valuable information for design and operation of engineering works. The differences between the conclusions among the investigators who have looked at geochronologic problems even in a single region indicate that much remains to be done and should indicate to the oncoming students how many interesting problems lie everywhere at hand.

REFERENCES

Antevs, E., 1955, Geologic-climatic dating in the West: *Am. Antiquity*, v. 20, no. 4, pp. 317–335.

Bryan, K., 1925, The Papago country, Arizona: *U. S. Geol. Survey, Water-Supply Paper* 499, 436 pp.

Bryan, K., 1941, Correlation of Sandia Cave with the glacial chronology: *Smithsonian Misc. Coll.*, v. 99, no. 23, pp. 45–64.

Bryan, K., and Albritton, C. C., 1943, Soil phenomena as evidence of climatic changes: *Am. Journ. Sci.*, v. 241, no. 8, pp. 469–490.

Bryan, K., 1948, Los suelos complejos y fósiles de la altiplanicie de Mexico en relación a los cambios climáticos: *Boletín de la Soc. Geol. Mexicana*, T. XIII, pp. 1–20.

Bryan, K., 1925, Date of channel trenching (arroyo cutting) in the arid Southwest: *Sci.* v. LXII, no. 1607, pp. 338–344.

Cooperrider, C. K., and Hendricks, B. A., 1937, Soil erosion and stream flow on range and forest lands of the upper Rio Grande watershed in relation to land resources and human welfare: *U. S. Dept. Agric. Tech. Bull. No. 567*, 88 pp.

Davis, M. B., and Goodlett, J. C., 1960, Comparison of the present vegetation with pollen-spectra in surface samples from Brownington Pond, Vermont: *Ecology*, v. 41, no. 2.

Erdtmann, G., 1954, An introduction to pollen analysis, 2nd rev. printing: Chronica Botanica, Waltham, Mass. 239 pp.

Glock, W. S., 1955, Tree growth and rainfall: *Am. Geophys. Union Trans.*, v. 36, no. 2, pp. 315–318.

Hunt, C. B., and Sokoloff, U. P., 1950, Pre-Wisconsin soil in the Rocky Mountain Region, a progress report: *U. S. Geol. Survey Prof. Paper* 221-G.

Hurst, H. E., 1950, Long-term storage capacity of reservoirs: *Am. Soc. Civil Engr. Proc.*, v. 76, Separate No. 11.

Hurst, H. E., 1952, The Nile: Constable, London, 326 pp.

Judson, S., 1953, Geology of the San Fan Site, eastern New Mexico: *Smithsonian Misc. Collection*, v. 121, no. 1, pp. 1–70.

Leopold, Estella B., 1958, Some aspects of late glacial climate in eastern United States: *Verhandlungen der IV Intern. Tagung der Quartärbotaniker*, Geobotanischas Inst. Rübel in Zürich. Heft, 54, pp. 80–85.

Leopold, L. B., 1951a, Rainfall frequency: an aspect of climatic variation: *Am. Geophys. Union Trans.*, v. 32, no. 3, pp. 347–357.

Leopold, L. B., 1951b, Vegetation of Southwestern watersheds in the nineteenth century: *Geog. Rev.*, v. XLI, no. 2, pp. 295–316.

Leopold, L. B., and Snyder, C. T., 1951, Alluvial fills near Gallup, New Mexico: *U. S. Geol. Survey Water-Supply Paper* 1110-A, pp. 1–19.

Leopold, L. B., and Miller, J. P., 1954, A post-glacial chronology for some alluvial valleys in Wyoming: *U. S. Geol. Survey Water-Supply Paper* 1261, 90 pp.

Martin, P. S., et al., 1961. The last 10,000 years: Geochronology Laboratories, Univ. of Arizona, Tucson, 119 pp.

Miller, J. P., and Wendorf, F., 1958, Alluvial chronology of the Tesuque Valley, New Mexico: *Journ. Geol.*, v. 66, no. 2, pp. 177–194.

Muller, J., 1959, Palynology of recent Orinoco delta and shelf sediments: Reports of the Orinoco Shelf Expedition, *Micropaleontology*, v. 5, no. 1, pp. 1–68.

Phipps, R. L., 1961, Analysis of 5 year dendrometer data obtained within 3 deciduous forest communities of Neotomo: *Neotomo Ecological and Bioclimatic Lab. Research Circ. 105, Ohio Agric. Exp. Sta.*

Schulman, E., 1956, Dendroclimatic changes in semi-arid America: Univ. of Ariz. Press, Tucson, 142 pp.

Simpson, J. H., 1850, Journal of a military reconnaissance from Santa Fe, New Mexico, to the Navajo country in 1849: 31st Congr., 1st sess., Senate ex. doc., no. 64.

Thornthwaite, C. W., Sharpe, C. F. S., and Dosch, E. F., 1942, Climate and accelerated erosion in the arid and semi-arid southwest, with special reference to the Polacca Wash Drainage Basin, Arizona: *U. S. Dept. of Agric. Tech. Bull.*, No. 808, 134 pp.

Chapter 10

Drainage Pattern Evolution

Development of Rill Systems

The relations of discharge parameters to drainage area were described in connection with the geometry of the drainage basin. Thus far, primary consideration has been given to stable conditions unaffected by evolutionary changes. Such changes will be considered here.

Consider initially the smallest channels, ephemeral rills, those channels which carry water only during storms and which most of the time are dry. If we observe the pattern of rills developing on a newly bladed road cut, the first thing that strikes us is the parallel pattern which is initially developed, and the fact that rills bite into the predetermined slope without materially altering the gradient. The head of a rill system may not extend all the way to the watershed divide. A zone near the divide may remain essentially unchanneled. In this zone the depth of overland flow is insufficient to develop an erosive force equal to the forces of cohesion tending to hold the soil particles in place.

The processes by which parallel shoestring rills on a fresh surface become integrated into a drainage net are cross-grading and micropiracy—the robbing of a small channel's drainage system by a larger channel. Micropiracy, as explained by Horton (1945, p. 335) is a consequence of the overtopping and breaking down of ridges that initially separate adjacent rills. Two parallel and adjacent rills usually will differ slightly in elevation and depth, the longer being slightly deeper or at a lower elevation or both, compared with the shorter. For where overland flow occurs on a strip of unit width, the depth of flow and thus the ability to erode increases with distance from the drainage divide.

During a heavy storm, when overland flow occurs to a depth sufficient

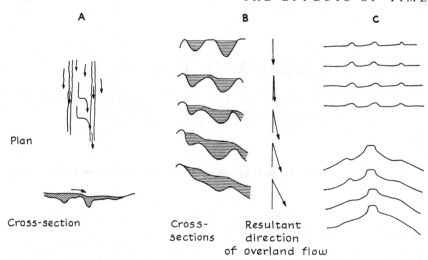

Plan

Cross-section

Figure 10-1.   *Development of rills on hillslopes and enlargement of rills by*
*cross-grading. [After Horton.] **A.** Overland flow over topping*
divide separating adjacent rills; arrows indicate direction of
*flow. **B.** Successive stages of cross-grading of rills. **C.** Con-*
tour lines on rilled surface (top) contour lines of the same
surface (bottom) after cross-grading.

not only to fill the adjacent rills but to overtop the intervening divide, the
water will tend to develop a lateral component of motion toward the slightly
lower rill, and this component across the inter-rill divide will erode it away,
as indicated in Fig. 10-1, **A.** As water is drawn away from one rill by the
lower elevation of an adjacent one, the latter will acquire by such micropiracy even more erosive ability at the expense of the former. The inter-
rill divides will progressively be broken down, and a component of flow
toward the master or main rill will progressively increase, as indicated in
Fig. 10-1, **B.** The development of this component across the main gradient
of the general surface was called by Horton "cross grading."

As the figure indicates, cross-grading gradually obliterates all the original
rill features, and there results a relatively smooth area of overland flow.
The contour lines on the rilled surface and those after cross-grading are
suggested in Fig. 10-1, **C.** The V-shaped re-entrant in the pattern of contours indicates the presence of a component of flow across the main gradient
of the hillslope as a result of the cross-grading. It is this new component of
slope that allows a new system of cross-grading and the development of a
drainage net.

The process illustrated in Fig. 10-1 is meant to apply to a rill system rather than a network of creeks or rivers. This qualification is added because the micropiracy or overtopping of inter-rill divides involves processes that can operate quite rapidly and can be observed. Processes operating in large river basins are believed to involve piracy and capture on a larger scale but they are more difficult to observe. The actual overtopping of interstream divides, or cross-grading, occurs but seldom, however, between channels larger than rills, save where rivers debouch from high mountains onto flatter aggradational plains or weaker rocks below.

In Fig. 10-2, **A,** a series of new parallel rills are sketched, the longest of which is *ab.* By overtopping of divides during overland flow this longest and therefore deepest rill drains water from adjacent rills, which are then

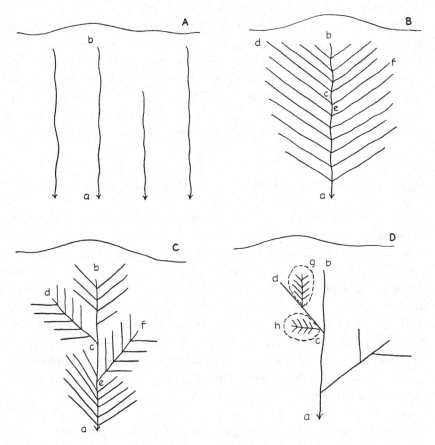

Figure 10-2.　*Development of tributary channels and a drainage net by cross-grading. [Partly after Horton.]*

obliterated by cross-grading. Figure 10-2, **B,** indicates a single rill *ab,* and the diagonal lines leading toward that rill suggest the direction of overland flow after cross-grading. Along the lines *cd* and *ef* the lengths of overland flow are supposedly long enough to attain sufficient depth of overland flow to cause cutting of new rills along those paths.

As the rills *cd* and *ef* deepen they similarly draw water from adjacent areas and cause cross-grading toward themselves, resulting in directions of overland flow suggested by the sets of parallel lines in Fig. 10-2, **C.** In turn, these new directions of overland flow cause cross-grading toward the positions of the longest length of overland flow and there develops a new pair of rills in positions *h* and *g* in Fig. 10-2, **D.**

This is the essence of Horton's argument to explain the geometric series by which number and lengths of streams of different orders occur. The sketches in Fig. 10-2 (and those in Horton's discussion of the process) suggest that in this system of rills there are more streams of order $n - 1$ than of order n. Though Horton's brilliant work brought out both the quantitative relations of the geometric series and a general argument in explanation, a somewhat more complete picture of tendencies guiding drainage-net development appears to be provided by the concept of greatest probability, derived from an analogy with entropy.

Effect of Longitudinal Profile on Tributary Junctions

A key element in drainage-net development and its configuration is that, within limits, the larger of two adjacent stream channels has a tendency to pirate or rob the smaller one of its drainage system. This is relatively easy to visualize in a newly developed system of small rills, owing to the fact that the greater the distance from the watershed divide, the larger will be the amount of water accumulating from a strip of area having unit width. By "a larger amount of water," in this instance, we mean a larger discharge, on the assumption that under a rain of uniform intensity and of sufficient duration each unit of area is contributing the same volume of surface runoff per unit of time. This does not apply to areas even of moderate size, but may hold for areas of a few square yards or even a few acres.

In the rill system, the greater the discharge, the greater the depth of flow and rate of rill cutting. At contiguous points in the basins of two adjacent rills, the longer rill will be slightly lower in elevation, and thus water will tend to flow toward that lower elevation when overland flow overtops the inter-rill divide.

In larger stream channel systems an analogous relation exists. One of the elements of the Horton system of drainage-net analysis is the relation of channel slope (stream gradient) to stream order. Channel slope decreases exponentially in relation to stream order; channels of larger order have flatter gradients. It would be expected from this generalization that in areas of uniform lithology drainage basins of the same size would be comparable in number and in the lengths of streams of various orders. Further, in similar basins at the same distance from the watershed divide along the principal stream, the channel gradient of the streams would be equal. Even with the variability expected among natural drainage basins, Hack (1957, Figs. 16 and 44) has shown that this holds true. It follows, then, that if in an area of uniform lithology two adjacent stream channels originating at the same elevation differ in length, the longer will lie at a lower elevation. This is shown in Fig. 10-3, where streams *ab* and *cd* are comparable except in length and thus in drainage area. The longitudinal profiles are exactly the same (da' ≡ ba) and originate at the same elevation. But because *cd* is longer than *ab*, *c* lies at a lower elevation than *a*. If, then, for any reason, the divide between the streams were broken down, water would flow from the smaller toward the larger basin, and the longer or larger stream would tend to capture the drainage area of the smaller.

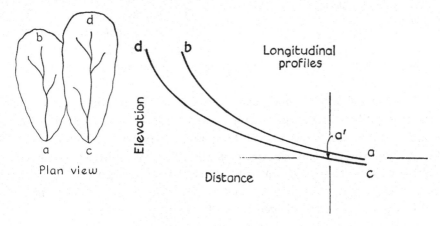

Figure 10-3. *A usual relation of longitudinal profile to stream length in adjacent basins. Profile da' is identical to ba, rising at the same elevation, but since dc is longer, c lies at a lower elevation than a.*

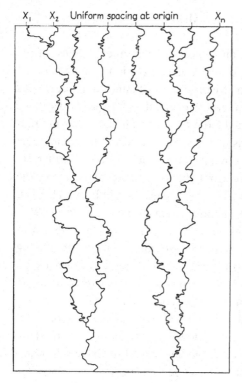

Figure 10-4.

Portion of random walk model of rill or stream network.

Probability and the Drainage Network

In the development of a drainage network there are many alternative ways in which rills and channels could develop and unit tributaries meet with their respective master streams. The length of overland flow, the drainage density, the relief, and many other factors that describe the drainage net would depend greatly on the character of the rocks or other materials on which the network is developing, on previous history, and on other parameters.

But because there are a very large number of rills, minor channels, and small tributary creeks, it might logically be supposed that in such a large population a mean state might be described. The mean state, or the most probable arrangement within the system, would be that one most likely to be encountered on the average in actual landscapes.

Random walks, as we have seen, provide a method by which a model of the most probable state can actually be developed, for it provides a mathematical model which can include the possibility of wide variation among individual examples and one which, on the average, will also yield a picture of the most probable state under the constraints postulated.

When precipitation falls on a uniformly sloping plane, an incipient set of rills would develop, oriented generally downhill. The cross-grading that would result and the micropiracy of incipient rills would, as Horton implied, be a matter of chance until the rills deepened sufficiently to become themselves master channels, or master rills. The randomness in the first stages of cross-grading might be approximated by the following model.

Consider a set of initial points $x_1, x_2, \cdots, x_n$ (Fig. 10-4) on a line, equidistant from one another at some particular spacing. Let random walks originate at each of these points, and in each unit of time let each random walk proceed from the initial line a unit distance. Let it be specified, however, that each walk may move forward, left, or right at any angle, but may not move backward. The accumulation of moves will produce in time sufficient accumulative departures from the orthogonal to the direction of the original line of points that some paths might meet. After such a junction only one walk proceeds forward, just as when two stream tributaries join and a single stream proceeds onward.

An example of the postulated model is shown in Fig. 10-4, which contains only part of the graphical construction actually made in the trial pictured.

The distance any individual random walk will proceed on the average before it meets another can be described by a statistical model called "the gambler's ruin," in which one may compute the most probable number of plays before one player loses all his capital to his adversary. If the capital of each is equal at the beginning of the game and if the size of each wager remains the same, the number of plays expected is one-fourth of the square of the total capital in the game, or $N \propto D^2$, where N is the number of plays and D is the total capital.

By analogy, the number of steps of adjacent random walks before one walk is absorbed by or meets another should depend on the square of initial separation distance. But because a walk may deviate either left or right from its general direction, it is less confined than the gambling model, so the meeting may occur as quickly as the first power of the initial separation distance, or $N \propto D$ (Leopold and Langbein, 1962, p. A14).

A symmetrical model of pairs of streams joining and resultant second-order streams joining would make the mean spacing of the second-order channels twice that of the first-order, and the mean spacing of third-order four times that of the first-order channels. This geometry is described by an equation, $D = d2^{R-1}$, where D is mean distance between streams of order R and d is mean distance between first-order streams.

Combining the equations yields the result that the average stream length from one junction to the next varies with the mean interfluve distance in the following form, where L is stream length between junctions and R is stream order:

$$L \propto 4^R$$

or

$$L \propto 2^R,$$

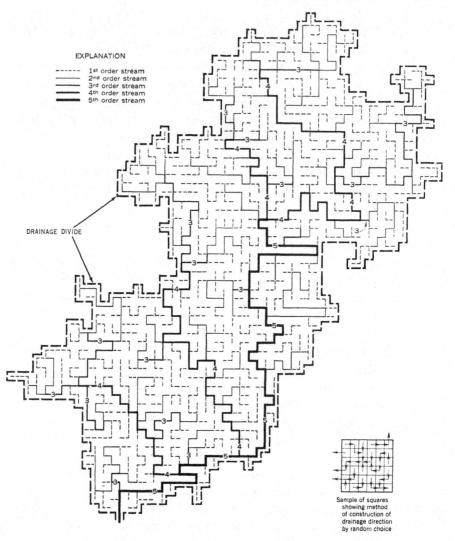

Figure 10-5.   *Sample of random-walk drainage network developed on rectangular graph paper.*

depending upon whether the length varies as the square or the first power of the distance between streams of a given order.

These equations, derived from the mathematical description of the random-walk model, suggest that the logarithms of stream length would vary linearly as functions of stream order and that the mean ratio between lengths of streams differing in order by 1 will be between 2 and 4. This result, derived from considerations of the most probable state, is in agreement with the actual configuration of most channel networks in nature.

Another type of test using random-walk techniques was also developed, using a sheet of rectangular cross-section paper. Each square is assumed to represent a unit area in a developing drainage net, and each square is to be drained. An arrow will be drawn from the center of each square on the graph paper to the center of an adjacent square in a direction specified by pure chance, the direction being decided by a table of random numbers or by some other scheme for producing random choices. Four equal possibilities are postulated for the direction of drainage from each unit square— the four cardinal directions. There is an equal chance, in other words, that the arrow indicating direction of drainage from a unit square will lead off in any of the four directions. This is subject only to the condition that no internal sinks can be developed.

On such a drainage grid the arrows connect, and a stream network is thus generated, an example of which is shown in Fig. 10-5. Divides are developed and the streams join so as to create drainage areas of various

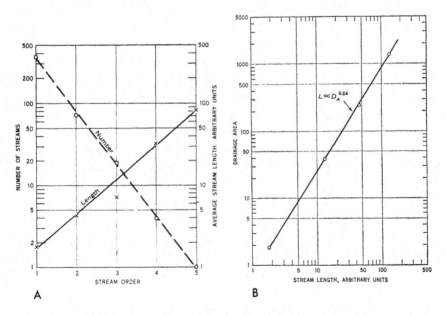

A   B

Figure 10-6.   *Relations among stream characteristics for drainage net of Fig. 9-5 developed by random walks.* **A.** *Relation of number and average lengths of streams to stream order.* **B.** *Relation of drainage area to stream length.*

sizes. Such a random-walk pattern may be considered to represent an example of a network in a structurally and lithologically homogeneous area and one in which no differential erosion is accorded to streams of different sizes. The drainage net so developed has statistical properties nearly identical to the properties of natural drainage nets; that is, logarithms of the number of streams of given orders and logarithms of the lengths of given streams are linearly related to stream order itself. This is demonstrated by Fig. 10-6, which refers to the drainage net developed by the random walk pictured in Fig. 10-5. Furthermore, in the random-walk model, stream length is related to the 0.64 power of drainage area, as shown in Fig. 10-6, **B,** a common value for the relationship between length and area in actual streams (Chapter 5).

It appears, then, that the logarithmic relations between stream length and numbers of streams with order number is the most probable condition.

Evolution of the Drainage Net

Two different approaches to an initial consideration of the drainage net have been introduced. The rational approach to drainage evolution—suggested by Horton and described here—is based on processes of cross-grading, micropiracy, and effects of overland flow. This indicates that a reasonable drainage net can be built up by applying two principles: (1) when the potential for erosion exceeds the resistance to erosion, a new channel will form or an old one will be extended, and (2) the initial cross-grading between rills is random. Assuming that erosion is a function of angle and length of slope, the general argument about cross-grading leads to the development of a drainage net in which, at least qualitatively, the geometric proportions followed the geometrical properties of natural drainage basins (Horton, 1945, p. 346 et seq.).

It is important to recognize that although the geometrical properties of the derived and natural drainage nets are quite similar, the scales are quite different. Even first-order basins in many humid regions are well beyond the incipient cross-grading stage; in all likelihood there are few areas that, on a regional scale, provide a clean slate on which the channel network can develop unhindered or unconstrained by differences in rock type, structure, or initial topographic irregularities. Nevertheless, the density and distribution of stream channels on a large or regional scale does fulfill the physical conditions postulated for the ideal case. The network transports debris derived from the hillslopes without continuously changing its form, and there is a relationship between precipitation, contributing drainage area, and size of first-order stream channels (see Chapter 5).

It has also been shown that these same geometric properties can be "generated" by applying a random process, the random walk, to trace out the paths of lines originating at points along the drainage divide. We have referred to the resulting drainage net as a "most probable configuration." From the relation of channel number to channel length for a large number of drainage basins it can also be concluded that, in general, large basins are geometrically similar to small basins with regard to relations of orders, numbers of streams, and stream lengths. By this we mean that as scale changes both the number of channels and their lengths change proportionately, so that the relation of channel number to channel length shows little scatter. However, the relation of stream length to drainage area changes uniformly but not as the square root of drainage area. The length of a drainage basin increases somewhat faster than the mean width, so basins get relatively longer and narrower with increase in size.

Although both Horton's approach and the random walk give an answer that agrees with observation—that is, the drainage systems so generated have many characteristics in common with real drainage systems—neither tells us whether, with the passage of time, drainage basins do in fact evolve randomly, following the postulated principles of erosion. It is also true that the agreement between model and nature is better between statistical or numerical relations than it is between spatial distributions or patterns.

To state this in another way, even if one were confident that drainage networks in nature represented the most probable state and the most probable state can be generated by several types of mathematical models, it does not follow that in nature real drainage nets are generated through time in the manner of the model. Although they provide fundamental general principles with which to begin, neither the rational nor the random walk descriptions indicate how drainage nets develop and change through time. It is seldom in nature that an area of any size is at any moment unrilled or undissected. In nearly all cases there are some inherited characteristics which will influence subsequent development or change and, furthermore, it is rare indeed that lithology and structure are so simple that homogeneity is characteristic of any large area.

A river or a drainage basin might best be considered to have a heritage rather than an origin. It is like an organic form, the product of a continuous evolutionary line through time.

The most probable state always exists, barring rapid or exceptional

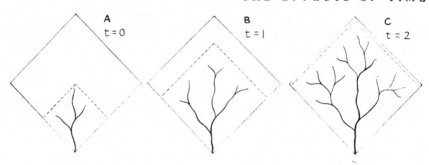

Figure 10-7.   *A hypothetical development of drainage net with time; the basin order increases from 2nd order, **A**, to 4th order, **C**.*

changes. This most probable state must always satisfy physical require-ments or laws and will therefore be adjusted or altered by the demands of physical inhomogeneities. It is useful, then, to inspect examples where either heritage or inhomogeneities are partially known or can be minimized.

The geometrical similarity shown by the relation of number and length implies that if a small drainage system begins to develop on a large area ($t = 0$ in Fig. 10-7, **A**), it will ramify in such a way that both channel length and channel number increase (Fig. 10-7, **A** to **C**). Considering the total area, *xy,* within the square basin, absolute frequency and drainage density both increase with the passage of time from $t = 0$ to $t = 2$, but if we look only at the area within the dashed lines, we see that each successive basin is geometrically similar to the one preceding it in time.

If we use the concept of stream order, the initial stream as drawn is of order 2, the next of order 3, and the last of order 4. Melton (1958, p. 49) described such a growth model, based on the following logic. Assume that all drainage basins have evolved through time. The innumerable drainage basins whose geometric properties form the statistical basis for the laws of drainage-basin morphology must, in fact, include basins in many stages of development. Yet the statistical data describing these varied drainage basins show no systematic variation which we should expect if different absolute ages were associated with different drainage textures or densities. Melton found (1958, p. 50) that channel frequency for a unit area was, in fact, related not only to channel length but also to basin relief according to the equation

$$N = 0.8147 \frac{L^{1.75}}{(\sqrt{A_d})^{1.5} Re^{0.25}},$$

where N is the number of channels, L the length of channels, A_d the area, and Re the total basin relief. Considered as a model of the changes in the drainage net over time where relief remains constant through uplift, the equation implies that the length of the channel system increases simply through an increase in the number of channels. Where relief is reduced, $Re = 1.899/L$; that is, length is inversely proportional to relief. In both cases, however, it is inferred that transformations of drainage nets over time look like existing differences in space.

One of the standard criteria long used by geologists to distinguish glacial tills of different ages is, in fact, the degree of drainage "development." It is postulated that the greater density and the more integrated the drainage pattern, the older the till sheet. The degree of weathering, the superposition of till sheets in stratigraphic sections, and carbon-14 dating correlated with the degree of drainage development have all substantiated this relationship. Thus drainage patterns developed on tills of successive

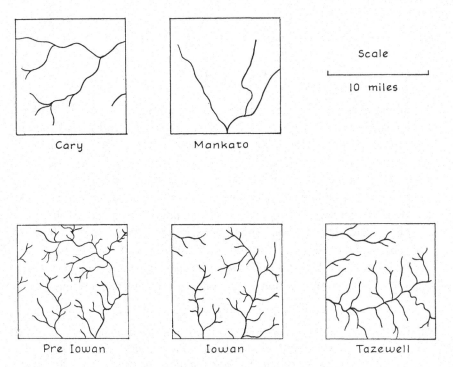

Cary   Mankato

Scale

10 miles

Pre Iowan   Iowan   Tazewell

Figure 10-8.   *Sample of drainage net on glacial till in Iowa of different age.*
[After Ruhe, 1952.]

ages can provide an example of progressive changes with increasing time, although variations in initial relief, lithology, and sedimentary composition make cautious comparisons advisable.

Samples of drainage patterns developed on glacial drift of different ages in Iowa are shown in Fig. 10-8 (Ruhe, 1952). For each sample unit area of the five drift sheets, Table 10-1 gives the approximate age in years,

Table 10-1.   Drainage density and other characteristics on five glacial till sheets that differ in age.

Characteristics	Pre-Iowan	Iowan	Tazewell	Cary	Mankato
Approximate age (B.P.)	40,000	≧40,000	17,000	15,000	13,000
Drainage density	89/100	78/100	77/100	31/100	25/100
Number of first-order tributaries	40	34	22	6	2
Number of second-order tributaries	14	11	6	2	1
Number of third-order tributaries	4	3	2	1	0
Number of fourth-order tributaries	1	1	0	0	0
Total	59	49	30	9	3

the drainage density, and the number of tributaries of successive orders. We recognize that we do not have perfect lithologic identity, and thus to some extent may be substituting space for time; nevertheless, the data do appear to follow the established geologic principle that, expressed numerically, indicates that both drainage density and stream frequency increase with the passage of time. The data are presented in Fig. 10-9, **A,** as a relation between drainage density and time. The graph implies that the rate of drainage development is most rapid in the first 20,000 years and subsequently levels off to a slower rate. Unfortunately, possible variations in soils and relief, combined with the fact that the dating is not sufficiently precise, do not permit us to state with confidence any more than

Figure 10-9.   *Relation of drainage characteristics to time on glacial till sheets in Iowa. [Data from maps of Ruhe, 1952.] **A.** Drainage density and stream length as related to time. **B.** Stream frequency and drainage density as related to time. **C.** Number of first order tributaries, and total number of tributaries as related to time.*

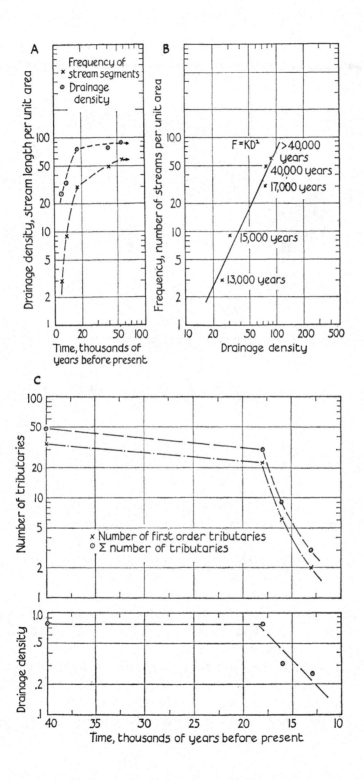

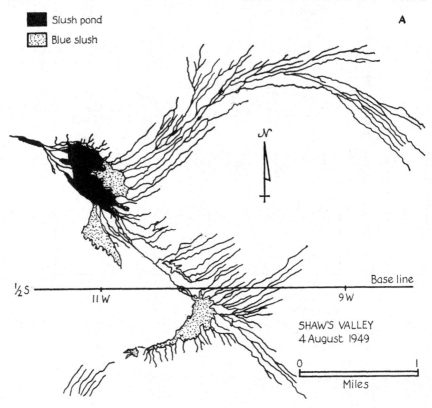

Figure 10-10. **A.** *Drainage pattern of meltwater streams on glacier, Shaw's Valley, Greenland Ice Cap, [From Holmes, 1955.] August 4, 1949.*

that channel length and number both increase with time. The leveling off in the rate of increase may reflect the establishment of equilibrium and the filling up of the basin area by stream channels.

Figure 10-9, **B,** from Ruhe's maps, shows that the number of streams per unit area and drainage density (total stream length per unit area) increase systematically with time. In Fig. 10-9, **C,** drainage density, number of first-order tributaries, and total number of tributaries are plotted on semi-logarithmic paper as functions of time.

It is interesting to note that the relation between channel number and channel length over time follows the geometric laws which earlier described the relation of number to length for drainage basins of different characteristics in different places. This supports the hypothesis that the spatial distribution is in fact a "growth" model.

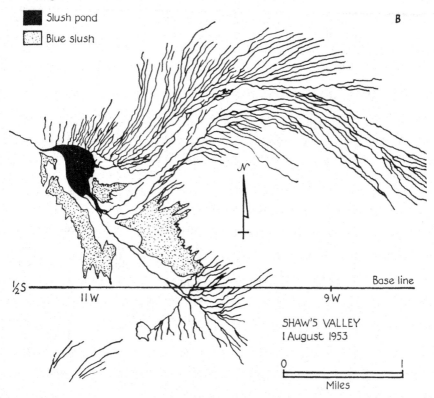

■ Slush pond

▨ Blue slush

B

½ S ———— Base line

II W   9 W

SHAW'S VALLEY
I August 1953

0 ————————— I
Miles

Figure 10-10. **B.** *Drainage pattern of meltwater streams on same glacier as* **A,** *August 1, 1953. [From Holmes, 1955.]*

Unfortunately there are few examples from different environments which illustrate how drainage patterns change with time. Where severe overgrazing or noxious gases from industrial plants have denuded hillslopes of vegetation, accelerated erosion has created new channels and both new and old channels have extended themselves in length.

The network of channels on a portion of the Greenland ice cap show an increase in a 4-year period of about 100% in channel length, and hence in drainage density, accompanied by an increase in the number of channels within the confines of major river valleys and divides on the ice (Fig. 10-10, **A** and **B**).

As the examples indicate, perpetual transformation of the overall composition of the drainage net does not appear to occur. Development may initially be quite rapid, but the change in pattern after this period is exceedingly slow and the geometrical properties appear then to represent a kind of equilibrium condition appropriate to the available processes, erodibility, and structure of the region. Some aspects of this concept of landform evolution are also considered in Chapter 5.

Modes of Drainage Extension

Three modes or physical processes of drainage extension can be cited: (1) surface runoff; (2) subsurface seepage; and (3) headward erosion of scarps or headcuts. Under conditions of surface runoff new first-order channels form where declivity, length of slope, and rainfall intensity create runoff capable of producing shear stresses that exceed the threshold of erosion of the soil. When this happens, the surface cover is broken and rills develop. This process is common on bare roadcuts and on badly managed farmlands. Under such conditions a rill channel may open up anywhere on a slope where local conditions are favorable, and the openings so formed may proceed downslope to already established drainage lines or upslope toward the divide.

Less obvious processes of incipient channel extension have recently been described by Bunting (1961) on moorlands in England. Detailed mapping of the subsurface soil mantle reveals seepage lines containing fine eluviated material and relatively deep layers of humus. These seepage lines have no topographic expression and are distinguishable at the surface only by their moist appearance at certain times of the year. The presence of moisture, and its slow downward movement evident in the soil profile, promotes continuous corrosion of the bedrock (Fig. 10-11, **A**).

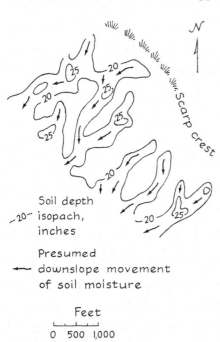

Soil depth
_20~ isopach,
inches

Presumed
← downslope movement
of soil moisture

Feet
0   500 1,000

Figure 10-11.

Subsurface seepage lines and their relation to surficial channels, in moors of Scotland. [After Bunting, 1961.] A gritstone backslope of cuesta, Bottom Moor, Matlock; subsurface seepage zones outlined by solid lines; direction of moisture movement shown by arrows.

In some areas the seepage lines connect directly with the perennial first-order drainage channels. Thus seepage promotes both a lowering of the surface of the bedrock by corrosion and an extension of the drainage network. Because water availability is limited at the crests of divides, continuous lowering of the adjacent hillslopes may leave isolated remnants of rock (tors) perched upon the drainage divides.

A third process of drainage extension—the headward retreat of vertical or nearly vertical channel scarps—has been described in many regions. A vertical face or drop in the bed of a stream channel, often called a headcut, may be created in a variety of ways. Faulting, lowering of the mainstream at the junction of a tributary, local concentrations of inflow, or topographic and stratigraphic irregularities may cause incipient gullies and headwalls to form. Once formed, a headcut may migrate upvalley, thus extending the drainage net within the drainage area. The mechanics of the process of headward migration of scarps is considered more fully in connection with channel degradation (Chapter 11). Gullies may extend themselves headward by basal sapping, aided on occasion by moisture or weaker strata at the base of the scarp, by concentrated erosion by surface flow in a plunge pool, or by combinations of these mechanisms as yet poorly understood. Not infrequently gully formation is an aspect of accelerated erosion produced by changes in land use or climate. Such changes may produce increased runoff, greater seepage, increased erodibility through reduced binding of the soil, or many other changes and combinations of effects.

Headward migration of channel scarps may increase the drainage density. In many cases the drainage net might consist of swales without marked channels, prior to trenching by gullies. Such swales are not unlike the subsurface seepage lines mentioned above, and the difference in composition of the drainage net before and after migration of the headcuts is in part a function of the way in which lines or channels in the network are defined. Where gullies form on slopes, however, these may constitute more obvious new channels, although the locus of formation may have been a modest depression or swale. In some regions the principal means of increasing the network of drainage channels may be through headward migration of gullies.

The rate of retreat of a gully headcut appears to be related to both runoff and moisture content of the soil. Repeated cycles in the southwestern United States show clearly that this process may create ramified drainage patterns comparable to those apparently created by other processes elsewhere.

Descriptions of erosional processes on slopes reveal that although specific

processes may be dominant in some climatic and physiographic environments, a variety of processes are usually operative in every region. One
should expect, therefore, in most drainage basins, that surface runoff,
seepage, and headcut retreat all participate in developing the drainage system. Where moisture is deficient, seepage may occur only after rain, as
it does in the free face of headcuts in semiarid regions. Where landslides
are frequent, they may intermittently obliterate evidence of stream channels
until surface runoff is again sufficient to trench the surface of the debris
provided by the slides. On a very much smaller scale, for example, Schumm
(1956) observed that rills formed on the surface of a fill at Perth Amboy,
New Jersey, during the spring, summer, and fall, and were obliterated each
winter by frost action.

Limits of Drainage Development

At present there is insufficient evidence to tie drainage evolution over
long periods of time to the detailed operation of specific processes. However, the operation of several basic mechanisms is apparent in the landscape. Where the surface of land in pasture, for example, is burned,
trampled, or otherwise destroyed, resistance to erosion is lowered, rills form
by surface runoff, and these in turn grow into gullies. Intrenchment of
gullies may produce steep walls. At points of weakness, where seepage
moisture is concentrated, portions of the wall may begin to retreat headward. These processes have been observed in virtually every environment
on the globe. All contribute to the development of the drainage system.

Drainage development, of course, cannot be divorced from hillslope
development. Assuming a constant drainage area, it is clear that as dissection proceeds and channel numbers and lengths increase, and if the
crests of divides are not lowered, slopes will increase in steepness. As
divides are lowered, slopes may decline in steepness, although changes in
the declivity of slopes over very long periods of time may be exceedingly
slow and will vary depending upon the relative rate of crest and channel
lowering.

One may well ask, is there a limit to drainage development? If so, what
is it like? The answer lies in the way one looks at or defines the dynamic
system and also in the time scale considered. An open system in steady
state may, in one sense, be constant in form, though energy is constantly
put into the system and degraded to heat within it. Such a view is both
theoretic and limiting. The utility of the concept of the open system in

steady state is that in nature close approximations to the steady state do occur. At the same time, however, landforms do evolve or change gradually through time while maintaining a steady-state form. It was pointed out in connection with the concept of entropy or the most probable distribution or form that the steady state is closely approximated by the average of a number of examples in a given population. For example, in a number of similar drainage basins in the same climate and on the same materials, the average characteristics of the group may describe very closely the steady state of an open system.

With random walks, furthermore, only a relatively small number of trials were necessary to obtain average results that closely approximated the average of a large number of trials. This finding suggests, even though it does not prove, that random processes operating within physical and hydraulic constraints develop rather quickly the same characteristics that obtain after a much longer period of time.

Thus there is plenty of play or room within the concept of the steady state of an open system for the gradual evolution and slow change of average conditions through time, maintaining all the while a close approximation to the condition of dynamic equilibrium with conditions existing at any point in time. Given topographic variability or relief and erosional processes operating over geologic time, uplift and denudation are not in perfect equilibrium. It should be expected that relief will change and in association with it hillslope and channel form. These changes, however, may be exceedingly slow and much overshadowed at any point in time by differences attributable to variations in processes or structure. Thus it is not only possible but profitable in the analysis of the origin and behavior of landforms to consider landscape as a system representing a steady state or dynamic equilibrium.

Drainage patterns and hillslopes show a high degree of adjustment of form and pattern to the tasks of handling variable quality and quantities of debris under diverse environmental conditions. This suggests simply that in the long run, adjustment of form to process in the landscape must be relatively rapid, and further, that the rate of change of many forms, including the drainage net, is slow save under unusual circumstances, such as man creates. As illustrated by the studies of Hack and Goodlett (1960) in the Shenandoah Valley, catastrophic rainfall and floods can create new stream channels with each catastrophe.

The number of stream channels does not appear to be ever increasing. Rather, some channels must be erased or filled as others develop, and the resultant drainage net continues to provide channels sufficient in frequency and size to carry off the hillslope debris and runoff.

The adjustment of form and pattern to structure and process is readily attained and changes slowly. Minor deviations from this adjusted or quasi-equilibrium condition are difficult to recognize. As statistical studies show, many parameters used to describe the profile and drainage pattern are relatively insensitive to departures from the average or most probable form.

REFERENCES

Bunting, B. T., 1961, The role of seepage moisture in soil formation, slope development, and stream initiation: *Am. Journ. Sci.,* v. 259, pp. 503–518.

Hack, J. T., 1957, Studies of longitudinal stream profiles in Virginia and Maryland: *U. S. Geol. Survey Prof. Paper* 294-B, 94 pp.

Hack, J. T., and Goodlett, J. C., 1960, Geomorphology and forest ecology of a mountain region in the Central Appalachians: *U. S. Geol. Survey Prof. Paper* 347.

Holmes, G. W., 1955, Morphology and hydrology of the Mint Julep area, Southwest Greenland: Mint Julep Repts., pt. II, Arctic, Desert, Tropic Information Center, U. S. Air University Publ. A-104-B, 50 pp.

Horton, R. E., 1945, Erosional development of streams and their drainage basins; hydrophysical approach to quantitative morphology: *Geol. Soc. Am. Bull.,* v. 56, pp. 275–370.

Leopold, L. B., and Langbein, W. B., 1962, The concept of entropy in landscape evolution: *U. S. Geol. Survey Prof. Paper* 500-A.

Melton, M. A., 1958, Geometric properties of mature drainage systems and their representation in an E_4 phase space: *Journ. Geol.,* v. 66, pp. 35–54.

Ruhe, R. V., 1952, Topographic discontinuities of the Des Moines lobe: *Am. Journ. Sci.,* v. 250, pp. 46–56.

Schumm, S. A., 1956, Evolution of drainage systems and slopes in badlands at Perth Amboy, New Jersey: *Geol. Soc. Am. Bull.,* v. 67, pp. 597–646.

Chapter 11 Channel Changes with Time

Channel Aggradation and the Accumulation of Valley Alluvium

Sediment deposition produces many constructional landforms. Among depositional processes alluviation, or subaerial sedimentation by rivers, is of major interest to geomorphologists and is the subject of a large literature. Statement of the controlling condition is relatively simple: alluviation occurs when the production of debris exceeds the amount that can be carried away by the processes of transportation. There are a variety of conditions leading to production of sediment in large quantities and to modification of regimens of sediment transport in large and small stream channels. The geomorphologist is concerned with these conditions and with the kinds of channels which produce different types of deposits, the rates at which these deposits accumulate, and the stratigraphic or depositional characteristics that can be used to interpret the conditions during deposition.

It would be desirable to know both the rate at which the surface of a valley fill increases in elevation and the rate of accumulation of sediment in a valley in terms of volume or weight per unit of time from a unit drainage area. These two types of data must be derived by different techniques and give results of varying accuracy and usefulness.

Some data on the rate at which the surface of a valley fill rises are available from a variety of sources. It is sometimes possible to obtain information on the year a fence was built by a rancher across a small tributary valley, and in those instances where sediment deposition has buried the fence there have been several generations of successive fences. Three generations of fences in several valleys in the Cheyenne River Basin, Wyoming, imply aggradation of a tributary valley totaling as much as 6 feet in 30 or 40 years. In the same river basin in Sioux County, Nebraska, a cottonwood tree approximately 60 years old was found buried by 8 feet of alluvium, and in the same period subsequent cutting eroded a channel slightly below the original roots of the tree (Schumm and Hadley, 1957, p. 170). Historical records, of course, provide unparalleled data on aggradation—if the record can be read. Classic among these is the record of the Nile, which had been building up its bed and flood plain at a rate of about 0.03 foot per year in the vicinity of Karnak and Memphis (Lyons, 1906, pp. 315–317). This rate is measurable only because gages, temples, and statuary provide a record of thousands of years. It may at first glance seem slow, but it is $\frac{1}{8}$ the rate of accumulation represented by the generations of fence posts, a rate which is considered rapid.

Like the Nile, the Tigris-Euphrates valley furnishes a record of irrigation agriculture extending back 6,000 years. The patterns of development at two different stages were worked out in detail and plotted on maps by Jacobsen and Adams (1958). Under both ancient and modern Mesopotamian conditions a clear distinction between canals and rivers is generally meaningless or impossible. Silt banks produced by cleaning canals are major topographic features; they extend for great distances and tower over all but the highest mounds marking the ruins of ancient towns and cities. Archeological soundings indicate an average accumulation of roughly 33 feet of silt in some alluvial areas during the last 5000 years. Values for the Indus River at Mohenjo-Daro are of the same order of magnitude. A part of this sediment in Mesopotamia is doubtless the result of severe floods, but much of it has come from cleaning canals and from sediment carried to the fields by irrigation waters. During the last 4500 years extension of the Mesopotamian delta has shifted the shore of the Persian Gulf approximately 180 miles to the southeast, a rate of delta building considerably greater than the present figure of 1 mile in 70 years. The degree of human influence on this process cannot be precisely defined.

There are cases, of course, where the degree of human influence is overriding. Man's activities have not been considered separately, however, because as a rule the geomorphic effects produced by man are the same as those produced without him. Usually man simply changes the magnitude of certain variables in the system. These in turn produce responses, perhaps only acceleration or deceleration, in the fundamental geomorphic processes. The appropriate principles are not abrogated. Not infrequently these principles are best illustrated where man has had a hand in molding the result.

The history of hydraulic mining in the Sierra Nevada, California, not only illustrates the effect of man on the landforms of a region but also provides a good example of aggradation as a result of increasing sediment yield without compensating increases in flow. In the early days of the gold rush only a small amount of dirt was disturbed, as most of the work was done by laborers with pick and shovel. As more efficient methods were developed, water power was substituted for manpower and vast quantities of earth were handled in separating the gold from the placer deposits in which it was found. Hydraulic mining increased steadily until 1884, when a series of injunctions brought by residents of downstream areas halted the entire operation. At the height of hydraulic mining it is estimated that scores of millions of cubic yards of earth were moved each year. Apart from the considerable topographic changes rendered directly by the mining, the principal effects were those on the streams, which resulted from over-loading with detritus and led to extensive aggradation over broad areas.

The general trend of such deposition is shown by the graphs of low-water records of the Yuba River at Marysville and the Sacramento River at Sacramento (Fig. 11-1) in the period 1849–1913. Gilbert (1917, pp. 46, 50) estimated that 2,375,000,000 cubic yards of sediment were moved

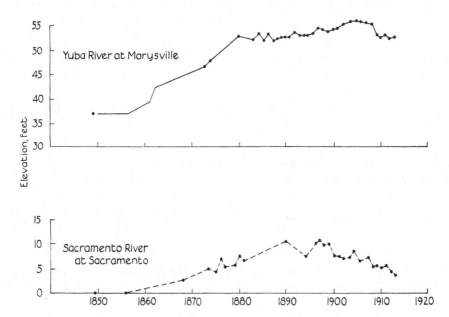

Figure 11-1.　*Fluctuations of low-water level due mainly to deposition of mining debris on stream beds [After Gilbert, 1913.]*

by mining operations to the San Francisco Bay system in the period 1850–1914, of which 1,675,000,000 cubic yards, or 62%, was considered to be mining debris. About 800,000,000 cubic yards were deposited in rivers and valleys above tidewater, virtually all of which was mining debris. Figure 11-1 shows that the bed of the Yuba River rose about 0.33 foot per year and the Sacramento 0.25 foot per year during this period.

Similar accelerated aggradation has been measured where sheet erosion from agricultural lands has resulted in aggradation in small valley bottoms at rates of 0.12 foot per year or more (Happ, Rittenhouse, and Dobson, 1940, p. 21). Severe gully erosion on occasion also provides the sediment for valley aggradation—that is, where channel cutting occurs in the uplands and aggradation in the valley bottoms. The contribution of gullies relative to sheet erosion, however, is usually small and decreases with increasing drainage area (Miller, Woodburn, Turner, 1962). Often deposition below gullies is also local and intermittent and related to upvalley migration of headcuts.

The question of simultaneity of valley filling in mainstream and tributaries throughout a basin is of geomorphic interest. The modern record suggests that when sheet erosion dominates on the drainage basin channel filling may be ubiquitous. Locally, where gullying is dominant, tributaries (gullies) may be eroding and the valleys filling immediately downstream. Aggradation of the Rio Grande Valley upstream from Elephant Butte Reservoir in New Mexico has been affected by several different factors and provides a particularly good example of the variety of possible controls of aggradation. During the period 1895–1935 the low-water (defined as 2,000 cfs) surface elevation at San Marcial increased 13 feet (Fig. 11-2). A widely held opinion attributes most of this aggradation to direct influences of the dam and reservoir a short distance downstream in which storage began in 1915. According to this view deposition at the head of the pool has affected the river upstream as far as San Marcial, a distance ranging from 5 to 30 miles, depending on the level of reservoir. As a matter of fact, however, the influence of the reservoir cannot be demonstrated in this case.

From Fig. 11-2, it is apparent that the rate of aggradation prior to construction of the dam was as great as afterward. Furthermore, the big jumps in the curve coincide with years of maximum runoff or great floods. During the period 1880–1915 diversions for irrigation reduced the natural flow of the Rio Grande at San Marcial by about one-half. At the same time accelerated erosion by tributary streams choked the Rio Grande Valley

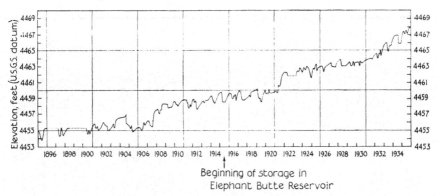

Figure 11-2. *Change with time of elevation of low-water surface of Rio Grande opposite San Marcial.*

with a sediment burden greatly in excess of amounts prevailing for several centuries previously. Between 1885 and 1929 nearly 400,000 acre-feet of sediment was contributed by the Rio Puerco, which joins the Rio Grande about 50 miles upstream from San Marcial. Still another factor which has affected aggradation in the San Marcial area is the presence of a railroad embankment built in 1880 and trending diagonally across the valley. Together with associated dikes, this structure behaves as a dam during flood periods and also prevents the river from distributing its sediment burden over its flood plain. All these influences, both natural and unnatural, combined to modify conditions in this relatively short reach of river. The controlling factor was not aggradation above the reservoir pool.

Passing from these direct observations of river and valley aggradation to the geomorphic aspects of the problem poses additional problems. Despite the interest in the subject, extension of the engineering observations to the geologic evidence is difficult, due to many still-missing links.

The outward characteristics of a river channel, whether large or small, are usually not diagnostic of whether a channel is in the process of aggrading or degrading. Many statements in the geologic literature imply that a braided stream is an overloaded one and that braided streams in general are in the process of aggradation. It is true that braiding is a pattern often associated with aggradation, but braided channels may represent an equilibrium pattern in the transport of the available discharge and load. The braiding pattern by itself is not prima facie evidence of aggradation, nor is the lack of braiding any proof that the channel is not aggrading. The Middle Rio Grande, for example, has been aggrading in modern

Figure 11-3.

Channel of the Rio Grande, near Bernalillo, New Mexico. Through many miles downstream from this location the river has an appearance similar to this and is aggrading, a fact not revealed by visual inspection. Bars of silt and fine sand are typical of this river at low flow. View is upstream; arrow shows location of gaging station.

times. It averages 270 feet in width, has an average slope of about 0.0009, an irregular silty bed, and low but definite banks. Looking at the channel of the Rio Grande, Fig. 11-3, one can not tell that it is gradually aggrading. The same might be said for the Nile.

There are many valleys in the western United States which in late-Pleistocene and post-Pleistocene time have been alluviated with a thick section of very uniform silt. The character of this valley alluvium can be seen in many places because in the recent epicycle streams have degraded and the alluvium is exposed in gully walls to considerable depth. Uniform silt exposed on the wall of the present channel and making up a major part of the valley fill is surprisingly ubiquitous and characterizes valleys of very different sizes. Flood-plain silt associated with large tributaries to the Ganges in Bihar, India (Fig. 11-4) is similar to that of the Powder River, Wyoming, a much smaller stream, and like that of even smaller tributaries of the Rio Chaco, New Mexico (compare Fig. 11-24).

The present stream may or may not be transporting material comparable to that it was transporting at the time it was depositing the large volumes of alluvium. There are many streams scattered throughout the semiarid area which presently flow over gravel or cobble bed, but in the period since the last glaciation, or about 10,000 years B.P., the progenitor of the present stream was depositing uniform silt practically devoid of gravel across a wide valley. Examples are Clear Creek, a tributary of Powder River, in the vicinity of Leiter, Wyoming, and the Popo Agie, a tributary of the Wind River near Hudson, Wyoming. The Popo Agie filled a valley nearly a mile wide with uniform silt to a depth averaging at least 20 feet. The present stream has a gravel bed. It displays a meandering habit which has typified the river, certainly for the several hundred years it must have taken to build the present flood plain; the scars of the meandering pattern are evidence of the maintenance of its present character during the most recent history. The question is, what was the regimen of the river and what did it look like at the time it was filling its valley with uniform silty alluvium?

In his studies of the relationship of the texture of bank material to

Figure 11-4.

Uniform silt of valley alluvium of the Burhi Gandak River, Bihar, India.
A levee is under construction, the silt being carried by hand in baskets
transferred from one man to another. The silt is too uniform for a good
embankment and finer material for a core-wall was hauled several miles
to this place.

channel pattern, Schumm (1960, p. 29) indicates that it is possible to separate aggrading from degrading reaches. Aggrading channels, he suggests, tend to have a higher width-depth ratio for a given value of weighted mean percent silt-clay than do degrading channels. If this method of separation proves on further investigation to be even reasonably dependable, it will be a valuable tool. Nevertheless, caution is urged in designating a given reach as aggrading or degrading on the basis of its general appearance.

Studies of the alluvial features of Wyoming valleys indicate that during the period when the major valleys were aggrading, aggradation also typified all of the larger tributaries and many minor ones and extended far up the hillside swales. This generalization applies to the great areas of plains or uplands of moderate relief and does not involve mountain valleys. The authors dug trenches through valley alluvium in small headwater draws, and have extended by shovel the cross section exposed in the headcuts of discontinuous gullies in the uplands.

Even headwater valleys only 5 to 15 feet wide are generally filled with alluvium to the depth of several feet over the bedrock. This alluvial mantle in the valleys of eastern Wyoming was graded to the highest of the postglacial silt terraces. The surface of the alluvium in the small tributary swales passes in a smooth unbroken curve to merge with the surface of the highest silty terrace.

The terraces of the master streams can be traced directly in the many tributaries of moderate size, indicating that deposition of alluvium in the valleys of the master streams was accompanied by alluviation in tributaries, even in the small ephemeral ones.

This implies that during the period of alluviation, when silt was being deposited in many valleys and tributary swales in this part of the semiarid west, the source of the silty material from hillslopes and hilltops was not related to erosion of tributary valleys now constituting the drainage network. Rather, this silt must have been derived from the unconsolidated shales and silt stones constituting bedrock over large areas by the processes of mass movement and sheet erosion on upland slopes. This conclusion from geologic evidence is in general agreement with observations of modern erosion and sedimentation patterns discussed previously (pp. 237, 349).

The same kind of evidence can be seen in other parts of the southwest. Within Mesa Verde National Park in southwestern Colorado there are many swales and valleys on top of the large plateau or mesa, the upper levels of which are underlain by sandstone. These valleys are separated from any base-level control of large master streams, for they terminate in nearly vertical cliffs where they reach the plateau scarp. These small isolated valleys exhibit alluvial fills and terraces, indicating an alternation of alluviation and degradation. At least one of the periods of alluviation was progressing during, and continued after, the Indian occupation of the 9th to 13th centuries A.D. Thus changes in rainfall-runoff relations associated with climatic variations cause even tributaries near the watershed divide to aggrade and degrade, and thus need not depend on changes of base level.

The geologist not acquainted with these conditions might well imagine that alluviation of a master valley is coincident with deepening and extension of tributary channels and rills near the headwater divides. This is not the condition in southwestern United States. There it appears to be generally true that tributary valleys alluviate simultaneously with main valleys. The source of the alluvium must be the hillslopes of the headwaters, and it was furnished not by rill erosion but by sheet erosion and probably by mass wasting.

Some evidence for rates of increase of surface or bed elevation in valley fills has been cited. Under the circumstances just described, what is the rate of accumulation of valley fills? The silty terraces in the valley of the Powder River, Wyoming, were tentatively dated on the basis of reasoning about a paleosol and general stratigraphic relations. The high terrace, called the Kaycee, stands on the average 25 to 35 feet above the present stream. The low terrace, called the Lightning, stands on the average 5 to 7 feet above the present stream.

Estimates of the volume of these alluvial fills and tentative assignment of ages allowed a computation of rate of sediment production from the tributary basins during the time these two alluvial fills were deposited. These computed rates of sediment production were compared with measured rates for modern streams which are not aggrading, in the midwestern and southwestern United States. A similar computation was made for the alluvial deposits of the Tesuque Valley, New Mexico. The latter is more trustworthy, because dates were established both by carbon-14 and by the stratigraphic position of datable pottery and other artifacts. Both the computations indicate that during alluviation under the alternating climate of the recent past in the semiarid United States, rates of sediment production were of the same order of magnitude as the rates observed in modern streams. It may also be inferred that perhaps these streams, when aggrading, appeared much as they do today.

Apparently the tendency for the maintenance of quasi-equilibrium in stream channels is sufficiently pervasive that only slight deviations, if sustained for long enough periods of time, may account for aggradational features of considerable magnitude, but the deviation from equilibrium conditions necessary for the construction of such depositional features cannot be recognized or identified by any criteria now available. Only by measurement over time can the net direction of river change be determined, at least in the usual instance. This applies particularly to channels which are aggrading. Stratigraphic studies of alluvial sequences all seem to indicate that large-scale alluviation in valley systems results from processes which act relatively slowly. But erosional features, such as valley-trenching, appear to occur more rapidly and may by the same time standards be considered episodic or catastrophic.

Degradation: Headcuts and Gullies

Degradation or lowering of the channel bed may occur as a result of structural or climatic changes associated with differences in topography, vegetation, and lithology. The extension of the drainage net by headward migration of scarps or headcuts is a significant process by which the channel bed is lowered, thereby creating a terrace.

The behavior of the knickpoint or the point at which the abrupt break in slope occurs is governed by the discharge regimen and by the structure and composition of the bed and bank materials of the river. Experimental studies indicate that in noncohesive materials an oversteepened slope will be reduced or flattened at a rate proportional to the rate of sediment transport, transport being a function of particle size and flow (Brush and Wolman, 1960, p. 68). The local maximum rate of transport and the accompanying flattening of slope occur at the knickpoint because at this point the shear stress—expressed as a function of the depth-slope product—is greatest. Progressive reduction of gradient in a laboratory study is shown in Fig. 11-5. Morisawa (1960, p. 1932) studied the gradients of natural streams in coarse unconsolidated gravel where they were oversteepened by normal faulting during the Hebgen earthquake of 1959 in Montana. Where flow was sufficient, oversteepened gradients also flattened with time. Meander cutoffs show a similar decrease in slope with the passage of time.

Experimental studies in homogeneous cohesive materials indicate that under certain conditions a knickpoint will retreat, maintaining a vertical face. The vertical face is the classical "headcut." For a given depth of flow, if the initial height, H, of a vertical face or abrupt drop in a channel bed is such that critical flow is attained at the knickpoint and a "plunge

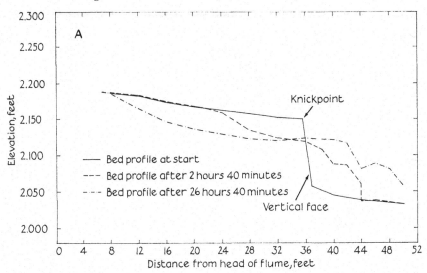

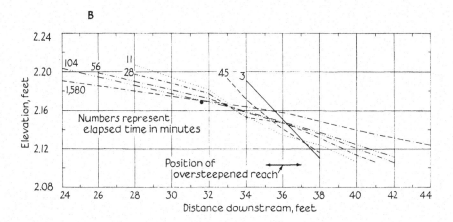

Figure 11-5. *Progressive flattening of gradients of knickpoints in labora-*
tory channels in noncohesive sediments. **A.** *Changes occur-*
ring in the longitudinal profile of the bed of a model channel
with time. **B.** *Changes in water-surface slope for the model*
channel.

pool" or hydraulic jump occurs at the base of the face, then the face will
retreat upstream. The face will be maintained, however, only if (1) the
material making up the bed at the knickpoint has a resistance to shear
stress greater than the stress provided by the flow, and (2) flow is sufficient
to transport the eroded material from the base of the face. In the model

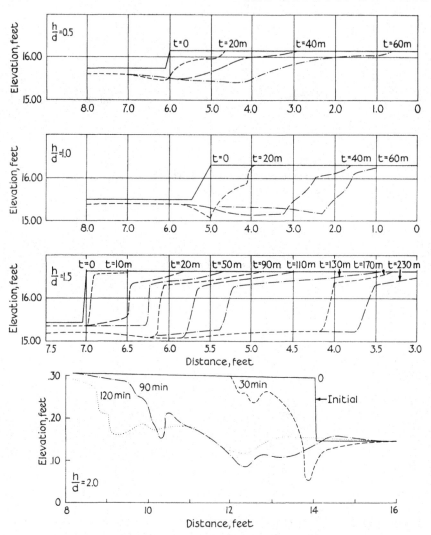

Figure 11-6.   *Up-valley migration of headcuts in laboratory channels in cohesive materials dependent upon ratio of flow depth to initial height of free face.*

studies (Fig. 11-6), where the ratio H/d between the height of the face, H, and depth, d, was large, for a constant slope and given material, migration of the vertical face occurred. Where H/d was less than or approximately equal to 1.0, the knickpoint was obliterated by flattening of the slope, as in the noncohesive materials.

Preliminary data suggest that the conditions necessary for maintenance of the headcut or vertical face can be schematically described by a diagram such as Fig. 11-7. For a channel on a given slope flowing at a depth d, upstream from the headcut, vertical retreat will occur if the ratio H/d is such that the flow passes through critical depth, $d_c = v/\sqrt{g}$, provided that at the knickpoint $d_c s$ (proportional to the shear stress) is less than τ_c/γ for the bed material.

Thus in Fig. 11-7, for a depth d_1, critical-flow depth occurs when H/d_1 is d_c, as shown on the figure. For materials which erode at the value of τ_c/γ, given by lines such as $\overline{OB}$ orthogonal to the lines of constant slope, headcuts will retreat, maintaining a vertical face if the value $d_1 s_2$ lies above the line $\overline{d_c AB}$, that is, where $d_1 s$ is less than τ_c/γ for values of d_1 equal to or less than d_c.

Knickpoints along the Thames River valley, cut into the relatively homogeneous London clay (Zeuner, 1959), appear to be not unlike those produced in the laboratory. Along the Thames, however, the height of the face seems to have decreased as the knickpoints migrated upstream. In general, the behavior of headcuts in nature is far more complex than in the laboratory examples. Field observations indicate that retreat of the face may be brought about by seepage or by pressure from water seeping out of the face following rains that raise the water table above the foot of the face. Climatic factors such as rain and frost may also increase the erodibility of the sediments. A plunge pool resulting from surface flow may contribute both to erosion at the base of the face and to downstream removal of the eroded sediments. Where the strata beneath the channel are layered, as at Niagara Falls, a vertical face may be maintained by sapping of weaker strata under-

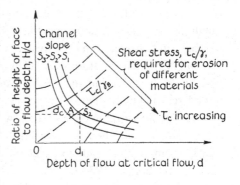

Figure 11-7.

Schematic diagram of migration of headcuts in cohesive materials controlled by relative height of face, depth of flow, and critical shear stress required for erosion of the bed material. Headcut migration will take place if at critical depth, d_c, the intersection of d_c and the available channel slope, s, lies above the orthogonals defining erodibility or critical shear stress of the available material.

lying more resistant beds. Some of the more important climatic and physical aspects of gullies and headcuts are described below.

Various studies by soil conservationists have shown that gully development in many areas is the principal way in which sediment is produced by erosive processes. Considering the importance of gully development and headward extension, surprisingly little is known, at least quantitatively, about the mechanics of the process.

Sapping at the base is undoubtedly one of the more effective processes in headward extension. Many gullies, particularly in relatively uniform material, advance headward, maintaining vertical cliffs at the head itself. In others, slumping of the vertical wall leaves a jumble of blocks at the base of the vertical face. To what extent the occurrence of such masses of slumped material at the base of the vertical face inhibits the process of sapping at the base is unknown.

Sapping is intensified and in many cases probably caused by the emergence of moisture that infiltrated the unchanneled swale above the headcut and moved under gravity to the base of the vertical wall. The moisture produces a pore pressure tending to slough the supporting base of the gully head; perhaps more importantly, moisture in materials containing clay weakens the supporting foundation at the base of the vertical cut, both in tensile and in compressive stress. In the gullies of the Piedmont of South Carolina, a stratum of relatively friable material occurs near the base of the headcut, but the top stratum is tough and resistant. This leads to undercutting by the emergence of groundwater and, although the foundation of the finer-grained material is weakened, the load of the overlying stratum is undiminished because large blocks of the upper material caves in. This combination leads to particularly rapid headward erosion (Ireland, Sharpe, and Eargle, 1939).

Undercutting of the vertical headcut by plunge-pool action during storm flow is presumably greatly aided by slumping of the moistened head wall after the storm flow ends. In extensive study of gullies in alluvial valleys we have but seldom experienced collapse of dry gully walls, nor is gully wall collapse characteristic of those short periods of storm discharge when water is actively flowing in the gully trench. But the rate and extent of local slumping and collapse in the hours following a storm flow, or on the day following the storm, is very marked.

The development of pipes or tunnel-like holes is an important element in the headward extension of many vertical cuts. Near gully walls these pipes often extend into the ungullied alluvium for distances of several hundred feet. Even at such distances depressions several feet in diameter can often be seen. Near the gully wall in silty alluvium of uniform texture the ground surface may appear pitted, and the collapsed tunnels may coalesce to form tributary gully heads. Piping is presumably often initiated through the occurrence of rodent burrows. However, the existence of piping is so prevalent that the action of rodents could hardly be considered a principal cause.

Piping undoubtedly represents zones where water has passed from the ungullied floor of the swale down to the base of the vertical gully wall along which the water moved not only salts in solution but also fine-grained sediment. Seepage movement of fine particles (eluviation) is known not only in the vertical movement of clay particles during soil profile development, but laterally as a sediment transport phenomenon. Ruxton (1958) noted surface deposits of eluviated clay near the edges of small pediments in the Sudan.

Gullies have been observed both in the middle west of the United States and in Canada, where the flat slopes of the land surface that has been gullied give no hint as to the reason for headcut formation. Rubey (1928), in investigating such an occurrence, postulated that the movement of water through the fine-grained material actually transported fine particles away from their original place of deposition and left voids. These voids then concentrated water, accelerating the process of soil removal and resulting eventually in the formation of collapsed holes from which full-fledged gullies developed.

The existence of vertical walls unsupported at the base can lead to fracture planes—which tend to be oriented either at 45°, sloping upward from the base of the vertical cliff—or to arcuate shear planes that begin nearly horizontally at the base of the cliff and curve upward to meet the surface in a nearly vertical orientation. Along the ungullied floor of a swale above, cracks are often evident, curving around gully headcuts or paralleling the gully wall. These often occur en echelon, oriented approximately parallel to the vertical gully wall.

In our experience in semiarid regions, the erosive action of water flowing over the vertical face of a gully head generally is not among the most active agents of headward progression.

Quantitative data on rates of gully development are scarce. A famous valley trench, Rio Puerco in New Mexico, is about 150 miles long and, in 1928, averaged 28 feet deep and 285 feet wide. Probably 95% of this was eroded between 1885 and 1928. In 42 years 395,000 acre-feet of sediment were contributed to the master stream, the Rio Grande (Bryan, 1928, p. 279).

Some details of movement of individual headcuts are presented in Table 11-1. The gullies observed in South Carolina are typical in that the head-

Table 11-1. Sample data on rate of erosion of gully headcuts, near Spartanburg, South Carolina. [From Ireland, Sharpe, and Eargle, 1939.]

Name of Gully	Drainage Area (acres)	Time Interval (days)	Rain in Interval (inches)	Erosion of Headcut (feet)	
				Length	Width
Layton	1.15	119	18	Slight erosion	
		2	4 to 6	15	12
Walden	0.3	978	?	50	
		2	6	8	5
Littlejohn, A	6.1	72		5	
		8	Soaking rain	4	
		10		1	
Littlejohn, B4		72		5	
		3	Soaking rain	3	
Cox, B	1.46	100		14	
		39		3	
		19		3	
Cox, A	1.68	100		6	
		188		0	

cuts remain unchanged for months or even years at a time and then, when heavy rains moisten the soil deeply, the head may erode several feet per day for short periods of time.

The Discontinuous Gully

A common characteristic of new gully systems is discontinuity along the length of the developing channel. A discontinuous gully system is characterized by a vertical headcut, a channel immediately below the headcut which often is slightly deeper than it is wide, and a decreasing depth of the channel downstream. Where the plane of the gully floor intersects the more steeply sloping plane of the original valley floor, the gully walls have decreased to zero in height and a fan occurs. The fan is characterized by some semblance of a concentration of water often along the topographic high of the cone-shaped deposit of debris. As a result, the position of this watercourse shifts frequently and, as in the case of other similar alluvial fans, gives the symmetry to the flat cone of deposition. Such a gully, having a vertical headcut at the upstream end and a fan downstream, is usually only one of a series occurring unsystematically along the length of the valley being gullied. Schumm and Hadley (1957) have described the discontinuous gully as a semicyclic phenomenon in which alluviation on a valley floor develops locally a gradient too steep to be stable; subsequently it erodes. They suggested that the fan formed at the mouth of a discontinuous gully has a local slope steeper than average for the valley; thus the fan is one of the features which may form the locus of a new gully, owing merely to the local gradient.

But *where* gully heads begin in the system of discontinuous gullies does not appear to be the most important aspect of the phenomenon. More important, in our opinion, is a hydraulic mechanism to explain the flat gradient of the gully floor relative to the steeper gradient of the original valley floor. The intersection of these two planes of different slope gives the discontinuous gully its particular character. All workers who have studied the problem seem to agree that through time there tends to be a deepening of each gullied segment and that this deepening alone would account for some increase in the length of the gullied segment. But it is usual for increased deepening to be accompanied by an increase in the gradient of the gully floor, for when a system of discontinuous gullies finally coalesces and a single uninterrupted channel is formed, the end result is a gully whose bed has a gradient practically identical to the gradient of the original valley floor. This end result indicates that there is a progressive steepening of the gully bed.

On an ungullied valley, a local weakening of vegetation—from grazing or trampling by stock, fire, or an exceptional storm—may allow an initial scarplet or small basin to form by erosion. The subsequent storms cause the head of this initial erosion feature to progress upvalley, and the excavated debris splays out at the downstream toe in the form of a low fan. The initial short channel terminates upstream in a vertical headcut. The concentration of water in the flumelike trench reduces the channel storage that had formerly existed when the same water could have spread out over a wide swale; consequently, from a storm of given size, the peak discharge passing through the channel is greater than would have been experienced on the ungullied valley floor. This increased peak discharge is accompanied by greater velocity and cutting power, and the initial gully advances with sufficient rapidity that growth of vegetation in the intervals between storms cannot stabilize the gully.

This process may occur simultaneously at various points along the length of the swale, probably concentrated at local zones of higher than average surface slope. The condition postulated is illustrated in part 2 of Fig. 11-8.

The plunge pool, where present, is always seen to be dug deeper than the level of the floor of the discontinuous gully just downstream from the plunge pool. The floor of a discontinuous gully is composed of a layer of newly deposited material, which must overlie the undisturbed alluvium. The bed of most discontinuous gullies, then, is a depositional rather than an erosional feature. Erosion in the plunge pool tends to deepen the channel faster than to widen it. At an early stage of gully development the channel has considerable depth but a restricted width. At the same time it is forming

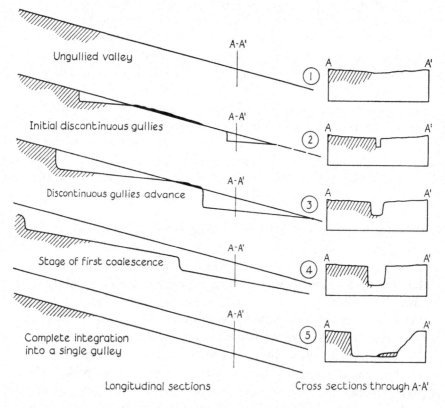

Figure 11-8.   *Stages of development of an arroyo from discontinuous gullies.*

a deposit just downstream from the plunge pool and thus is currently forming its bed slope, under conditions in which slope can be adjusted with relative rapidity as compared with channel width. Under the conditions that prevail in ordinary rivers, the reverse is true; width may adjust rapidly during floods, but slope adjusts only slowly.

Hydraulic considerations indicate that where roughness is constant a relatively large depth at any given velocity requires a small value of slope. As widening progresses during a later stage in gully development, indicated in parts 3 and 4 of Fig. 11-8, to carry a particular sediment load at a given discharge requires an increased shear as the width-depth ratio increases. To achieve the larger shear with decreasing depth requires an increase in slope. Thus, it appears that a low gradient of the channel bed should characterize the early and narrow stage of the discontinuous gully; slope should be expected to increase as the channel widens.

In summary, mutual accommodation causes the small width of the initial discontinuous gully to be associated with a gradient flatter than the original valley floor and, as a result, the gully is discontinuous downstream. As the process continues through time, a stage will be reached when the headcut of the downstream gully meets the toe of the one upstream and the two discontinuous gullies coalesce. As both width and depth continue to become greater, the channel reaches a size sufficient to contain—without overflow on the valley floor—the largest discharge that the basin upstream can produce. At this stage the effect of the large channel storage that characterized the flat-floored alluvial valley in its ungullied condition has been lost, and the discharge of a given reach increases greatly.

Figure 11-9 shows the longitudinal profile of a system of discontinuous gullies near Santa Fe, New Mexico. The gradient of the gully floor in the several discontinuous channel segments is clearly less steep than the average slope of the valley floor. In the example pictured, the gradient of the gully floor at Station 2500 was 0.014, a much smaller figure than the original gradient of the valley floor, 0.028.

Though the discontinuous gully is a common feature in the alluvial valleys of the western United States, the period of erosion which was initiated in that region in the last half of the 19th century has made mature gully systems even more ubiquitous. One characteristic of gullies throughout a large region is the depth to which the gully cuts. Detailed studies of the history of erosion and sedimentation made in various parts of the western United States have demonstrated that in each successive period of valley-trenching the gully in an individual valley was cut to approxi-

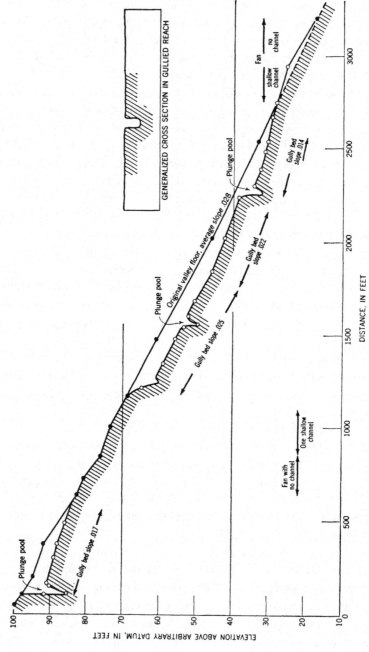

Figure 11-9. *Profile of a discontinuous gully, Arroyo Falta, near Las Dos, 6 miles northwest of Santa Fe, New Mexico.*

mately the same depth. In many valleys this depth was limited by bedrock on the valley floor or by a hard bed, often gravel, associated with some previous period of valley deposition. But in the post-Pleistocene period the several occurrences of gully erosion in the erosion-deposition sequence were similar in the depth to which the gullies cut in successive erosion cycles. We may therefore assume that in a period of valley-trenching the depth of gullying is limited generally by the existence of beds in the valley materials that are relatively resistant to erosion.

Degradation of Channels as a Result of Changes in Hydrologic Regimen

A mean or average elevation of the bed of a river channel is maintained over a period of time because the inflow of sediment to the reach equals the outflow. This balance is determined by the ratio of sediment and water derived from the drainage basin—a function of runoff, vegetation, and lithology or soil type. Gross average figures indicate that the maximum sediment yield can be expected under rainfall-runoff conditions comparable to those found in the short grassland or desert scrub areas of the western United States.

Extrapolation from this meager data suggests that in areas of low rainfall and sparse vegetation an increase in precipitation and runoff would produce an increased sediment yield to the stream channels. But beyond the maximum point of sediment yield, in regions of higher rainfall, an increase in precipitation would provide an increase in vegetation and hence a decrease in sediment yield. To the extent that an increased sediment yield from the watershed produces aggradation in the stream channel and a decreased sediment yield produces degradation, the effect of a change in climate would depend in this ideal circumstance upon the condition of the watershed prior to the change in climate. In terms of Fig. 3-7 (Chapter 3), if there was a climatic change toward increasing runoff, aggradation would be expected if the region experienced effective precipitation less than that corresponding to maximum sediment yield, and degradation would be expected if the region was above the maximum point.

The evidence of humid flora and fauna found in alluvial fills indicates that aggradation was associated with conditions of greater moisture (Hack, 1942). Arroyo-cutting generally seems to be associated with evidence of relative aridity. The data are not sufficient to test the hypothesis regarding a maximum point of sediment yield but conflict in the evidence precludes

Table 11-2.   Degradation below selected dams.

River, Dam, Location	Drainage Area (sq. mi.)	Period of Record	Average Lowering of Bed at All Sections (ft.)	Average Degradation per Year of Record (ft./yr.)
Missouri, Fort Peck, Montana	57,725	1936–1950	1.01	0.067
Missouri, Garrison, North Dakota	181,400	1949–1957	0.65	0.093
Arkansas, John Martin, Colorado	18,933	1943–1951	0.33	0.041
North Platte, Guarnsey, Wyoming	16,200	1927–1957	2.00	0.066
Red, Denison, Oklahoma-Texas	38,291	1942–1948	1.63	0.102
North Canadian, Canton, Oklahoma	12,483	1947–1959	1.82	0.150
South Canadian, Conchas, New Mexico	7,350	1935–1942	2.65[a]	0.378
Salt Fork Arkansas, Great Salt Plains, Oklahoma	3,200	1936–1945	1.0	0.125
Smoky Hill, Kanopolis, Kansas	7,860	1946–1961	0.76	0.051
Rio Grande, Elephant Butte, New Mexico	28,900	1917–1932	1.63	0.102[b]

[a] Apparent degradation small from 1942–1960.
[b] Below Perche diversion dam.

any unequivocal statement as to whether aggradation results from a particular direction of climatic shift.

Perhaps the best example of the effect of hydrologic changes on channel behavior is provided by the evidence of degradation below dams on many large and small rivers throughout the United States. Large dams in particular represent major changes that would be analogous to change in the climatic regimen of the rivers below them (see Table 11-2).

The purpose of a dam is to provide storage of flow during periods of high water, for use during periods of low flow. Hydrologically speaking, the dam tends to even out the duration curve, lowering the peak stages and increasing the base flow. This effect is shown in Table 11-3, which gives the reductions or increases in flow at selected intervals of the duration curve for rivers in the western United States.

In addition to creating changes in flow, dams with large storage capacity trap virtually 95 to 99% of the sediment that previously passed through the reach in which the dam is located. Thus clear water is released below the dam in place of the sediment-laden flows that existed prior to construction. The combination of clear water and changing flow regimen leads to erosion of the channel and lowering or degradation of the bed of the channel below the dam. These effects are clearly observable, for example, on the Red River below Denison Dam.

Measurements have shown that sixteen years after closure of the Denison Dam approximately 35,000 acre-feet of sediment have been removed from the Red River channel over a reach extending 100 miles downstream from the dam. This removal represents both a lowering of the bed and erosion of the banks of the channel. Degradation of the bed of the Red and other rivers is shown in Fig. 11-10. Within the first 10 miles below the dam the bed has been lowered on the order of 5 to 7 feet. Maximum degradation below a number of dams is of this same order, although values of over 10 feet have occurred locally. As shown by Table 11-2, the average rate of degradation has been of the order of 0.1 foot per year over periods of about ten to fifteen years following closure of the dams.

Reduction of the rate of degradation and cessation of the process can be brought about in several ways. Because degradation results in flattening of the slope in the vicinity of the dam, the slope may become so flat that the force required to transport the available materials is no longer provided by the flow. The reduction of floods by storage decreases the stages attained and hence the competence of the transporting stream. If the bed contains a mixture of particle sizes the river may be able to transport the

Table 11-3. Changes in flow before and after closure of dams.

River, Dam, Location	Change in Flow (cfs) at Specified Duration (expressed as percent of time)			
	90% of Time (329 days per year)		0.2% of Time (1 day per year)	
	Before	After	Before	After
Red, Denison, Texas	500	1200	200,000	90,000
Republican, Harlan County, Nebraska	85	8	12,000	4,000
Smoky Hill, Kanopolis, Kansas	20	36	11,000	4,300
North Canadian, Canton, Oklahoma	68	31	10,000	6,000
Salt Fork Arkansas, Great Salt Plains, Oklahoma	1[a]	16	18,000	15,000
Wolf Creek, Fort Supply, Oklahoma	<10	<10	3,600	3,500[b]

[a] Approximate.
[b] Short record.

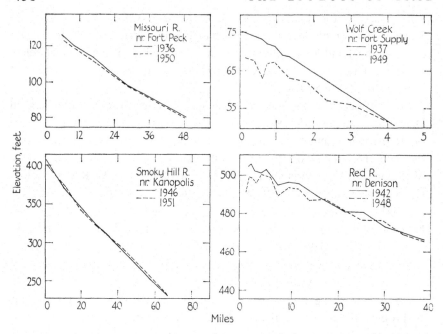

Figure 11-10. *Examples of degradation below dams on selected rivers in the western United States. Rates and amounts of degradation are variable depending upon flow, bed material size, location of bedrock, vegetation, and bank condition.*

finer sizes but not the larger, and the gradual winnowing of the fines will leave an armor of coarse material that cannot be transported. Armoring of the bed has taken place, for example, on the Red River below Denison Dam and on the Missouri River below Fort Peck Dam. In addition to changes in slope, the bed configuration may be altered as a result of changes in discharge and load, and these changes will affect the roughness of the bottom and the capacity for sediment transport. The hydraulic roughness may also be influenced by armoring, by growth of vegetation, by the appearance of bedrock, and so on. If tributaries enter carrying coarse material, they may provide an independent source of armoring material. Before dam construction, these were transported by the main stream during times of flood. The reduction of floods reduces the ability of the stream to rework the tributary debris. This is the case in some reaches of the Rio Grande River below Elephant Butte Dam.

Each of the above mechanisms may limit the rate or the amount of lowering of the bed. The limits of lateral erosion, although the same, may require a longer period for realization. The sixteen years of record on the Red River suggests that erosion may continue for long distances and over long periods of time. The control of this process lies in the sediment budget of the reach; so long as the supply of appropriate sediments is deficient, the flow may continue to make up the deficiency. Approximately 386 million tons of sediment have been deposited in Texoma Reservoir behind Denison Dam. Of this about 20%, or 77 million tons, is sand. Sediment removal by erosion downstream from the dam amounts to about 67 million tons. These data and a similar example from the North Canadian River below Canton, Oklahoma, indicate that the amount of sand deposited in the reservoir is roughly equivalent to what is eroded from the channel bed and banks in the reach downstream from the dam.

These changes in regimen associated with construction of dams are certainly heroic. It is difficult to conceive of a climatic change producing a change in vegetation on the watershed such that 95% of the sediment would no longer be delivered to the stream system.

Rates of valley cutting may, under special circumstances, be estimated crudely from data obtained in studies of terrace correlation. Datable artifacts such as pottery may furnish the essential information. Such facts seem to be provided by the study of alluvial fills in Tesuque Valley, New Mexico (Miller and Wendorf, 1958), in which both pottery and C14 dates were obtained. There it appears that over a period of 150 years the annual rate of degradation must have been about 0.13 feet per year, a figure comparable in magnitude to rates observed below dams.

The difficulty in such estimates comes from the fact that it is the alluvial fills which contain the datable objects. Degradation is known only to have taken place sometime between the youngest deposits of an older terrace and the oldest deposits in a younger alluvial fill.

The example provided by degradation below dams clearly indicates that the recognized control exercised by the ratio of sediment yield to water yield does operate, and at a relatively rapid rate. The evidence also indicates, however, that changes in vegetation in valley bottoms may be associated with changes in flow characteristics. Raising the base flow and eliminating floods appears to be associated in some instances with increasing density of vegetation on flood plains of semiarid regions. Large floods periodically devastate natural channels. If these floods occur after dry periods, they apparently change the channel to a great degree, probably because of the sparseness of the vegetation. Moderate climatic conditions, however, foster the development of vegetation and the vegetation serves to enhance the development of a confined channel and flood plain. Schumm (1963), in a study of historic changes on the Red River, presents suggestive evidence for this process, and it is corroborated by evidence on the development of vegetation along the Republican River below Harlan County Dam in Nebraska. The subtleties of these relationships between climate, vegetation, and sediment yield—both on the watershed and within the channel—are not yet clearly understood. But the mechanism by which degradation takes place and the hydraulic controls of the amount of degradation are suggested by the degradation below dams.

River Terraces

Degradational forces on an uplifted land mass tend to reduce it through time. Rivers as well as hillslopes tend to downcut gradually, while maintaining certain relationships along the river length. But the rate of downcutting is generally slow enough to allow processes causing lateral movement of the channel to operate. This results in the formation of flood plains in the valleys of most rivers and creeks. These flood plains vary in width, depending on the size of the river, the relative rates of downcutting, and the hardness or resistance of the rock material in the valley walls. Flood plains may occur in the valleys of creeks or torrents even a few feet wide but are generally absent along the most headwater tributaries, presumably because downcutting is sufficiently rapid that time is not sufficient for lateral movement of any magnitude.

In humid (as opposed to arid) climates, flood plains tend to be absent in the most headwater channels but appear at the point where flow in the channel changes from ephemeral to perennial—that is, where groundwater enters the channel in sufficient quantity to sustain flow through nonstorm periods. The reason for this apparent coincidence of flood-plain formation, however rudimentary, and the entrance of groundwater sustaining flow perennially, can only be inferred, for no detailed studies of this matter have, to our knowledge, been made. It seems reasonable, though, that perennial flow is influential in promoting rock weathering along the stream margin and sloughing into the channel, thus promoting lateral deposition and erosion along the small stream.

During any time period, then, when climatic characteristics remain approximately constant—and in the absence of uplift or change of base level—downcutting is slow enough that lateral swinging of the channel can usually make the valley wider than the channel itself. However, the elevation of the channel can be changed episodically, owing to alteration of tectonic and climatic factors.

In such a circumstance the flood-plain level previously associated with the stream is abandoned, either by downcutting or by aggradation. During downcutting the previous flood plain is dissected, and portions may remain as continuous benches bordering the river or, more often, as remnants of flat or nearly flat spurs jutting into the river valley. This sequence of events is pictured in **A** and **B**, Fig. 11-11. A different sequence of events that results in the same surface geometry is shown in **C, D,** and **E.**

For present purposes we will define a terrace as an abandoned flood plain. The terrace is composed of two parts, the scarp and the stair tread above and behind it. The term "terrace" is usually applied to both the scarp

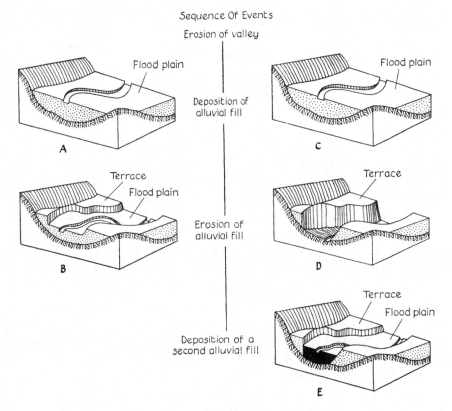

Figure 11-11. *Block diagrams illustrating the stages in development of a terrace. Two sequences of events leading to the same surface geometry are shown in diagrams **A, B,** and **C, D, E,** respectively.*

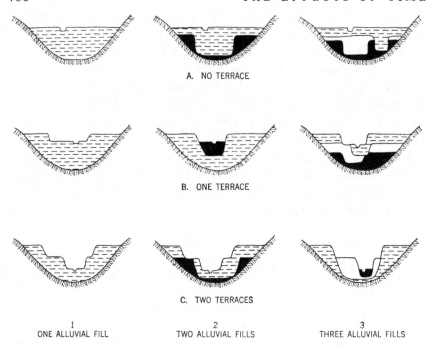

A. NO TERRACE

B. ONE TERRACE

C. TWO TERRACES

| 1 | 2 | 3 |
| ONE ALLUVIAL FILL | TWO ALLUVIAL FILLS | THREE ALLUVIAL FILLS |

Figure 11-12. *Examples of valley cross sections showing some possible stratigraphic relations in valley alluvium.* **A.** *No terrace.* **B.** *One terrace.* **C.** *Two terraces.*

and the flat tread—that is, to the whole feature of the landscape. (Terrace is also not infrequently used to include or to mean the deposit itself, when alluvium rather than bedrock underlies the tread and riser. This deposit, however, should more properly be referred to as a fill, alluvial fill, or alluvial deposit, in order to differentiate it from the topographic form.)

A distinction is usually made between terraces cut on bedrock, which are called "strath terraces," and those comprising former floors of alluvial valleys. The terms "cut terrace" and "fill terrace" have also been used by many writers, including us. Nevertheless, they have been found confusing to both students and investigators and should probably be abandoned.

The flood plain refers to the surface being constructed by the existing stream. It is not a terrace, though it is the surface of a formation composed of alluvial material. Alluvial fill, in the present context, is a deposit of unconsolidated or partially consolidated river-laid material in a stream valley, and is a single stratigraphic unit.

The point to be stressed is that a terrace is an abandoned surface not related to the present stream. The sequence of events leading to the observed features in the field may include several periods of alluvial deposition, and thus several alluvial fills. If incision and aggradation occur repeatedly, it is possible to develop any number of terraces. Depending on the magnitude and sequence of the deposition and erosion, any number of fills or different stratigraphic units could be deposited. Several possible combinations are sketched in Fig. 11-12.

It will be noted that several alluvial fills can comprise the valley sediments, even when no evidence of a terrace exists. The three sketches of Fig. 11-12, **A,** show some examples.

The stratigraphic relation of the alluvial deposits may be designated as "inset" or "overlapping." Figure 11-13, **A,** shows a situation in which the later of two fills (solid black) is inset in the trench cut in the earlier fill. In **B,** the later alluvial fill overlaps the earlier one; that is, the second fill was of sufficient volume to overflow the valley cut into earlier alluvium. These relations have been used as a basis for a classification scheme of alluvial valleys (Leopold and Miller, 1954, pp. 4–6).

The alluvial terraces just described demonstrate the principle that any alluvial terrace is a former level of the flood plain of a river. This generalization applies both to a strath terrace cut on bedrock but veneered with a layer of flood-plain alluvium, and to a terrace which is underlain by alluvium rather than rock. Cotton (1948) subdivides terraces into several categories: (a) rock terrace (strath terrace); (b) valley-plain terrace, surviving parts of former continuous valley flood plain; (c) slip-off slope terrace, formed by brief halts during vertical corrosion of a stream; (d) rock-defended terrace; and other categories. For present purposes these details do not add anything to the generalization that a terrace is an abandoned flood plain.

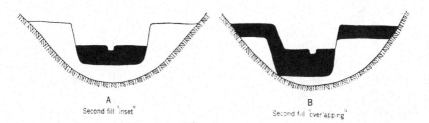

A
Second fill "inset"

B
Second fill "overlapping"

Figure 11-13. *Valley cross sections illustrating "inset" and "overlapping" relations of two alluvial fills.* **A.** *Second fill inset.* **B.** *Second fill overlapping.*

More important than subdividing terraces is the problem faced in the field of identifying and then interpreting a terrace form. Portions of a flood plain near the valley sides are subject to local deposition resulting from erosion of the valley sides by local wash, tributary rills, and mass movement. As a result there is generally a tendency near the valley margins for the surface of a flood plain to slope slightly from the valley edges toward the valley axis. Colluvium, or material washed onto the flood plain from the valley sides, tends to interfinger with the valley alluvium deposited by the main stream during valley aggradation. In some cases, of course, colluvium may actually be preponderant in the valley fill. The edges of a flood plain are often blurred, therefore, and merge gradually with valley-side colluvium. The contact can occasionally be seen when a trench across the valley exposes both the alluvium and valley-side material. In the valley of Beaverdam Run, Cambria County, Pennsylvania, for example, colluvium interfingers with valley alluvium on the west wall of the valley but spreads thickly over the east side of the valley, comprising an important part of the valley fill (Fig. 11-14).

The exposure at Beaverdam Run was excavated for a dam foundation. Lattman (1960) differentiated colluvium from river alluvium by the presence in the colluvium of slabs of bedrock typical of the local valley sides. In places these slabs dipped back into the slope, indicating that local slumping was one of the processes by which colluvium was contributed to the valley floor.

As a result of lateral or cross-valley slope of a terrace remnant, the exact level of the river at the time the flood plain in question was built is not always possible to ascertain in the field. The usual practice is to project toward the valley center the plane best representing the terrace surface. If the terraces are in pairs—that is, remnants on each side of the valley—

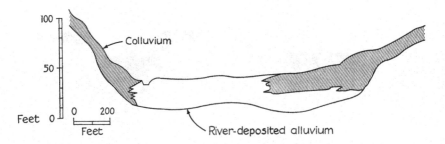

Figure 11-14.   *Cross section of valley of Beaverdam Run, Cambria County, Pennsylvania. [After Lattman, 1960.]*

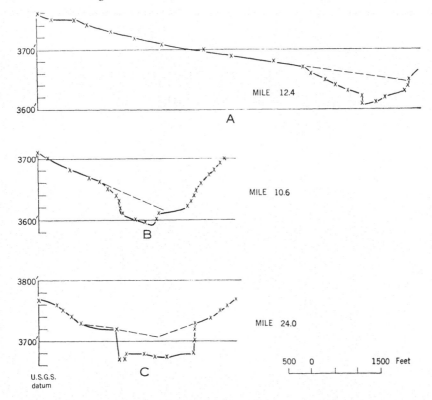

Figure 11-15. *Cross sections of the valley of the Powder River near Arvada, Wyoming.*

planes on each are projected to an intersection. Where the plane of a single terrace intersects the far valley side, or where the two planes from opposite sides intersect, is considered the lowest elevation at which the river could have been flowing when it formed the flood plain that later became the terrace surface. A typical example of such a projection is shown in Fig. 11-15.

An example of the distribution of the heights of terrace remnants above a present river is provided by the study by Bryan and Ray (1940) of the valley of the Cache la Poudre River, Colorado. Figure 11-16 shows a plot of the position of terrace remnants along the river. The variation in height of these remnants even on a supposed single or correlative surface can be seen in the figure. The investigators identified six terrace levels, the upper two of which did not extend headward beyond the most downstream moraine. As explained in their report, at least terrace No. 5, 90 feet above

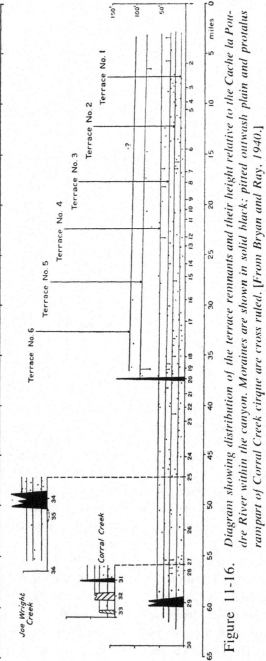

Figure 11-16. *Diagram showing distribution of the terrace remnants and their height relative to the Cache la Poudre River within the canyon. Moraines are shown in solid black; pitted outwash plain and protalus rampart of Corral Creek cirque are cross ruled. [From Bryan and Ray, 1940.]*

the river, ended at a well-defined moraine and was considered by them to be correlative with the advance of the ice associated with that moraine.

It is observed that the flood plains of present rivers tend to be narrow or even absent in narrow portions of the river valley and at the same time steeper than the average gradient of the river downstream, as is the river itself. Thus, in tracing terraces upstream into mountain valleys it can be expected that remnants may be scarce in canyons or narrow segments of the valley.

That a river should be steeper through a narrow portion of a valley is reasonable in that the presence of the constriction usually is a result of an outcropping of relatively hard rock or of coarse debris, as in a moraine that blocks a valley. Because the flood plain tends to be related to river stages of particular recurrence interval, the downvalley gradient of the flood plain tends to follow the gradient of the river itself.

Terrace Sequences and Correlation of Terrace Remnants

The most important criterion for correlating the terrace remnants or topographic benches in a valley is based on the premise that during a period of relative stability the stream cut a surface or built a flood plain, either of which was more or less continuous along the valley. When the stream cut down below this surface, thus forming a terrace, downcutting and lateral swinging of the main stream partly or completely eliminated the original surface. The degree of destruction of the original surface may vary, and thus the terrace remnants may be relatively continuous along the valley or they may remain only as isolated patches. They may occur at the same height on both sides of the valley, and thus be paired, or the remnants may occur only occasionally along the valley, sometimes on one side and sometimes on the other.

The continuity of a given surface along the valley and an associated tendency for terrace remnants to occur at uniform height above the present stream is a primary criterion for correlation (see Fig. 11-17). In contradistinction, a stream downcutting progressively will leave along the valley sides isolated flat spurs; these may not be paired, and they are of irregular height because they never were portions of a single continuous surface. Irregularities in elevation, of course, may also be produced by surficial erosion or by deposition of colluvium.

Continuity, then, is a basic criterion of terrace correlation. Relative

A

B

Figure 11-17.

A. *Terraces bordering Oraibi Wash, Arizona. Broad upper level under-lain by alluvium and lower terrace underlain by a thin layer of allu-vium and sand underneath which is bedrock, the horizontal bedding of which can be seen in banks of inner channel cut.* [Photograph by R. H. Hadley.]

B. *Terraces along Wind River near Dubois, Wyoming. In left center the bedrock underlying the high terrace is exposed, showing that the flat surface was cut across the structure of dipping beds. White barn in right middle ground stands on an alluvial terrace, and trees in mid-dle ground are, in part, on present river flood plain.*

elevation is another, but one which must be used with caution, as Frye and Leonard (1954) suggest.

Where several terrace levels exist in a valley, it is all too easy to postulate a correlation based on the relative position of a given remnant above the present stream. For example, if terrace remnants stand at three fairly well-defined heights, those at the intermediate height might seem obviously to represent the same surface. When the number of terrace heights is only two or three, this assumption might be valid, but without independent evidence it may nevertheless be dangerous. And when, as in some valleys, there seems to be a dozen or so levels, the assumption that remnants representing the tenth level at different places are correlative is liable to be in error.

Because both the principles and pitfalls of correlation are major concerns of geology and have been described voluminously elsewhere (Krumbein and Sloss, 1963, p. 332), they are not repeated in detail here. Because of a tendency to correlate terrace remnants solely on the basis of the longitudinal profile and their relative elevations, however, a few useful stratigraphic and physiographic criteria are enumerated. The criteria for correlation of terrace remnants may include: stratigraphic discontinuities between terrace fills, differences in particle size and sorting and in primary sedimentary structures, fossil fauna and flora, artifacts ranging from Indian pottery to tobacco tins, buried soils or paleosols, and frost features. In addition, physiographic relations to other landforms—the bounding valley sides and adjacent hills, glacial moraines, old lake beds—may serve as key elements in unraveling a terrace sequence.

Distinguishing between the lowest terrace and the flood plain may pose special problems. Because the surface of each is irregular, a clear distinction may not be evident. The colluvial slope, irregularity of the flood-plain surface, and a possible low terrace are shown in Fig. 11-18, a topographic map of the valley of Watts Branch near Rockville, Maryland.

A colluvial slope (A) splays out over a low terrace (B) which stands at a height only $\frac{1}{5}$ the channel depth above a modern point bar (C). Despite observation of this stream over a period of 10 years, it is still not clear whether the relation of B to C is indeed that of terrace to flood plain or whether both are portions of the same flood plain. The occurrence of several benches or berms along a small stream is not uncommon in the eastern United States. Where several are present, it is often difficult to pick out that level which represents the flood plain. The flood plain is not static but is undergoing frequent modification.

To the extent that the evidence presented in Chapter 7 is applicable,

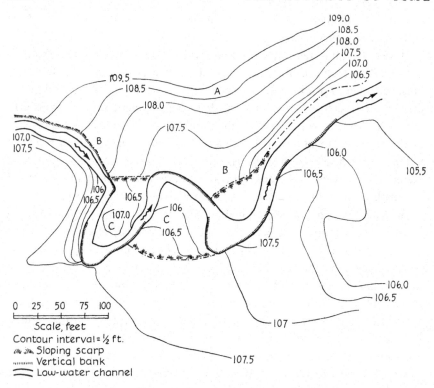

Figure 11-18. *Topographic map and cross section of Watts Branch, near Rockville, Maryland, showing colluvial slope and possible flood-plain and terrace levels.*

some terraces may be distinguished from the flood plain if we define the flood plain as that surface which is attained or just overtopped by floods with a recurrence interval, on the average, of 1 to 2 years.

Paleosols

Much of geomorphology is stratigraphic geology. Terrace correlation is all too often based on a mere assertion of contemporaneity, unsupported by detailed—much less quantitative—study of stratigraphy. Marker horizons are an integral part of terrace correlation, and identifiable features which may be used for markers of key beds include not only fossils but artifacts, fossil or buried soils (paleosols), pollen, or ash horizons. Soils as marker horizons have proved of exceptional value, but like other markers they may be misused or inadequately treated. Particularly needed are criteria or tests to demonstrate that a supposed soil is indeed valid and to differentiate one soil from another.

Surficial weathering in most climates tends to develop zonation in the form of a soil profile, whose characteristics and intensity reflect parent material, climate, and time. The abandonment of a flood plain by stream incision results in a terrace, and the same shift of stream regimen ordinarily is reflected to some extent in hillslopes or other surfaces even far removed from the master streams. Such changes often alter conditions of runoff and drainage and may then induce a shift in the character of the local soil. Climatic change, often the cause of shift in stream regimen and thus the cause of terrace formation, in itself tends to alter soil-forming processes.

As a result of such changes, a soil previously developing in response to original conditions is subjected to alteration over a period of time. The new conditions may gradually efface the original character of the soil, but before the effacement is complete some evidence of former conditions may persist and permit interpretation. Particularly, concentrations of iron compounds or calcium carbonate in nodules or solid masses may remain long after other evidence of the original soil-forming conditions have been obliterated. Distinctive features of this kind may provide not only a marker horizon but the key to inferences about conditions of climate and weathering in times past.

The soil characteristic of a given set of conditions may be partially preserved by burial as a result of sedimentation. When seen in an outcrop, the remnants may conspicuously be separated from the soil-forming processes characteristic of the present surface. But in speaking of buried or fossil soils, it must be recognized that when the soil-forming processes are removed, the zonation which remains cannot be considered a soil in the ordinary sense; by soil, we mean the dynamic complex of the microorganisms, vegetation, moisture, and minerals that maintain and develop the zonal character. What is preserved is partially determined by the time during which the profile was developing, as well as the intensity of the soil-forming processes. Preservation, of course, also is governed by subsequent events. A soil zone whitened and indurated by caliche accumulation subsequently covered by alluviation is common. This material may later be subject to erosion and there will remain only a calichefied zone of accumulation, representing the B horizon of a former soil, the A horizon of which has been eroded off.

Thus the opportunity for recognition in the outcrops depends in part on what Hunt and Sokoloff (1950) called the pedologic age, or relative maturity of the soil profile. As these authors point out, the chronologic or absolute age of an ancient soil refers to both the time and geologic date of its development, which is determined by its stratigraphic position.

A complex or polygenetic soil, as defined by Bryan and Albritton (1943, p. 478), is one which has experienced different soil-forming conditions— for example, a change from relatively arid to humid climate. A complex soil is postulated by them to explain a sequence observed at Praesum, Texas, in which the stratigraphic section was as follows:

DESCRIPTION	THICKNESS (inches)
(1) Humic soil, semicolumnar jointing	15
(2) Chestnut brown clay; cracks filled with alluvial clay	4
(3) Gray-brown sand; contains small veins of calcium carbonate, and solution holes	11
(4) Sand with angular rock fragments, strongly cemented with calcium carbonate	120

Bryan and Albritton interpret this sequence as indicating first an early relatively arid climate during which the main mass of caliche, including nodules, was formed as a pedocal, or an arid-land soil in which calcium carbonate accumulated in the B horizon. There followed a period of relative humidity in which caliche nodules were leached, leaving holes or cavities near the top of unit (4) and in unit (3). On alluvial flats there grew a good grass cover, which, together with some deposition, caused a soil to develop that contained more secondary clay and iron hydroxide than soils presently forming in the arid conditions of today. There followed another relatively recent change toward more arid conditions, during which the caliche films and veins in unit (3) were again deposited.

This is only one of several sequences of events which might be postulated to explain the observed soil characteristics. Bryan and Albritton discuss some of these alternatives. Our present purpose is only to give a sketch of the problem of interpretation of ancient soils, but one important point is that the thick layer of hard calcium carbonate present in unit (4) of the Praesum locality is a soil feature which could not have been formed under present conditions. Such a stratigraphic member can be used not only as a marker horizon but also as an indicator of conditions that existed in the geologic past.

The pre-Wisconsin soil studied by Hunt and Sokoloff (1950) illustrates particularly well the kind of field evidence used to establish the existence of an extended period of soil formation under conditions not present in the same area today. The general zonation observed is as follows:

DESCRIPTION[1]	THICKNESS (inches)
(a) Humic layer, dark gray or brown	2–3
(b) Light-colored lime-free layer	7–10
(c) Zone of some lime accumulation (present in areas of limy colluvium)	12–20
(d) Lime-free reddish-brown clay, structureless or prismatic; lower part may contain tiny veinlets of $CaCO_3$	20–50
(e) Lime-enriched parent material containing completely weathered residual stones	120±
(f) Comparatively fresh parent material; joints in upper part may contain $CaCO_3$	

[1] (a) to (c): soil related to present surface.

This thick layer of zonal ancient soil is much thicker than soils related to present surfaces. The paleosol is developed on a variety of parent materials and is present in patches over a large area in three states. It is often eroded at the top, so that the complete section is present only in some localities. In places it is covered with loess, containing extinct animals; in other places it is covered with materials deposited during times of the pluvial lakes in basins that are now desert, indicating a pre-Wisconsin age. It is associated with a distinct topographic unconformity; that is, it covers a surface of topographic relief much less rugged than the topography being carved at present.

Such variety of evidence strengthens the inference that this is an ancient soil of general and not merely local significance; thus it is a useful stratigraphic marker. For many supposed paleosols discussed in the literature, such variety of evidence is lacking. Dark zones or bands in alluvial materials often superficially resemble an ancient soil but are lacking in zonal development (evidence of leaching), continuity or geographic spread, and other characteristics which should be present if the postulated period of weathering and soil formation was general rather than local.

Because the accumulation of calcium carbonate is common in the B horizon of pedocalic soils of semiarid areas and because nodules and massive carbonates are resistant to later elimination by leaching, the presence of such zones of caliche accumulation is commonly reported in paleosols of western United States. But little is known about how fast and under what conditions $CaCO_3$ accumulates in soils. In some special localities soil scientists have been said to observe considerable accumulation of caliche in a period of 10 years, but the circumstances have not been published. If this is possible, one can imagine that it would be easy to mistake in outcrops local carbonate concentration having no value for correlation purposes. No doubt such misinterpretations have appeared in the literature and will continue to do so until reliable methods of identification, particularly quantitative ones, become available. A variety of evidence of different soils, as mentioned, should be sought in any study locality, and little reliance should be placed on any single type of evidence or limited number of outcrops. (The authors do not omit themselves from those to whom this suggestion is directed, for we have found by experience its necessity.)

Bands of whitish accumulation in alluvial materials are not always calcium carbonate. Chemical tests are often required to distinguish between caliche or calcium carbonate in veins and nodules, and gypsum, which may look similar. But the conditions for gypsum concentration in soils are even less clear than those for $CaCO_3$, and the interpretation of gypsum in sections where calcium carbonate might be expected is unclear. Also, a zone of whitish accumulation in a stratigraphic section of alluvial deposits may be caused by thin films of $CaCO_3$ covering soil particles, a condition which might well be local and requiring only a brief span of time to develop. Presumably hard nodules, tough blocky cementation and considerable thickness of $CaCO_3$ accumulation all tend to indicate a considerable span of time and some regional significance.

Figure 11-19 illustrates how a paleosol might appear in an outcrop of alluvium. The stratigraphic section accompanies the photograph. The example is associated with valley alluvium of the Powder River Basin, Wyoming.

Some techniques involving quantitative tests for analyzing possible paleosols have been used, though these at best only begin to fill the need. Figure 11-20, also associated with alluvial valley deposits in Wyoming, shows the variation of percentage of $CaCO_3$ in soil materials as a function of depth. The upper diagrams show that there is a concentration of carbonate, which increases with depth of burial by later alluviation of the Kaycee formation. This example illustrates a variety of evidence for the existence of a paleosol. The zone of carbonate accumulation is near the surface at the top of the hill but is progressively deeper as one moves toward the valley, as would be expected if the ancient soil developed on a topographic surface which later was altered by deposition of alluvium and colluvium. The carbonate concentration crosses several types of bedrock

Figure 11-19.

Paleosol exposed in valley alluvium bordering Powder River near Sussex, Wyoming.

Kaycee formation: Fine, tan, silty alluvium consisting primarily of quartz grains. Surface soil profile characterized by well-developed columnar structure to a depth of about 3 feet; no distinct humus zone near the surface; slightly calcareous from a depth of about 10 inches. Grades downward into Ucross formation . . . . *3–6 feet*

Ucross formation: Gray or white silt, with some gravel in the lower part, containing a high concentration of calcium carbonate interpreted as a paleosol. (Grades into next bed) . . . . *1–2 feet*

Gravel and sand, calcium carbonate crusts on the bottom sides of cobbles and pebbles. Gravel and sand, mostly unweathered, except in upper parts, but there are a few ironstained rocks. Grades downward into 2 to 3 feet of fine, gray sand containing clay lenses. Shows strongly developed red-brown layers due to iron concentration

7 to 10 feet

Unconformity.

Wasatch formation: Gray to light-green shale.
   (Miocene)

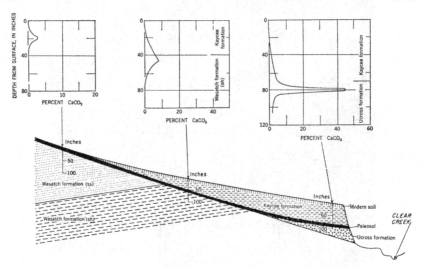

Figure 11-20.   *Diagram of stratigraphic relation of paleosol to alluvial fills. Changes in calcium carbonate content with depth at three positions on the Kaycee terrace varying in distance from the valley of Clear Creek near Buffalo, Wyoming. The depth of maximum calcium carbonate accumulation coincides with field evidence for the position of the paleosol.*

and therefore does not appear to be related merely to a particular parent material. Finally, the percentage concentration of carbonate is greater where the ancient surface was protected by an alluvial cover than where it remained at the surface near the top of the hill and was subjected to new soil-forming processes. The latter apparently tended to leach out the carbonate that had accumulated under the conditions associated with the development of the paleosol.

River Terraces and the Field Problem

Two fundamental or ultimate controls are responsible for terrace development. These are the tectonic and the climatic. Tectonic forces—such as faulting, folding, warping, or tilting—usually affect river systems through differential changes in gradient. Uplift may, in addition, place the watershed in a different climatic environment, or produce a change in the hillslopes of the river system, altering the delivery of sediments. These last might be called indirect rather than direct effects of tectonic controls.

Climatic controls are indirect. A change in climate may create a change in the hydrologic regimen of the river affecting the delivery of both water and sediment to the river system. As described in Chapter 2, sediment yield from a drainage basin is closely related to both vegetation and precipitation. Glaciers may also develop within a drainage basin or move into the basin as a result of climatic change. Lastly, on a worldwide scale, changes in sea level result from changes in climate. Thus, base level of rivers debouching into the sea is affected by eustatic changes in sea level.

Although the basic controls may be different, it is important to note that different "ultimate" causes may create identical conditions for the river system. For example, an abrupt break in gradient of the river profile or knickpoint, may be produced by (1) faulting, (2) lowering of sea level, or (3) a change in flow that causes a master stream to scour or degrade, thus steepening the profiles of tributary streams. Mechanically speaking, the adjustment required of the oversteepened river is the same in each case. The nature of these mechanical adjustments has been discussed earlier; here the primary concern is with the recognition of the history and field relations of the terrace remnants.

There is a general tendency for terraces and former valley floors to be nearly but not quite parallel to the longitudinal or downvalley profile of the modern river. It is observed that remnants of a single terrace usually stand at progressively lesser heights above the present stream as one proceeds downstream. This seems to be true of many alluvial terraces in unglaciated valleys and of glaciofluvial terraces; the latter surfaces of outwash plains or valley fills result from deposition downstream from a glacier.

Downstream convergence of terrace and present flood plain means, of course, that the former stream flowed on an average gradient somewhat steeper than the present one. Why this should be true in general is not obvious, but the fundamental question is the same whether the terrace profile diverges from or converges with the present stream profile. As pointed out in Chapter 7, one must deal with the complex of variables that control the slope or longitudinal profile of the stream. To the extent that terrace and river profiles differ, the balance of the controlling variables differed at the time of terrace formation.

In some situations the materials making up the valley fill underlying a terrace are coarser than materials presently being carried by the river. This may occur where the valley sediments underlying the terrace were brought into the valley by a glacier which has receded from the valley. Steepening of the outwash plain at the margin of the glacier is common. It may also be argued that the steeper gradient of the terrace was associated with coarser debris load than that carried by the present river, a debris load provided by different conditions on the watershed. Yet Hadley (1960) has found on Five Mile Creek, Wyoming, that the inset fills comprising the lower terraces, although coarser-grained than the oldest fill, nevertheless have about the same downstream gradient as the surface related to finer-grained material.

We have also noted that discharge is inversely related to slope (everything else being equal); hence another possible explanation of changed gradient may be that the discharge of the river that deposited the previous flood plain, now the terrace, differed from the discharge of the present river. The evidence usually indicates, particularly in glaciated regions, that the streams which built the flood plains that now are terraces had larger discharges than those of the modern rivers.

Where climatic changes result in the development and periodic advance and retreat of glaciers within a valley as the glacier advances, it pushes or entrains large amounts of sediment of heterogeneous character. The primary work of streams may be to redistribute the glacial debris downstream from the glacier front. If the water supply is limited, this incomplete transportation will lead to aggradation.

A subsequent climatic change producing melting and retreat of the glacier will alter the relative contribution of water and sediment. The additional quantities of water will not only rework the previously deposited glaciofluvial sediments but will also provide sufficient flow to trench the valley alluvium, thereby creating a terrace. Repetition of this sequence will lead to a sequence of terrace forms within a given valley. Correlation of these terraces with the appropriate glacial advances is best accomplished by tracing the terrace up to the end moraines of the glacial advance. Where upstream moraines are associated with terraces which then pass through moraines located downstream, it is clear that the downstream moraines and their associated terraces represent an older glacial advance. This relationship is shown schematically in Fig. 11-21, taken from the work of Holmes and Moss on the southwest flank of the Wind River Mountains in Wyoming, along Boulder Creek (1955, p. 639).

Climatic changes which lead to the development of glaciers also tend to produce changes in sea level. A stream heading in the mountains and debouching into the sea will at the same time be affected by the appearance of glaciers in the headwaters and by lowering of sea level at the mouth. In these circumstances aggradation of the valley will be taking place in the headwater portion, while in the lower reaches of the river the increased gradient produced by lowering of the sea level may lead to degradation.

This appears to have occurred in several river systems. Such a coincidence of valley filling and degradation would produce terrace levels whose profiles would intersect and cross the profile of the stream representing nonglacial conditions.

Where no glacier exists in the valley, climatic and accompanying eustatic changes may result in simultaneous deposition of alluvium upstream and marine transgression downstream. Butzer (1959) has correlated marine deposits in the delta area of the Nile River with alluvial materials derived as a result of climatic changes in the headwaters and tributaries of the Nile. Similar climatic changes and associated eustatic variations, again without the existence of valley glaciers, are postulated by Zeuner (1959, p. 356) for the Thames River in the vicinity of London.

In contrast to climatic changes which directly affect valley glaciation are those common climatic changes which differentially affect portions of the drainage basin and thus change the relation of sediment load to discharge. Such a differential effect on mountain and foothill areas has been offered to explain the relation of the terraces of Clear Creek, Wyoming, to those of its master stream, the Powder River. The present Powder River carries bed material composed of silt and fine sand. Its tributary has a bed of fine gravel. This difference appears to explain why, at points of equal discharge, Clear Creek flows at a steeper gradient than the Powder.

However, both streams are bordered by alluvial terraces that nearly

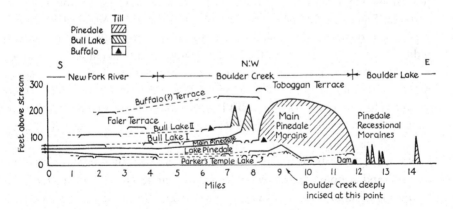

Figure 11-21. *The relation of an outwash terrace to a glacial moraine upstream illustrated by Boulder Creek, Wyoming. The two lower terraces pass through the main Pinedale moraine and are therefore presumed younger than that moraine. [After Holmes and Moss, 1955.]*

parallel the respective present streams, and the silty alluvium underlying the terraces in both is similar in texture. The terrace gradient is 16 feet per mile along Clear Creek and only 5.4 feet per mile along the Powder, at points where present discharge is comparable.

To explain these facts, it was postulated (Leopold and Miller, 1954, p. 64) that following the time when the terraces were active flood plains precipitation changed in such a way that the flow of Clear Creek decreased relative to that of the Powder River. It was assumed that this change was associated with a relatively larger runoff from the plains area where silty formations abound, compared with the mountain area from which discharge remained unchanged or decreased slightly. Such a climatic variation provides a reasonable mechanism of the river behavior and does appear meteorologically possible.

It seems quite likely that in many instances the slope of a former flood plain differed slightly from that of the present river, owing to changes in channel pattern (itself a product of discharge), discharge fluctuation, and sediment load. Braided streams and those having ill-defined channels are steeper than others of equal discharge but flowing in a single well-defined channel.

Tectonic movements provide a mechanism for producing a variety of relations between modern stream gradients and terrace profiles, including downstream convergence of terraces. Uplift, which may occur at any position along the longitudinal profile, will produce warping of the terrace surfaces. Figure 11-22 shows warping of a terrace of the Ventura River as it passes through the Coast Range of California. Deformation of the longitudinal trace of the terrace profile is due to faulting subsequent to the formation of the terrace. The river has continued to intrench itself at a more rapid rate than the rate of uplift, resulting in the humped or warped profile of the terrace. The timing and rate of uplift in some cases can be determined if the terrace chronology is known. Here again correlation of the terraces is made more difficult as a result of structural deformation.

No attempt has been made here to review all possible combinations of climatic and tectonic factors which produce variations in alluvial terrace landforms in nature. As in all natural phenomena, exceedingly complex relationships may be developed within any large drainage basin. The solution of problems of origin and correlation of terraces rests upon basic stratigraphic field evidence, which in the last analysis is the only tool available with which to verify hypotheses concerning origin. The examples outlined here are simply a sample illustrating a few of the many relationships that have been observed in the field, coupled with some hypotheses advanced to explain the origin of these field relations.

It is instructive to examine how field conditions may fortuitously provide a combination of different types of evidence for the interpretation of a sequence of geomorphic events and inferences about conditions prevailing during some of those events. Also, there is a considerable difference between concepts or possibilities conceived in an armchair or laboratory and their application to a particular set of conditions recognizable in the field. Geomorphology is, or should be, primarily a field study, and the field aspects cannot be eliminated by any growth in availability of maps, aerial photographs, or computers. These still are tools to assist in the pursuit of field investigations.

What is vitally needed for instruction and reference is more summary and collation of the types of evidence seen in the field, with discussion of how that evidence was used, together with general principles for unraveling the geologic history.

On a broad alluvial flat at the foot of the Lukachukai Mountains in

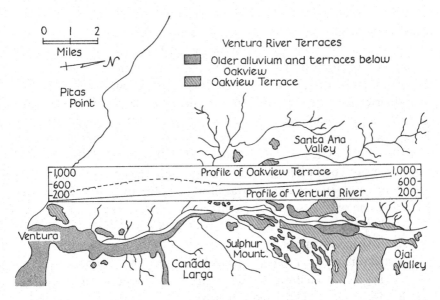

Figure 11-22. *Deformation of the profile of a terrace on the Ventura River as it passes through the Coast Range in California. Warping is produced by uplift of mountain area along faults roughly perpendicular to the course of the river. [Generalized after Putnam, 1942.]*

northwestern New Mexico are a series of arroyos trenching the alluvium. Near Mexican Springs (Nakaibito) these arroyos drain eastward and north into the Chaco River, a tributary to the San Juan. Interestingly, at the site of the former experiment station of the old Soil Erosion Service, a still needed and yet early-abandoned center of research on problems of erosion and land management, there are preserved in the stratigraphic record evidences of periods of erosion, sedimentation, and soil formation of interest to the student of land use and the effects of man.

On Mexican Springs wash, 18 miles north of Gallup and 4000 feet upstream from Highway 666, there is exposed in the arroyo wall stratigraphic relationships from which can be inferred a rather complete story of post-glacial geomorphic history. A sketch map and cross section of the locality are presented in Fig. 11-23. The photograph in Fig. 11-24 shows nearly the identical view diagramed in the cross section.

At this place the arroyo makes a tight S-shaped bend, and in doing so it has eroded laterally into one of the low mounds or prominences that stand slightly above the general flat of the alluvial plain, owing to local accumulations of wind-blown sand. These dunes are not active at present. Their original form has been modified by rainwash and the effects of vegetation, by which they are now at least temporarily stabilized.

The arroyo wall stands vertically nearly 30 feet at the point where the stream has impinged on the hillock. In the photograph and cross section, the central portion of the exposure consists of interbedded tan-brown sand and silt, irregularly cross-bedded and containing unconsolidated subrounded fine gravel, particularly near the base. To the left in this photo and section is exposed the cross section of a former stream meander and channel fill, which was cut in the main alluvial mass when the stream flowed at a level some 15 feet above its present bed. Before this level was abandoned by downcutting, the stream channel was filled in by layers of silt and clay, apparently as an oxbow isolated from the main stream except during periods of overflow, because the dish-shaped cross section was gradually flattened by successive increments of deposition.

The main alluvial deposit, the Nakaibito formation (Leopold and Snyder, 1951), rests unconformably on an irregular deposit of subangular and poorly sorted gravel, whose individual rocks are covered by a hard white caliche, which also permeates the interstices between them and consolidates the gravel. This gravel itself rests on an irregular and eroded surface of local bedrock, the Mesa Verde formation (Upper Cretaceous), as can be seen in Fig. 11-23.

Buried 14 feet below the surface of the main alluvial fill (Nakaibito formation), two potsherds were found in situ. These are comparable in age to others found on the surface of the wind-worked colluvium in the immediate vicinity. The design painted on these sherds dated them as being Pueblo I and II in age, 900–1100 A.D.

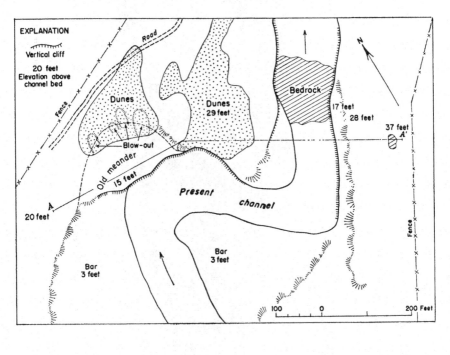

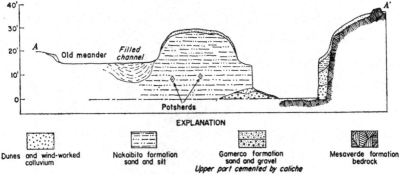

Figure 11-23. *Sketch map and cross section on Mexican Springs Wash, 18 miles north of Gallup, New Mexico. The visual aspect of the locality is shown in the photograph, Fig. 11-24.*

Figure 11-24.   *Photograph of alluvium exposed on gully wall of Mexican Springs Wash., diagramed by cross section on Fig. 11-23. The potsherds were found just to the right of the man. on a level with his head.*

The calichefied material (Gamerco formation) characteristically lies unconformably on bedrock and is itself eroded at the top, for it is unconformably overlain by the ubiquitous cover of silt and sand that makes up the main alluvium of the broad valley floors in the area. The relation of the Gamerco to over- and underlying materials, its existence in outcrops separated by at least 20 miles, and the existence of cementing caliche in alluvial materials of widely different texture all suggest that the calichefication was a general process occurring over a rather broad area and in different topographic situations. A soil-forming process would answer this specification and suggests soil formation under a condition of relative aridity. Because no such strong deposition and cementation by caliche is observed in the overlying materials of known recent but pre-Columbian age, it is surmised that this soil-forming period was different than the present and existed for a longer span of time than that represented by the overlying materials.

The Nakaibito formation implies a shift in conditions from those which resulted in calichefication and erosion of the underlying Gamerco formation, a shift presumably toward a more humid environment.

The relation between these bits of evidence suggests the following sequence of events, which generally appeared to be verified by stratigraphic relations elsewhere in the area.

The canyon erosion cycle (Pleistocene), during which the local valleys were eroded into bedrock, was followed by alluviation. In this early period of deposition, gravels from the nearby mountains were deposited directly on the local bedrock, at least in the valleys. This is the underlying gravel of the described locality assigned to the Gamerco formation, but elsewhere in the area the correlative material is silt rather than gravel. Recognition of contemporaneity rests on the calcium carbonate that occurs as coating and cement in the gravels and as hard nodules, veins, and tubes in silty members.

The deposition of the Nakaibito alluvium was in progress in the period 900–1100 A.D., and during that time Paleo-Indians camped in the vicinity and made pottery and flint objects. Deposition was ended by a period of channel erosion in which streams maintained courses generally similar to those observed today. This erosion at some places channeled the Nakaibito to bedrock or to the resistant Gamerco gravel but in other places the trenching was replaced by alluviation when the channels had eroded to only about half the depth of the present arroyo. Thus the level of the filled channel and meander scar in the left portion of the cross section (Fig. 11-23) and the photograph (Fig. 11-24) probably represents more than a chance preservation of a temporary position of the stream during downcutting, for it has a counterpart in evidence of channel trenching in the Nakaibito elsewhere.

It was during this interruption and period of alluviation that wind action concentrated sand in dunes on the general valley surface but extended down into the channels. The dunes became stabilized by vegetation. More recently the channels were trenched to bedrock or to the underlying resistant gravel.

These interpretations can be inferred from the site pictured and from those nearby. Other studies indicate that the most recent trenching began in the late 19th century and that the Gamerco formation is probably late Pleistocene. The existence of the calichefied paleosol is an important link in assigning an age to that member.

REFERENCES

Brush, L. M., Jr., and Wolman, M. G., 1960, Knickpoint behavior in noncohesive materials; a laboratory study: Geol. Soc. Am. Bull., v. 71, pp. 59–74.

Bryan, K., 1928, Historic evidence of changes in the channel of Rio Puerco, a tributary of the Rio Grande in New Mexico: Journ. Geol., v. 36, no. 3, pp. 265–282.

Bryan, K., and Ray, L. L., 1940, Geologic antiquity of the Linden meier site in Colorado: Smithsonian Misc. Coll., v. 99, no. 2, 76 pp.

Bryan, K., and Albritton, C. C., 1943, Soil phenomena as evidence of climatic changes: Am. Journ. Sci., v. 241, pp. 469–490.

Butzer, K. W., 1959, Contributions to the Pleistocene geology of the Nile valley: Erdkunde, bd. XIII, pp. 46–67.

Cotton, C. A., 1940, Classification and correlation of river terraces: Journ. Geomorphology, v. 3, pp. 27–37.

Eakin, H. M., and Brown, C. B., 1939, Silting of reservoirs: U. S. Dept. Agric. Tech. Bull. No. 524, 168 pp.

Fahnestock, R. K., 1963, Morphology and hydrology of a glacial stream: U. S. Geol. Survey Prof. Paper 422-A.

Frye, J. C., and Leonard, A. R., 1954, Some problems of alluvial terrace mapping: Am. Journ. Sci., v. 253, pp. 242–251.

Gilbert, G. K., 1917, Hydraulic-mining debris in the Sierra Nevada: U. S. Geol. Survey Prof. Paper 105.

Hack, J. T., 1942, The changing physical environment of the Hopi Indians of Arizona: Papers of the Peabody Museum of Am. Archeology and Ethnology, Harvard Univ., v. XXXV, no. 1.

Hadley, R. H., 1960, Recent sedimentation and erosional history of Fivemile Creek Fremont County, Wyoming: U. S. Geol. Survey Prof. Paper 352-A, pp. 1–16.

Happ, S. C., Rittenhouse, G., and Dobson, G. C., 1940, Some principles of accelerated stream and valley sedimentation: U. S. Dept. Agric. Tech. Bull. 695, 133 pp.

Holmes, G. W., and Moss, J. H., 1955, Pleistocene geology of the southwestern Wind River Mountains, Wyoming: *Geol. Soc. Am. Bull.*, v. 66, pp. 629–654.

Hunt, C. B., and Sokoloff, V. P., 1950, Pre-Wisconsin soil in the Rocky Mountain region, a progress report: *U. S. Geol. Survey Prof. Paper* 221-G.

Ireland, H. A., Sharpe, C. F. S., and Eargle, D. H., 1939, Principles of gully erosion in the Piedmont of South Carolina: *U. S. Dept. Agric. Tech. Bull.* no. 633, 143 pp.

Jacobsen, T., and Adams, R. M., 1958, Salt and silt in ancient Mesopotamian agriculture: *Science*, v. 128, pp. 1251–1258.

Krumbein, W. C., and Sloss, L. L., 1963, Stratigraphy and sedimentation (2nd ed.): Freeman, San Francisco, 660 pp.

Lattman, L. H., 1960, Cross section of a flood plain in a moist region of moderate relief: *Journ. Sed. Pet.*, v. 30, no. 2, pp. 275–282.

Leopold, L. B., and Snyder, C. T., 1951, Alluvial fills near Gallup, New Mexico: *U. S. Geol. Survey Water-Supply Paper* 1110-A.

Leopold, L. B., and Miller, J. P., 1954, A post-glacial chronology for some alluvial valleys in Wyoming: *U. S. Geol. Survey Water-Supply Paper* 1261.

Leopold, L. B., and Wolman, M. G., 1957, River channel patterns; braided, meandering, and straight: *U. S. Geol. Survey Prof. Paper* 282-B.

Lyons, H. G., 1906, The physiography of the River Nile and its Basin: Survey Dept., Cairo, Egypt, 411 pp.

Miller, C. R., Woodburn, R., and Turner, H. R., 1962, Upland gully sediment production: *Internatl. Geophys. Union*, Symposium of Land Erosion, Bari, Italy, 25 pp.

Miller, J. P., and Wendorf, F., 1958, Alluvial chronology of the Tesuque Valley, New Mex.: *Journ. Geol.*, v. 66, no. 2, pp. 177–194.

Morisawa, M., 1960, Erosion rates on Hebgen earthquake scarps, Montana (abstr.): *Geol. Soc. Am. Bull.*, v. 71, p. 1932.

Putnam, W. C., 1942, Geomorphology of the Ventura region, California: *Geol. Soc. Am. Bull.*, v. 53, pp. 691–754.

Rubey, W. W., 1928, Gullies in the Great Plains formed by sinking of the ground: *Am. Journ. Sci.*, v. XV, pp. 417–422.

Ruxton, B. P., 1958, Weathering and subsurface erosion in granite at the piedmont angle, Balos, Sudan: *Geol. Mag.*, v. 95, pp. 353–377.

Schumm, S. A., and Hadley, R. F., 1957, Arroyos and the semi-arid cycle of erosion: *Am. Journ. Sci.*, v. 255, pp. 161–174.

Schumm, S. A., 1960, The shape of alluvial channels in relation to sediment type: *U. S. Geol. Survey Prof. Paper* 352-B, pp. 17–30.

Schumm, S. A., 1963, Channel widening and flood plain construction, Cimarron River, southwest Kansas: *U. S. Geol. Survey Prof. Paper* 352.

Wentworth, C. K., 1928, Principles of stream erosion in Hawaii: *Journ. Geol.*, v. 36, pp. 385–410.

Zeuner, F. E., 1959, The Pleistocene period, its climate, chronology, and faunal succession: Hutchinson Scientific and Technical Publ., London, 447 pp.

Chapter 12 Evolution of Hillslopes

The hills are shadows, and they flow
From form to form, and nothing stands;
They melt like mist, the solid lands,
Like clouds they shape themselves and go.

LORD TENNYSON
In Memoriam

Divergent Views of Hillslope Evolution

The evolution through time of the form of hills has provided more lively discussion than perhaps any other subject in geomorphology. The range of geomorphic thought on this subject has been covered in several comprehensive papers, such as L. C. King's "Canons of landscape evolution," and Kirk Bryan's "The retreat of slopes." For this reason we will not attempt to resummarize these observations but will concentrate instead upon selected examples which appear to provide specific evidence of the way in which hillslopes evolve. Although, as noted in Chapter 7, pediments are landforms developed by stream erosion, we have included certain aspects of pediment development in this chapter because the form of the pediment is closely related to the retreat of hillslopes.

In connection with the examples from the field which will be mentioned here, it is particularly important to keep in mind that in studying the effect of time on the landscape, landforms in one locality have usually been compared with those in another. There may be good reason to believe that one form is sequential to the other, but in most studies space is substituted for time—with the exception of very short periods.

Even greater caution is required in the explanations of so-called theoretical or mathematical models. Because a slope form derived from a hypothetical model is analogous to that observed in nature, there is no guarantee that the conditions and processes assumed in the model are those which in fact produced the similar form in nature.

It was Penck's thesis (1953) that steep slopes retreat without loss in their inclination and that steepness disappears only because the upper part of the ridge is consumed. Gentle slopes replace steep slopes only when the gentle gradients below the ridge meet at the divide. Davis, however, stated that slopes flatten during the progress of the erosion cycle from youth through maturity to old age. The valley sides are reduced under weathering and surface creep to smaller and smaller angles. As the slopes flatten, the interfluves are lowered more rapidly than the river bed until there remains only a landform of faint relief, to which Davis assigned the name peneplain.

It is not necessary to reargue Penck's assertion that straight slopes furnish valid evidence of slow uplift approximately balanced with downward degradation and that convex slopes are expressions of rapid uplift which exceeds the rate of degradation. The evidence presented in Chapter 8 suggests that hillslope form is not uniquely determined by the rate of uplift. But to test or compare these two major concepts of slope evolution, one should consider Penck's third type of slope development, in which slope evolution begins after uplift is complete and base level is either stable or rising very gradually. This situation is basically similar to the assumption made by Davis, in which the cycle of erosion was assumed to begin after completion of a relatively rapid uplift.

Because these contrasting views have received so much attention in the literature, the contribution which seems most useful here is an illustrative summary of the types of evidence supporting these alternative views. Davis himself cited few examples of field conditions that can be considered to support his contention of gradual slope reduction. In his classic essay on the geographical cycle he is content to point out that "the higher parts of the interstream uplands, acted on only by the weather, without the concentration of water and streams, waste away much more slowly . . . relief is decreasing faster than at any other time, and the slope of the valley sides is becoming much gentler than before . . . the slopes become fainter and fainter, so that sometime after the latest stage . . . the region is only a rolling lowland, whatever may have been its original character" (Davis, 1954 ed., p. 255).

The field conditions which he used as illustrations of the progression through time under humid conditions indicate that he was considerably influenced by the forms observed in the Appalachian Mountains and in New England. The illustrative sketches in the essay on the rivers and valleys of Pennsylvania all happen to be examples of rounded hilltops (p. 428).

In a sketch comparing two valleys, one of which was affected by an early erosion cycle, a reduction of hillslope with time is implied. He maintained that the two valley profiles resembled, respectively, the gorge of the Rhine and the Frazer River Valley in British Columbia (p. 378). Davis, always a keen observer, was aware of the steep slopes typical of arid regions, but originally he considered the cycle of arid erosion merely a variant of the cycle of humid erosion. Later, however, he emphasized the "homologous" features of humid and semiarid regions (1930, pp. 19–20).

Interestingly, the field evidence used by Penck to support his hypotheses differs from that of Davis principally in the following respect. Penck's book is well illustrated with many photographs of landforms in various parts of the world. But in surprisingly few cases did Penck discuss in detail the characteristics of an individual field example and relate that example to his general argument. More often than not the photographs carry but the briefest reference to the generalized model so carefully developed in text and diagram. It is left to the reader to interpret the field illustrations in terms of the principles that Penck discussed.

To analyze the evidence bearing on the differing view of slope evolution, one must turn to studies containing observations of active processes and quantitative descriptions of field conditions. Perhaps some of the authors cited would not care to be categorized as quantitative geomorphologists, but their studies do provide quantitative evidence with which to test working hypotheses; they are, then, in the best sense of the word, quantitative geomorphologists.

Types of Evidence Supporting Some Postulates on Landscape Evolution

To provide a basis for analysis, samples of the evidence bearing on fundamental problems of the evolution of hillslopes are related here to specific questions or to specific generalizations that have been widely discussed by workers in the field.

1. "The steepness of slope of hills and mountains, inherent in the interrelations of the processes of arid erosion and the resistance of the rocks, is

maintained to the end of the erosion cycle . . ." (Bryan, 1940, p. 261).
As Bryan pointed out, the best proof of this generalization lies in the
existence of residuals of all sizes.

King (1948) describes the bornhardts (inselbergs) of semiarid South
Africa—isolated masses of native rock protruding from a relatively
smooth surface. These mounts are variously called inselbergs, koppies,
or tors. The important observation is that the bornhardts are not composed of different types or qualities of rock from those of the surrounding apron of bedrock (Fig. 12-1). He believes that the margins of these
isolated masses are, almost without exception, independent of petrologi-
cal boundaries. In a region comprising many hundreds of square miles
underlain by rock of a type suitable for the formation of these inselbergs,
the bornhardts occupy only a small portion of the area. These observations lead King to the conclusion that the surrounding plains are
erosional features and the bornhardts are erosional forms.

Figure 12-1.

*Kashawa Bornhardt on Raree Farm west of Umvukwes, Southern Rhodesia.
Both the bornhardt (inselberg) and the surrounding apron are granite bedrock.
[Photograph by L. C. King.]*

2. Slopes of hills are characteristic of the climate and the rock, and these slopes—once formed—persist in their inclination as they retreat. They disappear only when all the volume of rock above the encroaching foot slopes has been consumed. This applies in arid, tropical, and humid-temperate climates (Bryan, 1940, p. 266). Furthermore, rock-floored beveled surfaces (pediments) occur in all three climatic types.

Three examples are summarized below, in each of which slightly different types of evidence are used in differentiating the ages of features that are geometrically similar.

(a) The Hawaiian Islands are alike in that all are dome-shaped cones of basaltic rock. Studies of the geologic history, however, show clearly that the islands differ considerably in age and thus in the time during which subaerial erosion has been operative. Comparison of Maunalei Gulch on a young island, Lanai, and a valley of comparable size but greater age, on the southwest slope of the Koolau Range, Oahu, shows that each approaches a U shape with uniform steepness of their walls nearly to the top of the valley sides (Wentworth, 1928, pp. 402, 406).

(b) A quite different setting is provided by an example from North-amptonshire, England, pictured in Fig. 12-2, **A.** A small hill developed by erosion of a weak bedrock (clay) stands about a mile from a similar larger hill, also on clay but capped by a resistant bed of sandstone, the side slopes of which stand at 28°. The slopes developed on clay are similar in gradient, 11°, and straight in profile, merging with the lower part of the waning slope that progressively flattens. The small outlying hill appears to have maintained its constant angle of slope without flattening with age and appears to have reached its present position by parallel retreat.

(c) Moreover, nearby is another hill, shown in Fig. 12-2, **B,** display-ing a similar 11° slope that joins with a waning slope of progres-sively flatter gradient toward the River Cherwell. This waning slope passes without any break in gradient from the clay to a strong underlying limestone. These field relations lend support to the concept that in humid-temperate regions, as well as in arid, the gradient of hillslopes is appropriate to the acting processes and the rock characteristics, and does not alter progressively with age (Dury, 1959), provided the form is controlled by slope processes and not by processes acting at the base of the slope.

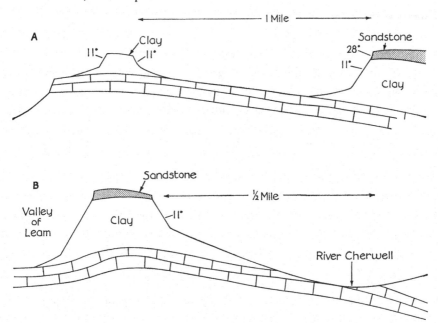

Figure 12-2. *Erosional slopes in Northamptonshire, England. [After Dury, 1959.]* **A.** *Comparison of two nearby hills composed of clayey bedrock, the larger capped by sandstone; note similarity of slope angles.* **B.** *Slope cutting two types of bedrock, clay and limestone, but with no break in slope angle.*

3. Many, if not indeed most, slope forms in the field are complicated by alteration in the intensity or types of processes acting upon the slopes during passage of time. When this occurs, the form or configuration of the slope will alter with time.

Delineation of the differential effects of various processes and the time over which each acts is uncommon. Occasionally conditions exist in the field where through time, because of peculiar circumstances, a portion of a hillslope is progressively removed from the action of some particular process. As a result, a series of features representing a progression in time is exhibited. Such a unique situation was recognized by Savigear (1952), who described a sequence of slope forms along a segment of the coast of South Wales.

At Gilman Point west of Laugharne, downcurrent from a re-entrant of the coast (Fig. 12-3), a sand spit has been desposited. Initially the south coast in the Taf Estuary was exposed throughout its length to wave action, which maintained cliffs at the seaward face. Gradual encroachment of wind-blown and marsh deposits between the developing spit and the original shoreline caused the sea to abandon the base of the cliff.

This removal of the influence of the sea has been progressive from west to east. Thus, the most westerly cliffs on the south facing shore are those which have undergone subaerial erosion without waves cutting at their base; those to the east are progressively younger, more recently abandoned by the sea. All slopes are formed on the Old Red Sandstone, a formation consisting of marls and sandstones containing occasional thin shale partings. Profiles *A*, *B*, and *C* (Fig. 12-3) east of the Coygan Promontory are those most recently abandoned by the sea; profiles *L*, *M*, and *N* are those which have undergone successively longer periods of erosion since their abandonment by the sea. The younger profiles (*A*, *B*, *C*) still retain portions of the cliff face at high vertical angles. These cliffs have declivities on the order of 80°; the angle itself is dependent upon the relation of the slope face to the dip and strike of shale partings in the Old Red Sandstone. The base of the slopes is concave or straight where talus has accumulated. Both the

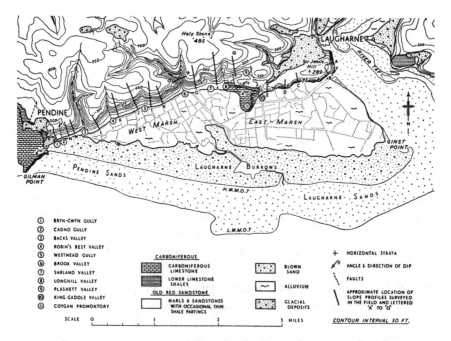

Figure 12-3.   *Coast between Pendine and Taf Estuary, South Wales. [After Savigear, 1956.] Cliffs to west are oldest, longest removed from wave action.*

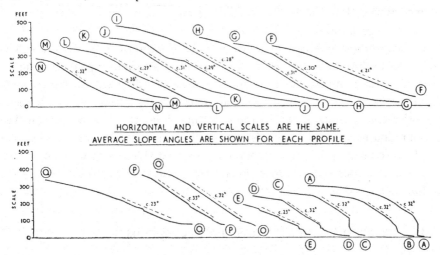

Figure 12-4.   *General forms of certain slopes of cliff area shown in Fig. 12-3. Profiles A, B, C are youngest and L, M, N oldest (longest protected from wave action).*

more recent slopes and the older ones (Fig. 12-4) have midslopes approximating 30°.

All the profiles demonstrate that the upper portions of the slope are broadly convex. Detailed measurements show that this broad convexity is actually made up of a number of convex and concave elements as well as a number of facets. Debris on slopes of more than 32° or 33° is discontinuous, but on slopes *N* through *F* the thickness of the debris generally increases downslope. The concavity of the basal slope is produced by the accumulation of debris which is lenticular in form, thinning upward both toward the convex midslope and toward the sea. Where material is still being removed from the lower limit of the convex midslope, as in profiles *A, B,* and *C,* slopes have angles of about 32°.

In contrast, on profiles *L, M,* and *N,* the average angles of the midslope are less than 32° (Fig. 12-4). To describe these differences, Savigear (1952) uses the words "unimpeded removal" and "impeded removal," and he concludes that unimpeded removal leads to parallel retreat of slopes and impeded removal to decline of slopes. On the foot slope the form is being altered by replacement.

To the extent that these progressive age distinctions are correct, this example permits a valid transposition of space and time. In this instance the evolution of the hillslopes under subaerial weathering is accompanied by a reduction in declivity as a result of the changing complex of processes during passage of time.

Savigear (1956) has suggested that the slope form is polyphase—that is, the result of successive phases or intervals of time interrupted by changes in base level. Where sea cliffs are subjected to a different combination of processes, the once vertical scarps are reduced in gradient to 32° and less, and declining gradient with passage of time is indicated. However, on the same coast facets and elements of the hillslopes, once established at angles of approximately 20° to 30°, maintain these equilibrium forms throughout their subsequent evolution. This appears to indicate that processes there have stabilized and remain constant.

Ultimate Forms: Pediments and Peneplains

Mountains in semiarid and arid regions are commonly bordered by smooth piedmont slopes, cut on rock and veneered with a thin layer of alluvium. These surfaces, called pediments, are usually characterized by a remarkably sharp break in slope where the pediment meets the mountain front (Fig. 12-5).

The areal and climatic distribution of pediments is not easily defined. Features which have been described as pediments are confined largely to arid or semiarid regions of the world. Surfaces beveled on bedrock and capped with gravel do occur in humid regions; Hack (1960) has described some in the central Appalachian Mountains. It is reasonable to suppose that they are less obvious than in arid regions because of the deeper, more mobile soil cover and the presence of luxuriant vegetation.

Although there are a number of variations in pediment form, the literature indicates reasonably good agreement on the following properties of pediments occurring in arid or semiarid regions.

Gradients reported range from 750 feet per mile (8°5′) to less than 10 feet per mile (0°7′), with a high percentage of examples between 200 and 300 feet per mile (2°11′ to 3°15′). Longitudinal profiles are gently concave upward and transverse profiles are generally flat and display irregularities of fine grain. Gradients of the adjacent slopes are commonly 20° to 35°, but, depending on the character of the rock and weathered mantle, may approach vertical or even overhang at the base of some residual rock masses.

The principal features of pediments in plan view are often related to the initial form of an escarpment, which may have had either an erosional or tectonic origin. Differences in rock resistance and relative sizes of stream channels also affect pediment form. Residual mountain masses may be islandlike as a result of junctions of pediment surfaces from opposite sides of a mountain, forming pediment passes through the mass.

Pediments are rock-cut surfaces. It is widely believed that during formation they are either bare of debris or veneered with such thickness as can be moved by floods. The size of debris ranges from boulders a few feet in diameter to fine clay; generally mean or median size decreases downslope. On the Sandia pediment, median diameter of channel material is more than twice that on interchannel areas. The presence of a very thick debris cover over a rock-cut surface has been interpreted as the result of aggradation after pediment formation, and in many places the alluvial cover appears to be in process of removal, suggesting that the pediment is an exhumed surface.

Most pediments have complex drainage patterns with channels commonly cut a few feet to a few tens of feet below the general level of the rock-

Figure 12-5.

Desert pediment at foot of Santan Mountains near Phoenix, Arizona. Note abrupt change in slope at base of mountain mass, inselberg, and network of channels. [Photograph by Howard T. Chapman.]

cut surface. This has been used as evidence of dissection of the pediment sometime after its formation. Another interpretation is that pediments are "born dissected" in that combinations of trenching by headward erosion of channels and lateral planation by some channels leads to isolation of remnants even during progressive degradation.

Pediments and pediment remnants at different levels in the same area reflect changes of local base level, climatic fluctuations, or tectonic changes. These diverse "controls" lead to a variety of topographic and stratigraphic relationships between mountain slope, pediment surface, and alluvial cover, including alluvial fans, surficial alluvial cover, and the distal or basinward alluvial deposits that accumulate where the pediment surface drains to a closed depression. (For description of diverse pediments, see Tuan, 1959.)

Flowing water is the dominant erosive agent, the effects of wind being minor and local. However, opinion still differs as to whether lateral planation by streams in well-defined channels, as opposed to sheetflooding coupled with back-wearing of slopes by weathering and rill-wash, is more effective in the formation of pediments. All of these processes do occur on pediments in different regions.

Despite general concurrence on the importance of flowing water to the origin of pediments, remarkably few observations on this subject are recorded, and no measurements are reported. Apparently the views of Davis (1938) on possible sources of information have prevailed. Regarding the relative effectiveness of lateral erosion and back-weathering of slopes, Davis (1938, p. 1367) wrote: "In attempting to settle this theoretical question, little can be accomplished by direct observation of the agencies concerned because they work so slowly and intermittently in the desert that their action can seldom be seen. Like the traveler who stood on a river bank in a rainy country, waiting for the river to run dry, an observer might spend years and years in the desert waiting for a sheetflood or a streamflood to appear. Recourse must therefore be had to a mental analysis of the problem. . . ."

In keeping with this somewhat introspective view we do find relatively little published literature describing movement of water on pediments (McGee, 1897; Berkey and Morris, 1927; Ives, 1936; King, 1953). All of these descriptions of sheetfloods are brief. King (1953, p. 733) states that sheets of water on the upper slopes of a pediment are in laminar flow and hence cannot cause erosion. It is questionable whether laminar flow does prevail on these slopes, but it is true that laminar as well as turbulent flow can transport sediment, as has been demonstrated in theory and by experiment

(Bagnold, 1954). McGee (1897) mentions being caught in a sheetflood more than 3 feet deep that had pronounced waves on the surface.

Information collected by the authors on the pediment at the base of the Sandia Mountains east of Albuquerque, New Mexico, demonstrates a very high drainage density and an elongate dendritic pattern. Furthermore, Horton's laws for relation of stream order to stream length and number of streams apply in this case. Whether the Sandia pediment is in the process of dissection is not known, however, for if a pediment is born dissected there is at present no way of judging what is normal and what is dissection unrelated to the ordinary process of pediment evolution. Additional data on channel pattern as well as channel and nonchannel flow on pediments would be valuable in differentiating among controls of these processes in different regions.

Sharp (1940), in his study of the Ruby-East Humboldt Range, Nevada, determined the angle through which streams turn after they leave the mountains and cross the pediment. For more than 200 major streams the median departure from a course normal to the mountain front is $11\frac{1}{2}°$. Less than 10% turn through an angle greater than $35°$. From this information Sharp concluded that approximately 40% of the mountain front retreat is caused by lateral planation of streams and the remainder by weathering and wash. Whatever the exact percentages, this study shows that lateral planation cannot account for the abrupt break in slope between mountain and pediment. A mountain front produced by lateral planation of streams should be serrate rather than straight.

There has been a tendency in the field of geomorphology to make a sharp division between the end forms produced by continuous subaerial denudation in semiarid and humid regions. The evidence presented in this chapter and earlier indicates that the forms of hillslopes may be identical in all climatic regions. There is no convincing evidence that certain major landforms are restricted to any given environment. However, there is a great deal of information to show that specific processes may vary in their relative importance in different environments. If specific erosional processes operate at significantly greater rates than others and if these processes are confined to specific climatic environments, then in a given length of time more erosion will have taken place in such a climatic region. This obvious statement does not, however, carry the corollary that the region so affected will of necessity be more subdued or have less local relief than another region of different structure or lithology or subject to different processes for the same length of time. These questions confront us when we try to establish the form that a land mass will have after being eroded for a long period of time.

There is a growing conviction that pediments occur in humid as well as arid environments. Indeed, King (1953) states that the pediment is the general or normal result of landscape evolution in all climates, and other forms are modifications or specialized cases. Hack's work shows that what have long been considered type examples of dissected peneplains may instead be examples of local pedimentation, which, as a result of expectable noncyclic operation of present processes, assume a form frequently ascribed to the peneplain.

If a pediment is a planed surface developed on bedrock by the action of rills, channels, and rainwash, it may be asked whether forms that answer this description in humid areas are not also pediments.

In Fig. 12-2 field relations were described in Northamptonshire, southern England, where rills, channels, and rainwash have developed a smooth surface on native rock, in that instance a weak clay. Even the steeper head-slopes are present in the sides of the clay hills, which have uniform gradient of 11°. To be sure, this is less steep than the mountain slopes of 25° to 35° common in arid regions, but the two appear analogous.

The Piedmont region of the Appalachian geomorphic province is underlain by greatly deformed and metamorphosed pre-Cambrian and Paleozoic rocks. It is what is commonly and aptly called "rolling topography." The rolling form transects the bedrock geology and thus is the product of erosion. If one constructs an imaginary surface over the real land, touching only the uplands, the imaginary surface developed is still rolling but more "subdued" than the actual surface and the relief is far less by virtue of the absence of dissection. When parallel profiles of the area are projected on a single plane, the composite or Barrellian profile idealizes or emphasizes dominant uplands at various elevations. It has been suggested that these profiles represent a "reconstructed" surface, once actually a real land surface which with minor modification has subsequently been uplifted and dissected. This hypothesis faces several hurdles. First, modification of the uplands by erosion subsequent to uplift cannot be neglected, as the evidence (Denny, 1956, p. 49) indicates that it may have been very great. Second, the "more subdued" relief of the imaginary surface cannot be used as evidence of a former land surface until there is additional evidence that such a land surface does actually result from long-continued erosion. This, of course, is the same problem encountered earlier in distinguishing polyphase and single-phase landscapes in connection with the origin of the hillslopes of Cornwall. The alternative hypothesis must be considered: the present landforms, including the degree of dissection, are explainable without resort to several cycles. That is, the result of long-continued erosion may be simply the rolling topography of the Piedmont today.

Hack (1960) has pointed out that the hypothetical surface produced by long-continued erosion and lateral planation, consisting of a plain interrupted by residual hills and veneered by waste, is not found in the landscape of the United States today. Instead, he points out, "the plainlands of the earth are either depositional surfaces like alluvial plains, deltas, drift plains, and coastal plains, or if they are erosion surfaces in humid areas, they are hilly with rounded divides and steep-walled valleys that have generally come to be described as 'maturely dissected peneplains.' " To avoid a genetic connotation, Hack prefers to call the Appalachian landscape "ridge and ravine" topography.

Hack postulates that present landforms in the Piedmont region of the Appalachians are adjusted to the geology and to the geomorphic processes operating upon them, a view based upon the following types of observations. In this region the particle size available from different rock types, the shapes of longitudinal profiles of stream channels, and the forms of divide areas are closely related. Steep profiles and coarse material are associated with mountain areas underlain by sandstone. Where these areas abut valleys underlain by shales and limestones, large fans are observed, composed of sediments brought down by streams carrying sandstone particles. Such gravel deposits overlie pediment surfaces cut on the shale below. Headward erosion by streams in the softer shale dissects the shale bedrock, and these flow at a flatter gradient than do streams heading in the harder sandstone. Steep sandstone-bearing streams on occasion spill over or are diverted into the streams heading in shale. Diversion or capture by the streams in shale leaves as terrace remnants large areas of sandstone gravels, overlying and protecting the shale. The remnants of planed bedrock (shale) covered with gravel resemble a dissected pediment.

Both the process and form described for the central Appalachians are similar to those described by Hunt, Averitt, and Miller (1956) in the Henry Mountains in southern Utah. In the latter, trachyte is the resistant lithology in contact with softer sandstones and shales. Despite marked climatic differences in southern Utah, causing the streams to lose discharge in the downstream direction, these landforms in the Henry Mountains and in the Appalachians are the same. In both regions the landforms are controlled by the spatial distribution of the different lithologies.

In the Appalachian highlands there are many extensive surfaces produced by the processes described briefly above. Extensive pedimentlike landscapes have long been mistaken for a formerly gravel-covered broad-valley stage or peneplain, now dissected. Actually the gravel-capped surfaces testify to the contrast in resistance between rocks of the mountains and of the valleys and are part of the equilibrium between the two (Hack, 1960, p. 94).

Because the hills are not everlasting it is safe to say that long-continued erosion at a constant base level will ultimately reduce relief. Where relief is reduced, the gradients of stream channels will be reduced. If one assumed a sufficiently long period of time of uninterrupted denudation without uplift, presumably the topography would become subdued.

Under some conditions a resultant land surface might approach the form of the peneplain of the classical geomorphic literature—that is, an extensive surface of moderate to low relief surmounted by monadnocks or minor hills composed of more resistant strata produced by subaerial erosion. There are limited areas in which varied lithologies are beveled by erosion, but the history of relief in these areas cannot be assumed and no large surfaces fitting this description are found at the earth's surface today. Some unconformities such as the Ep-Archaean or Ep-Algonkian traces exposed in the pre-Cambrian section of the Grand Canyon have some attributes of the classical peneplain. It is clear, however, that the existence at a time in the geologic past of an erosion surface of low relief near base level is an historical condition which must be proved; one cannot be inferred simply from an accordance of summits or subaerial origin of hill forms.

Models of Slope Evolution

Because no single process controls slope forms, as streamflow does the river channel, any true model of hillslope development will be complex. As an illustrative model of landform development the rate of erosion might be assumed to be a function of shear stress and the form of the land derived as a function of various lithologies and possible rates and loci of uplift. All manner of concave and convex forms and combinations of these shapes can be generated by assuming various combinations of erosion and uplift, and quite reasonable results can be obtained. But whatever model is used, the forms derived are not necessarily unique to the conditions assumed and to those only.

Whereas such a model involving erosion, uplift, and lithology is exceedingly general, several more restricted derivations can be made which appear to have certain counterparts in nature. Assuming a homogeneous rock, a fixed basal slope of low inclination, and a constant angle of repose, Bakker and LeHeux (1952) have shown that, under conditions of rectilinear retreat of slopes in the absence of deposition of a large volume of talus, the ultimate slope of a rock wall will be a rock surface with straight profile, sloping at an angle equal to the angle of repose of the rock debris. Where the slope retreat is parallel and waste material is allowed to accumulate slowly at the base of the slope, the talus will lie upon and protect a rock nucleus with convex profile.

Both derivations are based on a differential equation in which infinitesimal volumes of rock removed are equated to volumes of talus:

$$\text{Volume of rock removed} = (1 - c) \text{ volume of talus,}$$

where c is a constant. When the value of c becomes exceedingly small, then the rock surface beneath the veneer of waste assumes a nearly straight surface at the angle of repose of the rock debris. Where little or no talus accumulates, then the slope declines from the original rock face to the angle of repose and the slope form evolves by replacement as material accumulates. The landforms derived by these assumptions are comparable to many found in nature.

However, field observations indicate that on many slopes the combination of processes is not sufficiently well known to assert, for example, that a constant angle of repose will be maintained. The assumption of rectilinear retreat is just the factor that usually cannot be assumed in the field. It must indeed be proven. Thus, although comparable forms are evident in nature, it is likely that they represent combinations of processes involving differential rates of erosion on slope surfaces. The similarity of field and model does not prove that the assumption of uniform rate of removal is applicable in any particular case nor that the assumption of increasing rate of removal from base to crest, as implied by the model, is matched in nature.

The form of a hillslope may be markedly influenced by a river cutting at its base. This is perhaps the best illustration we have of base-level control of slopes. Scheidegger (1961) has shown what the effect of an undercutting river will be on a slope otherwise subject to a surficial rate of erosion, proportional to the angle of slope. An equilibrium or balance between the effect of the undercutting river and the erosional processes on the slope itself will result in a slope intermediate between the concave-convex forms and nearly vertical slopes.

Here again the derived curves are undoubtedly reasonable representations of slopes formed under similar controls in nature. Yet the previous discussion of profiles of slopes measured on the coast of Cornwall reveals that the actual history of such composite slopes may in fact be more complex.

No attempt has been made to apply to hillslopes concepts of equilibrium distribution of energy comparable to those used in deriving the most probable river channel forms and longitudinal profiles. However, the concept of steady state, grade, or equilibrium has been applied to hillslopes by Davis and many others, and it may well be that the so-called typical concave-convex slope profile does represent an equilibrium or most probable form in a statistical sense. The appropriate assumptions regarding the downhill changes in specific variables, however, would be more complex than in rivers inasmuch as both creep and rainwash should probably be included in the model. Application of principles of energy distribution to hillslopes cannot yield useful results until more is known about the processes acting on the slopes. Only now are there beginning the concentrated studies of hill processes needed.

It seems desirable to emphasize, in connection with models of hillslope development through time, the distinction between two kinds of mathematical formulations; both are useful in hydrologic and geomorphic analysis but can be expected to yield results of different levels of generality.

The rational approach to research in the physical sciences involves reasoning from basic physics in which relations between forces, motions, and states are expressed in rational form. The equations of motion are typically involved. The processes acting are described in terms of the Newtonian laws, involving principles such as the conservation of mass and energy and the equation of state. In this manner relations are derived from these basic tenets that may then be tested by experimental or field observations. Oceanography and meteorology are disciplines in which this classical approach to research has been widely used. The method seeks to develop hypotheses from physical laws that predict certain relations which can be tested by observation.

The engineering sciences also utilize mathematical nomenclature to express relations between observed phenomena, although usually in a deductive rather than an inductive way. Many models have been formulated in recent studies of slope evolution, but most involve assumptions about the processes rather than about the physics. Therefore, though mathematical, these formulations are of less general applicability than those derived by induction from physical laws. Owing to the long period during which the geological sciences centered around descriptive rather than mathematical statements, the difference in utility of various mathematical relations may not be as apparent to the geologist as to the physicist. Caution is needed in evaluating various mathematical models, particular attention being directed to the assumptions involved. Such models must be tested against field conditions, keeping in mind that mathematical and conceptual models are only steps in analysis and are in themselves neither proofs nor end products.

To summarize, theoretical models of hillslope evolution involving a degradation term proportional to the angle of slope provide reasonable resemblances of hillslopes found in many regions. These are primarily concave-convex. Superficially this erosional profile is not distinguishable, however, from profiles formed by erosion of the crest slope and deposition at the foot slope. In regions where weathering and weathering removal of rock waste unaided by running water are the principal mechanisms of denudation of slopes, the geometric models produce hillforms comparable to those found in nature. Like all such models, however, the mere agreement of natural form and model-derived form is not proof of the validity of the assumptions made in constructing the model. Such proof depends upon matching observation of process and form with the reasoning and assumptions of the model.

REFERENCES

Bagnold, R. A., 1954, Some flume experiments on large grains but little denser than the transporting fluid, and their implications: *Inst. Civil Engr. Proc.,* Paper No. 6041, pp. 174–205.

Bakker, J. P., and LeHeux, J. W. N., 1952, A remarkable new geomorphological law: *Koninkl. Nederl. Akad. van Wetenschappen (Amsterdam) Proc.,* ser. B, v. 55, no. 4, pp. 399–571.

Berkey, C. P., and Morris, F. K., 1927, Geology of Mongolia: Natural History of Central Asia, v. II, 449 pp.

Bryan, K., 1940, The retreat of slopes: *Assoc. Am. Geog. Annals,* v. 30, pp. 254–268.

Davis, W. M., 1902, Geographical essays: Ginn, Boston, 777 pp. (Reprinted 1954, Dover, New York.)

Davis, W. M., 1938, Sheet floods and stream floods: *Geol. Soc. Am. Bull.,* v. 49, pp. 1337–1416.

Denny, C. S., 1956, Surficial geology and geomorphology of Potter County, Pennsylvania: *U. S. Geol. Survey Prof. Paper 288.*

Dury, G. H., 1959, The face of the earth: Penguin Books, Baltimore, 223 pp.

Hack, J. T., 1960, Interpretation of erosional topography in humid temperate regions: *Am. Journ. Sci.,* Bradley Volume, v. 258A, pp. 80–97.

Hunt, C. B., Averitt, P., and Miller, R. L., 1953, Geology and geography of the Henry Mountains region, Utah: *U. S. Geol. Survey Prof. Paper* 288.

Ives, R. L., 1936, Desert floods in the Sonoyta Valley: *Am. Journ. Sci.,* v. 32, 5th series, pp. 349–360.

King, L. C., 1948, A theory of bornhardts: *Geog. Journ.,* v. 112, pp. 83–87.

King, L. C., 1953, Canons of landscape evolution: *Geol. Soc. Am. Bull.,* v. 64, pp. 721–752.

McGee, W. J., 1897, Sheetflood erosion: *Geol. Soc. Am. Bull.,* v. 8, pp. 87–112.

Penck, W., 1953, Morphological analysis of land forms: Macmillan, London, 429 pp. (Translation of 1927 ed.)

Savigear, R. A. G., 1952, Some observations on slope development in South Wales: *Inst. of British Geog. Trans.,* Publ. no. 18, pp. 31–52.

Savigear, R. A. G., 1956, Technique and terminology in the investigation of slope forms: Union Géogr. Internatl. Premier Rapport de la Comm. pour l'étude des versants, Rio de Janeiro, pp. 66–75.

Scheidegger, A. E., 1961, Theoretical geomorphology: Springer-Verlag, Berlin, 327 pp.

Sharp, R. P., 1940, Geomorphology of the Ruby-East Humboldt range, Nevada: *Geol. Soc. Am. Bull.,* v. 51, pp. 337–372.

Tuan, Yi-Fu, 1959, Pediments in southeastern Arizona: *Univ. of California Publ. in Geography,* v. 13, 163 pp.

Wentworth, C. K., 1928, Principles of stream erosion in Hawaii: *Journ. Geol.* v. 36, n. 5, pp. 385–410.

Conversion of Units and Equivalents

1 cu. ft. = 7.48 gallons

1 cfs = 449 gal. per minute = 0.646 mil. gal. per day (mpd)

1 cfs for 1 day = 1.98 acre-feet

1 inch per hour from 1 sq. mi. = 640 cfs

1 cubic meter per second = 35.3 cfs

1 inch per hour of runoff = 1 cfs from 1 acre

1000 ppm in 1 cfs = 3.46 tons per day (assuming sediment density 80 lbs. per cu. ft.)

1 acre foot = 325,851 gallons

At 70°F, properties of water include:

(ρ) Density = 1.94 slugs per cu. ft.

(γ) Specific weight = 62.3 lbs. per cu. ft.

(μ) Dynamic viscosity = 2.04×10^{-5} lb. sec. per sq. ft.

(v) Kinematic viscosity = 1.05×10^{-5} ft.^{2} per sec.

1 inch = 2.54 cm

1 mile = 1.61 km

Symbols and Nomenclature

A Cross-sectional area

A_d Drainage area

C Chezy coefficient

D Grain diameter; distance between streams (Chapter 10)

E Energy level or amount

F Force

G Sediment transport rate

H Heat quantity in thermal units; also height

I Dynamic transport rate

K Coefficient or constant

L Length; stream length

L_R Average length of streams of order R

M Mass; rank of number in an array (Chapter 3)

N Number of years of record (Chapter 3)
Number of streams (Chapter 5)

P Perpendicular force; precipitation (Chapter 8)

Q Discharge rate

R Hydraulic radius (hydraulic mean depth); order number of streams (Chapter 5)

Re Total basin relief

T Time unit; frictional stress opposing motion; absolute temperature; texture ratio (Chapter 5)

V Velocity of grain relative to bed

W Work; weight (Chapter 8)

F Froude number

K Von Kármán coefficient

R Reynolds number

H Height above base level

f Darcy-Weisbach resistance factor

a Coefficient; a reference level

b Change of width with discharge

c Flow coefficient; cohesion (Chapter 8); with overbar $\bar{c}$, denotes sediment concentration

d Depth; distance between first-order streams (Chapter 10)

e Efficiency factor; base of Napierian logarithms

f Change of depth with discharge; function of; a measure of curvature of slopes

g Acceleration of gravity

h Height

i Rate of transport per unit width of channel

j Rate of increase of suspended load with discharge

k Constant or coefficient; height of roughness element

l Length

m Change of velocity with discharge

n Manning coefficient of roughness; n' coefficient of Manning type

p Probability; wetted perimeter

q Discharge or flow rate per unit width

r Radius; rainfall supply rate (Chapter 8)

s Slope; shearing resistance (Chapter 8)

t Time

v Velocity of fluid

v_s Settling velocity of grain

w Width; unit weight (Chapter 8)

x Abscissa direction; slope distance (Chapter 8)

y Ordinate direction; exponent in relation of resistance to discharge (Chapter 6)

z Vertical direction; exponent in relation of slope to discharge (Chapter 6)

SUBSCRIPTS:

b With respect to bed load; when used with Q, refers to bankfull

c Critical value

e With respect to eroding force or resistance

m With respect to momentum

r_b Is bifurcation ratio; r_e is length ratio, ratio of average length of streams of given order to streams of next lower order

s With respect to suspended load; where used with v signifies settling velocity of grain

g With respect to the grain

$*$ With v or U, refers to shear velocity

1 Some particular value

OVERBAR, —: Average value

α Angle of repose; angle of slope (Chapter 8)

β Slope of channels, bed, or energy grade line

γ Specific weight

η Packing coefficient

μ Dynamic viscosity

ν Kinematic viscosity

ρ Mass density of fluid

σ Mass density of grain

τ Shear stress

ϕ An angle

ω Stream power per unit width

$\mathfrak{z}$ Total applied stress

Ω Total power

Author Index

Subject Index